T0132971

Triumph of the Expert

Ohio University Press
Series in Ecology and History

James L. A. Webb, Jr., Series Editor

Conrad Totman
The Green Archipelago: Forestry in Preindustrial Japan

Timo Myllyntaus and Mikko Saiku, eds.
Encountering the Past in Nature: Essays in Environmental History

James L. A. Webb, Jr.
Tropical Pioneers: Human Agency and Ecological
Change in the Highlands of Sri Lanka, 1800–1900

Stephen Dovers, Ruth Edgecombe, and Bill Guest, eds.
South Africa's Environmental History: Cases and Comparisons

David M. Anderson
Eroding the Commons: The Politics of Ecology in
Baringo, Kenya, 1890s–1963

William Beinart and JoAnn McGregor, eds.
Social History and African Environments

Michael L. Lewis
Inventing Global Ecology: Tracking the Biodiversity
Ideal in India, 1947–1997

Christopher A. Conte
Highland Sanctuary: Environmental History in
Tanzania's Usambara Mountains

Kate B. Showers
Imperial Gullies: Soil Erosion and Conservation in Lesotho

Franz-Josef Brüggemeier, Mark Cioc, and Thomas Zeller, eds.
How Green Were the Nazis? Nature, Environment, and
Nation in the Third Reich

Peter Thorsheim
Inventing Pollution: Coal, Smoke, and Culture in Britain since 1800

Joseph Morgan Hodge
Triumph of the Expert: Agrarian Doctrines of Development and the
Legacies of British Colonialism

Triumph of the Expert

Agrarian Doctrines of Development and the Legacies of British Colonialism

Joseph Morgan Hodge

OHIO UNIVERSITY PRESS

ATHENS

Illustrations

Figures

Plates

Following page 178

Tables

Acknowledgments

The genesis of this book can be traced back to 1989, when in my last year as a history major at the University of Waterloo, Canada, I took a senior seminar called "Canada and the Third World" and decided to write my term paper on the history of the Canadian International Development Agency (CIDA). I was so intrigued by the subject that I resolved to pursue a career in the field of international development, enrolling at the University of Guelph, Canada, where I completed a master's degree in comparative international development studies and sociology. My time at Guelph was challenging and exciting, but after three years of study I had grown increasingly critical of "development" as a set of national and international policy initiatives. I sensed, as many other scholars and practitioners argued at the time, that development theory was at a crossroads: the old theoretical paradigms and models no longer seemed to hold, yet no real alternatives had appeared. What was missing from the discussion was a solid understanding of the broader historical, political, and institutional context from which development had emerged as a pervasive set of ideas and practices. Most critics assumed that development as a global discourse began at the end of the Second World War with the emergence of the United States as the dominant world power. My sense of history told me that the origins and problems went much deeper and that perhaps the way out of the so-called impasse in development studies and theory was to examine those origins more carefully and thoughtfully.

This quest for historical perspective led me to Dr. Robert W. Shenton, a professor of history at Queen's University in Canada, who together with the late Dr. Michael Cowen, the former director of the Institute of Development Studies in Helsinki, was researching the history of development doctrines in Europe in the nineteenth century and subsequently in other regions of the world from Australia and Canada to Kenya. It was Shenton

EAVRO	East African Veterinary Research Organization
EMB	Empire Marketing Board
FAO	Food and Agriculture Organization of the United Nations
IIALC	International Institute of African Languages and Cultures
IMC	International Missionary Council
IMF	International Monetary Fund
ICTA	Imperial College of Tropical Agriculture
LNHO	League of Nations Health Organization
MRC	Medical Research Council
ODA	Overseas Development Administration
ODM	Ministry of Overseas Development
OFC	Overseas Food Corporation
TDRF	Tropical Diseases Research Fund
UN	United Nations
UNDP	United Nations Development Program
UNESCO	United Nations Educational, Scientific and Cultural Organization
WACRI	West African Cocoa Research Institute
WAMS	West African Medical Staff
WHO	World Health Organization
WIDWO	West Indies Development and Welfare Organization

Introduction

Expertise, Development, and the State at the Climax of Empire

> One thing is certain: world population is increasing today at a rate which threatens the future of mankind. Food production is limping behind the expansion of population . . . What is needed is a spectacular increase in food production on a scale that exceeds the increase of populations and thus makes possible a higher standard of living . . . The facts of world poverty and the danger of growing crisis by starvation should be sufficient to convince even the most selfish and insular of the need for an all-out crusade for world development. The dangers to world peace are equally obvious. The dragons' teeth of poverty and hunger inevitably produce violence, for hungry men are dangerous men.
>
> —*Harold Wilson, 1953*

IT HAS BEEN MORE THAN fifty years since Harold Wilson made his memorable appeal to the conscience of mankind for a war on world poverty.[1] Viewed through the "Cold War lens," Wilson's crusade can be seen as part of a new postcolonial strategy that gained international currency among Western policymakers in the 1950s and 1960s as they grappled with the pressing problem of transforming the newly emergent nations of the third world into productive, modern economies.[2] Today, the passing of the Cold War and the resurgence of neoliberal economics have exposed the apparently flawed and defunct principles of the development era and ushered in, in its place, a new age of globalization. A barrage of critical reassessments of contemporary development theory and policy has appeared in recent years, attesting to the sense of moral and intellectual impasse.[3] At the same time, historians and other social scientists have shown growing interest in unearthing the broader historical, political, and institutional context in which this pervasive set of ideas and practices was first set in motion. For many, that context began at the end of the Second World

War with the emergence of the United States as the dominant world power and the rise of anticolonial nationalist movements, which hastened the end of European rule in Africa, Asia, and elsewhere.[4]

For poststructuralist analysts in particular, the discourse of development is presented as the outcome of a specific historical conjuncture, formally inaugurated by President Harry Truman's Point Four Program in 1949, when the "discovery" of mass poverty in the third world came to occupy a prominent place in the minds of U.S. and other Western power elites.[5] The offer of technical and financial assistance as part of a new deal for the former colonies is invariably tied to the U.S.-led campaign to counteract communist influence in the newly emerging nations. "Development," as Arturo Escobar relates, "became the grand strategy for advancing [East-West] rivalry and, at the same time, the designs of industrial civilization. The confrontation between the United States and the Soviet Union thus lent legitimacy to the enterprise of modernization and development; to extend the sphere of political and cultural influence became an end in itself."[6] Despite its countless failures and the many twists and reincarnations of recent years, postdevelopment scholars argue that the dominance of Western knowledge and power have enabled the central tenets of this discourse to persist, seemingly impervious to criticism and incapable of meaningful reform. We are left with a paradigm that is totalizing and undifferentiated in scope, an all-encompassing knowledge-power regime seeking to impose an unwanted modernity on the rest of the world. Development as theory and practice, from this standpoint, began sometime in the decade following the Second World War, and from its inception it has always been and remains in essence a project synonymous with the globalizing of Western capitalism and the modernizing of non-Western cultures.

This book offers a different narrative, shifting our gaze to connections and debates that not too long ago were regarded in the literature on development as largely irrelevant. Building on several important studies, it examines the way development as a framework of ideas and practices emerged out of efforts to manage the social, economic, and ecological crises of the late colonial world.[7] The severe economic depression of the early 1930s and the rising social unrest in the form of strikes, riots, and disturbances that followed in its wake marked a critical turning point in the colonial encounter, setting off a far-reaching process of official rethinking and reform designed to forestall popular discontent and give a new lease on life and legitimacy to the imperial project. The concept of development, as Frederick Cooper and Randall Packard suggest, "became a framing device bringing together a range of interventionist policies and metropolitan finance with the explicit goal of raising colonial standards of living."[8] From this perspective, the postwar crusade to end world poverty represented not so much a novel proposal marking the dawn of a new age, as the zenith of decades, indeed centuries, of debate over the control and use of the natural and human resources of colonized regions.

More recent research on development has sought to extend the analysis further, arguing not only that many of the key themes and central concerns of the postwar development era have their roots in earlier colonial efforts but also that Western development models and ideologies should not be taken as self-evident, self-generated, hegemonic discourses imposed without check from above. Again, as Cooper and Packard remind us, "The appropriations, deflections, and challenges emerging within the overall construct of development—and the limits to them—deserve careful attention."[9] Some of the most innovative research of recent years has been by Africanists and other colonial historians, who stress that wider imperial ideologies and agendas were inevitably intertwined with, and often challenged, modified, and even molded by, the actual practice and locality of development.[10] Helen Tilley's work on the African Research Survey, for example, has shown how in the interwar period new ideas about development were emerging that paid careful attention to local conditions, needs, environments, and even local knowledge.[11] Such efforts drew heavily on the new sciences of ecology, nutrition, and social anthropology, as well as on decades of local experience that brought into question earlier assumptions regarding the economic potential of tropical Africa. In a similar vein, Suzanne Moon's study of scientific authority and development in the Netherlands East Indies has also shown how detailed, empirical knowledge and awareness of the variability of local conditions and indigenous agricultural practices became the basis of the Dutch approach to agricultural improvement and extension. This greater understanding and knowledge of place, she argues, guarded against indiscriminate colonial economic and environmental transformations.[12] Perhaps the most penetrating analysis so far is Monica van Beusekom's case study of the French Soudan's Office du Niger.[13] Colonial planners at the office, van Beusekom contends, were unable to effect the wholesale transformation of agriculture and rural society as they had initially intended when the project began in the 1920s, in large part because African settlers on the scheme failed to embrace their rationalist vision of intensive plow agriculture and crop rotations on irrigated plots. Instead, a complex process of interaction and negotiation between African farmers and project officials and experts gave rise to development practices that were often pragmatic and makeshift in nature and that diverged on a number of crucial issues from the social evolutionist model that underpinned the scheme's original plans. What is more, the failure of these earlier efforts led to a significant disjuncture in approach at the project level. After the Second World War, as she details, project managers began integrating local agricultural practices, such as abandoning crop rotations in favor of fallow periods and allowing nomadic pastoralists to graze their cattle on office lands during the dry season as a way of maintaining soil fertility. And although they still believed in the superiority of European science, they began to draw heavily on indigenous knowledge such as local soil classification systems to determine which crops and varieties were best suited to which soils.

The present study, as will become clear, owes a great deal to the insights of recent scholarship on the politics and role of local practices in shaping and altering the wider discourse of development, but it approaches the subject from a somewhat different angle. Rather than focusing on any specific development experience or particular colonial locality or region, it examines the emergence of and continuities and ruptures in colonial development doctrine through the lens of a distinct group of actors operating *within* the late imperial state—the network of specialist advisers, scientific researchers, and technical experts involved in colonial policy debates and project planning during the period from 1895 to 1960. I argue that the reevaluation of British imperial goals which began in the wake of the Depression stemmed in large measure from the increasing sensitivity and awareness of the complexities of local conditions articulated by technical officers and researchers working on the ground as part of the colonial professional departments. Their reassessment of local conditions and earlier colonial development models contributed to fundamental shifts, not only in practices at the project level but also in the broader theoretical assumptions and policy paradigms operating at the Colonial Office (CO) as well. Indeed, one of the main goals of the present study is to deepen our understanding of how colonial knowledge was produced and institutionalized, by examining the process through which the concerns and visions of practitioners operating on the peripheries of empire were filtered back up to and had an influence on policy debates in London.

It is, however, no less important to understand how the perspectives of metropolitan officials and experts, imbued with a particular reading of history and agendas emanating from wider imperial and international research currents, were in turn circulated outward. For it is equally clear that the shift in colonial development debates and agendas that began in the 1930s bore the stamp of preoccupations and practices radiating from the center, as well as the fault lines of world economic and political events such as the Depression and the Second World War. A kind of neo-Malthusian crisis narrative began to creep into official discourse as experts warned of land shortages, famine, desertification, and widespread social unrest.[14] The template and symbolism of earlier European and imperial historical experiences were overlaid upon a mounting political and social instability, as officials in London struggled to understand and assert control over the turmoil erupting in the colonies. I make the case that one of the key sets of actors facilitating this movement and exchange of ideas—elevating the concerns of local officers onto the imperial stage while at the same time filtering them through the metropolitan lens—were the specialist advisers and experts enlisted by the CO. These advisers helped promote colonial scientific cooperation and the greater exchange and synthesizing of colonial knowledge through collaboration with imperial research institutes and clearinghouses in Britain; through periodic visits and regional tours to different colonial territories as well as countries outside the empire; through encouraging collaboration among scientific practitioners in different colonies and regions working on related problems, and the coordination of local efforts through the setting up of regional research institutions; and through the organization of regional and pan-colonial conferences and attendance at various imperial and international scientific fora. In a very real sense, these experts played a critical role in the growing institutionalization and globalization of colonial scientific knowledge and authority in the 1930s, 1940s, and 1950s.

Although there already exists an extensive literature on the history of late British imperial policy, most of this scholarship concentrates on the role of prominent political leaders and senior administrative officials or on the political, administrative, and ideological background behind "high" metropolitan decision making at the climax of empire.[15] Without the inclusion of specialist advisers and experts and the role they played in the imperial project, our understanding of late-colonial environmental and development debates and their legacy remains incomplete. Imperial discourse and power were never static or hegemonic. Nor can we speak of a unifying and all-embracing colonial mission, as the work of an earlier scientific expertise to deal with the problems of capitalist production, resource management, and social order.[25]

In Britain, the new forward-looking doctrines became synonymous with Joseph Chamberlain's campaign to supplant the laissez-faire philosophy of nineteenth-century liberals with a more "constructive" and interventionist ideology of colonial development. As secretary of state for the colonies from 1895 to 1903, Chamberlain pressed for state-directed development and imperial assistance so that "those estates which belong to the British Crown may be developed for the benefit of their population and for the benefit of the greater population which is outside."[26] Under Chamberlain, the CO in London began to promote tropical medical and agricultural research and training through the appointment of medical and botanical advisers, the founding of the Liverpool and London Schools of Tropical Medicine, and support for the Royal Botanic Gardens at Kew.[27] Subsequent parliamentary ministers and imperial enthusiasts from across the political spectrum would invoke the "great estates" analogy, and over the next half-century the imperatives of colonial science and development would be firmly entrenched within the government's agenda.

Indeed, I make the case that the most striking feature of British colonialism in the twentieth century is the growing confidence it placed in the use of science and expertise, joined with the new bureaucratic capacities of the state, to develop the natural and human resources of the empire and manage the perceived problems and disorder generated by colonial rule. The possibilities of planned, rational state intervention, guided by the advice and impartiality of expert opinion, helped to reinvigorate and morally rearm the imperial mission in the late colonial epoch.[28] Fallout from the Great Depression cast doubt on that mission as escalating unrest and rioting exposed the deplorable working and living conditions that existed in many colonial territories, while the collapse of world commodity prices shattered confidence in free trade and export-oriented development strategies. The CO initiated a far-reaching process of policy reform, which led by the end of the decade to the passing of the Colonial Development and Welfare Act of 1940. The Second World War acted as a further catalyst for the revolution of metropolitan thought, creating new pressures for state planning and social engineering both on the metropolitan home front and in the colonies, which heightened the sense of urgency and demand for action.[29]

For the colonial empire, the new initiative created an immense need for new kinds of knowledge and scientific organization. A great pooling of British scientific and academic expertise, assembled by the CO, was matched by an equally dramatic expansion of the colonial technical departments in the field. Even before 1940, the CO had sought to widen the range and sources of technical knowledge and expertise available to the secretary of state and the area departments by gradually building up a network of permanent professional advisers and standing committees. The beginnings of the CO's advisory network stretch back to before the First World War, with Chamberlain, but it was in the interwar period that the expansion and consolidation of technical expertise began in earnest as new areas of advice were added in education, agriculture, animal health, fisheries, nutrition, and labor. In structure and composition, these advisers and committees acted as important models for the final and most substantial phase of growth, which began in the 1940s in preparation for the postwar development and welfare offensive. By the late 1940s, the CO had been reinforced by the inclusion of nearly two dozen principal advisers and consultants and an equal number of specialized advisory councils and committees, each with the authority to plan new initiatives and to co-opt the services of the country's leading experts in their respective fields (see tables 0.1 and 0.2).[30]

Table 0.1. Principal Colonial Office advisers and consultants

Advisers and Consultants	*Years Appointed*
Chief Medical Adviser (later Chief Medical Officer)	1897–1912, 1926–61
Agricultural Adviser	1929–61
Adviser on Animal Health	1931–32, 1940–61
Fisheries Adviser	1931–37, 1948–61
Labor Adviser	1938–61
Educational Adviser	1940–61
Forestry Adviser	1941–61
Secretary for Colonial Agricultural Research	1945–61
Director of Colonial Medical Research	1945–61
Adviser of Social Welfare	1947–61
Adviser on Co-operation	1947–61
Surveys Adviser and Director of Colonial Geodetic and Topographic Surveys	1948–61
Geological Adviser and Director of Colonial Geological Surveys	1948–61
Head of African Studies Branch	1949–61
Adviser on Drainage and Irrigation	1953–54, 1957–61
Colonial Building Research Liaison Officer and Housing Adviser	1953–54, 1957–61
Engineer-in-Chief, Crown Agents (Engineering Adviser)	1953–61

Source: Charles Jeffries, *The Colonial Office* (London: Allen and Unwin, 1956); Anne Thurston, *Records of the Colonial Office, Dominions Office, Commonwealth Relations Office and Commonwealth Office* (London: HMSO, 1995), chap. 8.

Table 0.2. Principal Colonial Office advisory committees and councils

Advisory Committees and Councils	*Years Appointed*
Colonial Survey (later Survey and Geophysical) Committee / Advisory Committee on Colonial Geology and Mineral Resources	1905–57
Colonial Veterinary Committee	1907–19
Advisory Medical and Sanitary Committee for Tropical Africa	1909–22
Colonial Research Committee (later Council)	1919–32, 1942–59
Colonial Advisory Medical and Sanitary Committee / Colonial Advisory Medical Committee	1922–61
Advisory Committee on Native Education in Tropical Africa	1923–28
Colonial Medical Research Committee	1927–30, 1945–60
Advisory Committee on Education in the Colonies	1929–61
Colonial Development Advisory Committee	1929–40
Colonial Advisory Council on Agriculture and Animal Health (and, later, Forestry)	1929–61
Central Welfare Coordinating Committee/Colonial Social Welfare Advisory Committee / Advisory Committee on Social Development in the Colonies	1941–61
Colonial Labor Advisory Committee	1942–61
Colonial Economic Advisory Committee / Colonial Economic and Development Council	1943–51
Colonial Fisheries Advisory Committee	1943–61
Colonial Social Science Research Council	1944–61
Colonial Agriculture, Animal Health and Forestry Research Committee	1945–61
Advisory Committee on Co-operation in the Colonies	1947–61
Colonial Economic Research Committee	1947–62

Source: Charles Jeffries, *The Colonial Office* (London: Allen and Unwin, 1956); Anne Thurston, *Records of the Colonial Office, Dominions Office, Commonwealth Relations Office and Commonwealth Office* (London: HMSO, 1995), chap. 9.

Paralleling these moves at the center was a related enlargement and coordination of the colonial technical and research services. Prior to the First World War, organization of the technical services was sporadic and carried out on a departmental and territorial basis, but during the interwar period technical recruits rose dramatically, accounting for more than 40 percent of overall appointments to the colonial service. A colonial research infrastructure began to take shape with the creation of the first in a planned network of central research stations that were to be linked to colonial resource departments as well as scientific bodies and research institutes in Britain. Organizational efforts were temporarily set back by the declines in recruitment, retrenchment of staff, and disruptions in applied research that followed on the heels of the Great Depression and the disruption of the Second World War, but as the war drew to a close, earlier models and recommendations were revived and reformulated into ten-year development plans, and there was a great surge in training of new technical candidates (see table 0.3). At the same time, research work intensified and concentrated in new or extended regional organizations and institutes. In many ways, then, late British colonial imperialism *was* an imperialism of science and knowledge, under which academic and scientific experts rose to positions of unparalleled triumph and authority.

This book follows the origins, course, and legacies of the strategic engagement between scientific and technical expertise, development, and the state at the climax of the British colonial empire. It focuses on the views of the CO's specialist advisers and councils, as well as those of scientific practitioners and researchers employed in the colonial technical services. While British colonial administrative structures made use of specialists trained in an extensive range of scientific and technical fields, this account is necessarily selective, concentrating primarily on agriculture—the most important colonial science in regard to environment and development—and to a lesser extent on tropical medicine and education, which, as we will see, came to be seen as the fields of imperial knowledge most closely and significantly connected with the new policies concerning colonial development and welfare. Further, the study focuses primarily on sub-Saharan Africa, although other regions, including South and Southeast Asia and the Caribbean, are taken into account as well. This concentration reflects not only the importance of tropical Africa to British

Table 0.3. Colonial Service recruitment, 1913–52

	1913–19	*1920–29*	*1930–39*	*1940–44*	*1945–52*
Administrative	190 (35%)	983 (25%)	610 (27%)	166 (22%)	1,934 (19%)
Scientific Expertise[a]	167 (31%)	1,720 (44%)	928 (42%)	311 (40%)	3,724 (37%)
Nonscientific Expertise[b]	32 (6%)	514 (13%)	161 (7%)	141 (18%)	2,689 (26%)
Other Appointments[c]	154 (28%)	725 (18%)	548 (24%)	154 (20%)	1,801 (18%)
Total	**543**	**3,942**	**2,247**	**772**	**10,148**

a. Includes appointments in Agriculture, Chemistry, Engineering, Fisheries, Forestry, Geology, Marine, Medicine, Meteorology, Surveying, Veterinary, and other/miscellaneous scientific posts.

b. Includes appointments predominantly in Education but also Architecture and Town Planning, Civil Aviation, Commerce and Industry, Cooperation, Dentistry, Development Officers, Economics, Labor, Mining, and Social Welfare.

c. Includes Auditing, Broadcasting, Finance and Customs, Legal, Personnel, Police, Prisons, Public Relations, Statistics, and other/miscellaneous posts.

Source: Anthony Kirk-Greene, *On Crown Service: A History of HM Colonial and Overseas Civil Services 1837–1997* (London: I. B. Tauris, 1999), 22, 24, 26, 37, 52–53.

imperial ambitions in the twentieth century but also the fact that many of the colonial technical services and advisory bodies established in this period were initially working on this region and only later extended to include the entire colonial empire.

As scholars of colonial history have begun to recognize, these metropolitan specialists and their technical counterparts in the field played decisive roles in the final decades of imperial rule, formulating policies and plans at Whitehall and setting the agenda within professional departments overseas. Academic and scientific advisers and technical officers became the principal conduit through which new ideas and influence flowed, as authority in these years moved away from the fabled district administrator who "knew his natives" to the specialist who "knew his science."[31] Crucial to the present study is the way in which they imagined and framed the problems gripping the colonies in the late 1930s, 1940s, and early 1950s, and how the analyses they produced often provided the rationale for administrative solutions that promoted external intervention and control over local resources and practices. But while the sway of scientific opinion in the 1940s and 1950s was unprecedented and extensive, it was not absolute. The rise and "triumph" of the experts was a great deal more ambiguous and fragmented than the sheer increase in their numbers would lead us to believe. By the early 1940s, the new battalion of experts and advisory bodies seemed poised to rule, busily pumping out memoranda and reports that would provide the blueprints for postwar reconstruction and reform. Their victory was short-lived, however, and to a large extent illusory. Indeed, one of the main questions arising out of the study of colonial expertise is why the vast increase in specialist personnel, imperial financing, and bureaucratic services that took place in the late 1940s and early 1950s failed to achieve the intended goals of the new development policies. For despite the increasingly ambitious tone of metropolitan planning initiatives, many of the resulting programs and schemes proved unsuccessful and at times even disastrous.

This disjuncture between metropolitan discourse and colonial practice has its roots in the structural constraints and material contradictions punctuating the colonial project. As studies of the political economy of the colonial state have shown, the inadequacies of colonial power in Africa and elsewhere forced regimes early on to retreat from their initially ambitious plans for the introduction of private property and wage-based labor relations.[32] Difficulties in securing an adequate supply of labor, in particular, proved insurmountable, sparking confrontations and clashes with preexisting indigenous societies and heated public debate back in Europe. By the First World War, colonial states had come to rely instead on alliances with "traditional" elites and local chiefs in order to maintain political order, collect taxes, and guarantee the preconditions for capitalist development. Africans were "containerized" into tribes and subjected to the power of the Native Authority or hierarchy of chiefs, while the colonial state for its part undertook to steer, and at times even block, the economic forces it had helped to set in motion in the first place.[33] This makeshift settlement began to unravel under the weight of the Great Depression, which had a profound effect throughout much of Africa, the West Indies, and elsewhere, as export prices for primary commodities collapsed. The collapse of exports triggered a sharp fall in overall wage employment and wage rates, which remained low even after world prices began to rebound. Real incomes and living standards for colonial wage earners plummeted even as the cost of food, imported goods, and urban rents rose sharply. In response, social unrest and strikes erupted throughout the colonial empire in the late 1930s. The Second World War exacerbated the crisis, generating new pressures for the mobilization of colonial resources for metropolitan needs, while at the same time requiring colonial governments to drastically reduce consumption by raising taxes and cutting imports.[34] Social unrest accelerated under the severities of wartime as a further wave of strikes and protests spread across the vital ports and communication centers of empire.

Britain's postwar economic troubles accentuated the tensions even further. Once again, governors in the tropical colonies were instructed to maximize production of food and raw materials to meet British as well as colonial needs. The urgency and scope of state intervention in colonial production and marketing has led historians to describe this moment as "the second colonial occupation."[35] The postwar colonial state may have been revitalized by the influx of new specialist personnel and the new emphasis on integrative planning and coordination of government services, but the inherent weakness and incongruities of colonial power remained. After the war, colonial authorities increasingly resorted to heavy-handed state measures and compulsory legislation to solve rural problems, inadvertently deepening social discontent and anticolonial resistance in many African territories.[36] The high-water mark of intervention came with the Fabian-inspired development offensive of 1947, when the imperial government attempted to circumvent private capital by establishing large-scale, state-managed projects for agricultural production on estates, while maintaining control over colonial commodity trade to, in effect, subsidize British consumers.[37] But in the absence of a functioning civil society to provide moral legitimacy, and with governing structures and habits that were deeply patriarchal, racialist, and authoritarian in nature, the capacity of the colonial state to deliver a metropolitan-style development and welfare program was critically and fatally circumscribed.[38]

The structural inadequacies and limitations of colonial power in the face of mounting social, economic, and ecological crisis form the backdrop to the bureaucratic tensions and policy debates analyzed in this book. Contradictions in colonial development doctrine reflected the factionalized and contradictory nature of colonial power in general. The conflicting pressures of metropolitan needs, on the one hand, and mounting material crises and social resistance on the other, generated bureaucratic confusion and rivalry among many officials and experts. These fissures had an impact on ongoing policy debates at both the local and imperial levels, often leading to strong criticism against many proposed development schemes, or to the formulation of new approaches that promised to undo the "mistakes" of earlier efforts. The influence of long-standing CO traditions remained strong in the 1930s and 1940s, especially among the older generation of senior officials, who viewed the growth of the advisory committees and subject departments with alarm, as undermining the importance of local knowledge and experience and the intimacy of personal communication that had once existed between governors and their contacts in the geographical departments.[39] Nor was there always agreement between political heads, permanent staff, and advisers, or even among advisers themselves, over the right course of action to take. Imperial administrators and politicians at the time used science and development as evidence of the progressive and altruistic virtues of empire, while public scientists and academics used empire as a way to advance individual careers and secure state funding for research and institutional network-building. But beyond the mutually beneficial partnership and unbridled optimism of science and empire, important disagreements and tensions existed over the direction and purpose of research and policy.

Outside Whitehall, there were divisions and conflicts, not only between the CO and local administrations but also between technical and administrative departments within colonial governments and between research specialists and general administrative officers within technical departments. Indeed, it has been argued in recent research that the failure of the postwar colonial development project stems to a very large degree from the internal conflicts surrounding scientific authority within the colonial state itself.[40] The transformative project envisioned by those in London required the specialist departments of the colonial state to take on a much more assertive and central position of authority than before, but this would have required a revolution in colonial government and the subordinating of district administration to the technical branches. Yet the political foundations of the colonial state were tied to the administrative service, with its close ties to local intermediaries, and could not be easily modified without raising serious political problems. In the end, the technical branches remained subservient to the administrative side.

This structural impasse was compounded by a divergence of views over the meaning of expertise itself. Although they shared many of the same assumptions and attitudes, generalist administrative officers tended to place greater value on local experience and the understanding of particular local knowledge, whereas technical officers considered the subordination of this knowledge to the principles of a universal science as necessary for the success of the new development initiatives. In many ways, the universalizing discourse of mathematical calculation and efficiency that characterized imperial science was at odds with the outlook of colonial administrators, whose own sense of expertise was premised on years of experience and knowledge of local conditions and cultures, as well as with the whole structure of local power, privilege, and customary rights upon which the legitimacy of imperial administration perched.[41] While the present account certainly lends support to such a distinction, it also suggests that matters were by no means so clear-cut. Many technical staff, it is true, were committed to introducing revolutionary changes and transformative technologies; others in the field urged a slower and more cautious approach, and were outspoken in their criticism of the ill-planned and hasty directives coming from political and metropolitan authorities. The evidence suggests that the gap between the rhetoric of science as an important tool of empire and the reality of science as practiced was often very wide indeed.

The overall picture that emerges is one of a colonial development discourse being formed through a complex interaction between various levels of colonial administration, each with its own perceptions of the nature of the problems which confronted it, as well as by the influences of broader international and imperial scientific networks and trends. As David Anderson perceptively observes, "It cannot simply be argued that events in the colonies forced a shift in policies, or that awakening of concern for [colonial] development in London prompted a new set of policies 'from above.' Both explanations are unsophisticated, and ignore the movement of ideas that went from colony to London and back again, that were modified by experience both within and outside the Empire, and that often resulted in reforms that went far beyond what was initially intended."[42] Out of the complex and dialectical intersection of ideas, expertise, and bureaucratic power emerged a collective imperial agenda torn by inconsistency, indecisiveness, and objectives pulling in divergent and often conflicting directions. The friction and paralysis of late colonialism, I argue, was to a very great extent the product, not only of structural limitations and administrative wrangling but of the contradictory and ambivalent aims of the colonial mission itself: a mission shaped as much by concerns about the relative decline of Britain and the apparent failure of progress in the colonies as it was by high-modernist belief in the inevitable march of human and material advancement. By the early 1940s, the old Chamberlainite doctrine, which had linked colonial development abroad to social reform at home in an effort to solve the problems of industrial decline and surplus population in Britain, had been substantially redefined by a new liberal, paternalist agenda, which looked to state intervention to deal with the problems of growing rural-urban migration, unemployment, and loss of productive resources—*not in Britain*, but in the colonies themselves. In the place of expansionary schemes to pry open the colonies for British investment, settlers, and access to commodities, with science directing the exploitation of colonial nature, metropolitan experts became increasingly preoccupied with the conservation of the empire's resources and the self-sufficiency and welfare of its rural communities. Like the earlier social-imperialist doctrines, imperial science was still about developing the empire, but it increasingly became "a voice of reason and restraint" and one of sustainable management for the long term.[43]

This period is significant, then, not only for the degree to which the imperial government was prepared to utilize the services of scientific experts but also for the changing rationale that lay behind its greater accommodation. Earlier colonial development debates had revolved around the central problem of population scarcity amid an apparent bounty of untapped wealth. This discourse of emptiness and underpopulation in many ways spoke to the fundamental problem faced by almost every regime in the early decades of colonial rule: that of exerting effective control over local labor. Indeed, in the 1920s, the shortage and inefficiency of manpower due to debilitating diseases and unsanitary conditions was considered one of the greatest obstacles to the development of the tropics, and its resolution through increased medical and sanitary intervention was seen as perhaps the most evident display of the utility of science for empire. Such optimism was tempered, however, by indications of a crisis of biological reproduction among native populations of many colonial territories.[44] In the late 1920s and early 1930s, concerns about depopulation and cultural "degeneration" were particularly strong in colonial medical and public health circles, whose members lobbied for the extension of mother and child welfare services in an effort to check high maternal and infant mortality rates. Alarm over high infant mortality rates also precipitated research on native health and diet, which concluded that malnutrition was a widespread and serious peril in many colonies and would require greater cooperation between agricultural, educational, and health services and experts to remedy. Nor were infant mortality and malnutrition the only social ills laid bare by colonial researchers between the wars. Initially, most colonial agricultural departments had concentrated on the introduction of new cash crops and the expansion of export agricultural production, but in the wake of the Depression these policies came under fire for encouraging "selfish individualism" and overtaxing the soil.[45] Such criticisms may well have been influenced by a series of droughts that occurred in the 1920s and 1930s over large areas of the United States, Australia, and sub-Saharan Africa. The dust storms and drought that swept across the Great Plains in the 1930s are well known, but similar droughts hit the center of the Australian continent and the savannah lands of sub-Saharan Africa as well, leading some observers to suggest that these areas, under the pressure of human settlement, were drying up and becoming uninhabitable.[46]

These problems and many others started to register more prominently in both local and imperial research currents and policy debates in the 1930s and 1940s, precipitating a widespread rethinking of native welfare and colonial land use, and, more broadly, of colonial development itself. Humanitarian lobbyists and social reformers argued that the "development of the people" should be the first priority of the imperial government, through the introduction of a more enlightened policy of colonial scientific trusteeship. "Native" education, broadly defined as the upbringing of the individual by the whole community, was the linchpin in this new concern for the human side of development. The new field of social anthropology was held up as a potential theoretical model for understanding and ameliorating the problem of "culture contact." At the same time, a new "systems approach" to ecological research was beginning to take shape through the influence of key imperial institutions such as Oxford University, which saw in ecology a potential administrative tool not only for the study of the environment but for the improved management of the empire's human and material resources as well.[47]

At the center of much of this new research was the appearance—or, rather, the perception—of a budding population problem. In contrast to the concern about underpopulation and depopulation which had initially inspired demand for more information and investigation, by the late 1930s and 1940s officials and experts at the CO in London were increasingly troubled by what they saw as a pending crisis of *surplus* population and unemployment.[48] The recognition of a relative surplus population in the colonial empire had much to do with the changing nature of the labor question. The old problem of labor shortage was transformed by the Depression into a new crisis of labor surplus, in the form of growing unemployment and underemployment, low wages, and widespread urban and rural immiserization. Alarms were increasingly sounded over what officials termed "detribalization," or, in other words, the breakdown of rural village communities and the unwanted drift of natives to towns and mining centers.

It is with this backdrop of growing deprivation and the threat of rural emigration in mind that we need to read the emerging consensus in favor of state-managed colonial development in the late 1930s, 1940s, and early 1950s. For many metropolitan officials and experts, the unfolding drama that began to grip the colonies had a familiar ring to it. The social problems arising from increasing urbanization and poverty in Britain and the specter of rural indebtedness and impoverishment in India cast particularly long shadows over imperial debate. As authorities began to contemplate and fear the possibility of an empire-wide environmental and population crisis that might lead to a breakdown in colonial order, they were motivated to envision increasingly bolder and more extensive schemes for state-directed intervention and social engineering. Many of the problems confronting colonial authorities were compared to metropolitan ills and judged to be amenable to metropolitan-type solutions. The state-centered approaches to welfare entitlements in vogue in Britain were considered by many reformers to be readily adapted to managing the social and ecological problems of the empire as well. But the fear of colonial disorder also produced indecisiveness and profound ambivalence over the aims of government policy. Officials and experts vacillated between the need to reassert order and stability, on the one hand, and the imperative of increasing productivity and efficiency on the other. Some wanted to use the state as the prime agent for transforming and modernizing rural communities, while others hoped to employ its power to bolster and strengthen them. Even those wishing to preserve traditional society could see no way of introducing the "improvements" they felt were necessary without unleashing centrifugal forces of individualism and class division that threatened to break it apart. The tensions inherent in the late colonial mission were further exacerbated by the ill-conceived and urgent nature of the postwar Fabian colonial offensive, which ultimately led the demise of British colonial rule altogether.

Although the late colonial development initiative may have failed on its own terms, it was not without its lasting effects. The most direct outcome, as others have stressed, was the heavy bias in favor of state-centered ideologies and development structures that remained largely intact even after the transfer of formal colonial power. Hand in hand with this was the depoliticization of poverty and power achieved by recasting social and economic problems as technical ones that could be fixed by rational planning and expert knowledge.[49] But perhaps even more enduring were the ideas and ideological assumptions these experts left behind in the wake of the colonial failures of the early 1950s. Colonial advisers and experts laid down a framework of ideas that championed an agrarian doctrine and vision of development that would become deeply embedded in international policies and institutions in the decades following the end of colonial rule. The pioneering studies and texts they produced became key reference works for subsequent generations of scientists and academic advisers. Many of the philosophical assumptions and apocalyptic narratives that have become part of the conventional wisdom and lexicon of contemporary development and environment policies and practices have their roots in these earlier colonial debates and doctrines.[50] In addition, in a very direct way, the scientists and academics who served the CO played an active role in establishing the network of international aid organizations that sprang up after the Second World War. Some of the most prominent metropolitan advisers would later go on to help establish and direct such UN specialist agencies as the Food and Agriculture Organization (FAO) and the United Nations Educational, Scientific and Cultural Organization (UNESCO). When their careers in the colonial service were brought to an abrupt end in the 1960s, many of the colonial scientific practitioners and technical officers would go on to become consultants working for the United Nations, the World Bank, the Commonwealth Development Corporation (CDC), the Overseas Development Administration (ODA), and other donor agencies.

Through this case study of colonial scientific and technical experts and the part they played in the development and environmental policy debates and agendas of the mid-twentieth century, this book seeks to understand the quandaries that led up to this important transformation in British imperial thought and practice, and to begin to piece together the intellectual and administrative legacies it left behind. While there is no denying the significance of Cold War geopolitics and American strategic interests in elevating the idea of development to the status of a hegemonic, global doctrine after 1945, it is important to realize the continuities that exist with these earlier doctrines and debates. In the context of findings that indicated a shift in demographic trends in favor of the poorer areas of the world, Western power elites and experts in the 1950s and 1960s were increasingly disturbed by the specter of colonial and third world poverty and population growth, which they imagined might lead to increasing conflict over global resources.[51] These regions were thought to be in the grips of a social revolution that had to be carefully guided and controlled through "positive" state intervention in order to forestall discontent. Viewed through the lens of imperial crisis, the impetus for the late colonial initiatives and the subsequent appeal for a war on global poverty reflected as much angst over an uncertain future as it did optimism regarding modernity as the destiny of humankind.

CHAPTER 1

Setting the Terms of the Debate

Science, the State, and the "New Imperialism"

IN ONE OF HIS FIRST major speeches after becoming secretary of state for the colonies in 1895, Joseph Chamberlain, the one-time Radical mayor from Birmingham, announced at Walsall that

> Great Britain, the little centre of a vaster Empire than the world has ever seen, owns great possessions in every part of the globe, and many of those possessions are still unexplored, entirely undeveloped. What would a great landlord do in a similar case with a great estate? We know perfectly well, if he had the money, he would expend some of it, at any rate, in improving the property, in making communications, in making outlets for the products of his land, and that, it seems to me, is what a wealthy country ought to do with regard to these territories which it is called upon to control and govern. That is why I am an advocate of the extension of Empire.[1]

By the time Chamberlain made this declaration he had become one of the greatest exponents of the fin-de-siècle drive for European overseas expansion and conquest in Africa and Asia. The seemingly limitless possibilities opened up by the recent penetration of the African continent and other tropical regions inspired many prominent explorers, philanthropists, businessmen, and political leaders in various European countries to urge their governments to, as King Leopold of Belgium put it, claim their piece of this magnificent African cake.

Chamberlain began his parliamentary career in the 1870s as a Liberal MP under William Gladstone and quickly rose to become leader of the party's Radical wing. The Radical program aimed to transform Britain into a popular middle-class democracy, completely unrestrained by tradition and free from the power and privilege of Anglicanism and the landed elite. But Chamberlain clashed with Gladstone over the question of Irish home rule. He and his supporters, known as Liberal Unionists for their support for the Act of Union between Britain and Ireland, ran as a separate party in the general election of 1886 and later defected to the Conservatives. A trip to North America in 1887 proved to be an epiphany for Chamberlain, opening his eyes to the ideal of a Greater Britain as a way of countering what he perceived to be the steady disintegration of the ties of empire. Upon his return, and for the rest of his life, Chamberlain reinvented himself as the most self-conscious and ardent imperial spokesman of his day, giving voice to the dream of a new imperial federation between the white colonies and Britain.

Chamberlain and his supporters went further, though, offering a bold, forward-looking policy for the more efficient use of the country's vast colonial possessions in order to revive the British economy, ailing under the effects of the late nineteenth-century depression and growing competition and protectionism from the United States and continental Europe. Under Lord Salisbury's Unionist coalition government,[2] from 1895 until 1903, Chamberlain gave shape to his imperial estates program, attempting to supplant the laissez-faire philosophy of nineteenth-century liberals with a new kind of state-directed, "constructive" imperialism. He called on the British government to provide the necessary financial and technical assistance for the extension of imperial communications, especially railways, and other infrastructural projects in order to facilitate the mobilization of the largely unexplored wealth of Britain's tropical colonies in Africa and the East.[3] He lobbied hard to convince the permanent staff at the Colonial Office (CO) of the utility of scientific knowledge and technical expertise, especially in the fields of tropical medical and tropical agricultural research and training.[4] He played a pivotal role in setting up the West India Royal Commission, which included botanical advisers from Kew and which recommended grants and subsidies to aid the declining sugar industry and to encourage economic diversification of the islands.[5]

By employing the agrarian metaphor of a great estate in need of development, Chamberlain was very consciously articulating language associated with a landed, conservative ethos in order to appease Tories who feared his Radical roots and previous association with urban politics.[6] This was, to some degree, a tactical move by Chamberlain to broaden his "business" view in a way that offered the Conservative-dominated Unionist Party a chance to refashion itself as the party of empire through a new and dynamic imperial policy. But his rhetoric also spoke to the continuing resonance of long-held assumptions and ideological tropes rooted in earlier moments of imperial crisis and expansion. There is a strong ideological bridge, as recent studies have shown, between the fin-de-siècle doctrines of empire-builders like Chamberlain and the earlier domestic and imperial policies of the age of William Pitt the Younger and Sir Joseph Banks.[7] Late eighteenth-century advocates of agrarian improvement also envisioned an alliance of science and the state that would work toward the dissemination of progress both in Britain and the empire. However, this continuity at the ideological level should not distract us from the fact that Chamberlain's proposal for an association of government and expertise to actively and intentionally develop the natural (and, later, human) resources of the colonies *did* indeed represent a watershed in the rationalization of the British Empire. It set the terms and parameters of political and philosophical debate for decades to come.

In recent years, historians have once again highlighted the importance of the cultural and ideological dimension of modern European colonialism.[8] Imperial ideologies were not simply window dressing or rhetorical flourishes meant to obscure the real material motives behind expansion and conquest. Understanding the way in which ideological discourses were created, disseminated, institutionalized, and recast is central to understanding how officials viewed themselves as rulers and how they perceived the peoples and lands over which they ruled. Such discourses defined the limits of what could and could not be done. Nor were these sets of meanings monolithic and unchanging. Recent scholarship has highlighted the tensions and constraints that afflicted imperial efforts to impose power on other peoples.[9] Although there was much continuity to imperial doctrine, the meaning of the language of imperialism evolved over time in response to changing circumstances both at home and abroad. It was also laden with paradox and ambivalence, reflecting the contradictory and often enigmatic nature of policy-making and colonial administration on the ground.

This chapter builds on the new colonial historiography, examining the origins and reiterations of earlier assumptions and ideologies of "improvement" and "development" both in Britain and the empire, as well as the significance of Chamberlain's crucial entrance into colonial affairs in the 1890s. I argue that Chamberlain's years at the CO represent a critical moment in the British Empire, marking the effective beginning of the story of the triumph of the expert. In other words, the significance of the "New Imperialism" lies not so much in the novelty of its rhetoric as in the way it tied older arguments and metaphors to a new vision of the imperial state, and in the way it bound the development imperative to the rising agency of scientific knowledge, technical expertise, and state planning.

From Empty Lands to Tropical Abundance: Ideologies of Improvement and Development in Britain and the Empire before 1890

The term "development," according to H. W. Arndt, first appeared in the lexicon of economic discourse around the mid-nineteenth century.[10] Before then, political economists and moral philosophers had spoken instead of the progress of nations toward a state of "opulence" and "improvement." It is indeed difficult to exaggerate the degree to which the language of improvement—that ultimate Georgian buzzword—reverberated in the minds, treatises, and speeches of British Enlightenment thinkers.[11] At the heart of the eighteenth-century idea of universal improvement was a belief that those who made the most efficient use of the material and human resources of the nation and its colonies had a right to control them. It implied a way of living and ecological use that exalted the benefits of fixed agriculture and husbandry, private property, and the production of commodities for the market.[12] We can trace the origins of the late eighteenth-century ideology of agrarian improvement back many centuries. It had long been associated with the reclamation of unused lands or wastes. This particular usage derived from the Anglo-Norman concept of waste land, which signified not only an uncultivated terrain such as the Fens but also land inhabited by few or no people.[13]

Prospective lands of colonization were also described as "empty" or "virgin," rich in abundant resources but void of inhabitants, or else sparsely populated by nomadic peoples who were too primitive to effectively harness the land's potential wealth. Arguments of this sort had been used to buttress European claims of the right of conquest and settlement since at least the earliest English settlements along the eastern seaboard of North America. Early settler claims rested initially on a mixture of religious and economic imperatives.[14] The task of cultivating the soil was seen as part of the extension of the Lord's kingdom, a providential sign of England's chosen status as the true defender of Godly reformation. Early colonists drew a parallel between "taming the waste," by establishing new gardens and enclosures, and pacifying the "spiritual wilderness" of the heathen by bringing him into the world of Christian civilization.[15] They also based their claims to the land on the Roman legal principle of *res nullius,* which defined all "empty things"—including unoccupied lands—as the common property of all, the fruits of which became the property of the first persons to make such terrain a "betterment" by mixing their labor with it. Perhaps the most well-known exposition of this kind comes from John Locke, who argued in his *Second Treatise of Government* that "whatsoever [a man] tilled and reaped, laid up and made use of, before it spoiled, that was his peculiar right; whatsoever he enclosed, and could feed, and make use of, the cattle and product was also his. But if either the grass of his inclosure rotted on the ground, or the fruit of his planting perished without gathering, and laying up, this part of the earth, notwithstanding his inclosure, was still to be looked on as waste, and might be the possession of any other." It's revealing that Locke chose to disparage, as the most fitting example of this lesson, the supposed misuse of nature by Native Americans, "who are rich in land, and poor in all the comforts of life; who nature having furnished as liberally as any other people, with the materials of plenty, i.e. a fruitful soil, apt to produce in abundance, what might serve for food, raiment, and delight; yet for *want of improving it by labour,* have not one hundredth part of the conveniencies we enjoy."[16] The image of America as a resource-rich yet sparsely populated wilderness, inhabited by peoples who were unwilling to improve the land or were indeed incapable of doing so, served to ideologically render Native Americans as savages, nomadic hunters with no natural right to the lands they occupied. They roamed over these vast tracts rather than actually inhabiting them. Their subjugation and displacement by those who would plant, enclose, and improve the land was thus grounded on both biblical authority and natural law theory.

The wedding of agrarian improvement to the imperial mission reached its apogee during the long struggle against Revolutionary France, when price" for land granted, the government could limit the quantity of land on the market and thereby compel laborers to work longer for wages before they could make the move to land proprietorship. The revenues that accrued to government from the sale of Crown land could be used in turn as an emigration fund to assist poor laborers to make the passage from Britain to the colonies.[23]

In Australia, then, it appeared that the inherent march of progress was being stymied by critical barriers that could be overcome so as to make sustained settlement possible only through the aid of government initiatives to bring people and capital from overseas and to construct railways and irrigation works. It is no historical accident, according to Arndt, that development in the transitive sense—as something that had to be *made to happen,* as opposed to something that occurs naturally—was first employed in Australian (and, to a lesser extent, Canadian) economic writings of the 1830s and 1840s.[24] It is here, Arndt asserts, that the term was first conceived to describe deliberate efforts by states and private commercial companies to steadily occupy and exploit a country's natural resources. But there was a flip side to the motif of "development before settlement," for as Wakefield's expositions make clear, the question of empty lands overseas was closely and consciously tied to the problem of surplus population in Europe itself. Indeed, as Michael Cowen and Robert Shenton insist, the origins of development as an intentional state practice may be equally, and for purposes of this study more usefully, located in the early nineteenth-century debate surrounding the Enlightenment belief in unlimited progress.[25] The origin of the debate is usually attributed to the writings of Thomas Malthus, who postulated that it was an "iron law of nature" for humanity to periodically succumb to misery and vice when population growth outstripped the finite necessities of life. Here, then, was a profoundly different understanding of the problem of population and its relationship to resources, one in which land was finite and people abundant.

In the tumultuous years of early industrial transformation in Britain—roughly from the end of the Napoleonic Wars until the 1840s—Malthus's prediction that *surplus* population would lead to unemployment and immiserization of the poor in the newly forged urban slums of Britain seemed to be amply confirmed. It was here, amid the fear that agrarian revolt and growing industrial unrest might lead to social anarchy and possibly even revolution, that Victorian liberal thinkers like John Stuart Mill first conceived of development as a means by which the state could impose order on unruly societal forces; that is, an idea of intentional development was formulated that attempted to reconcile progress with order. Central to this idea was the principle of trusteeship, whereby those with the knowledge of how to create order out of the social instability of rapid urbanization, poverty, and unemployment were to act as the stewards of humanity. It fell to the experts, to those who possessed scientific knowledge and understood the laws of social statics, and thus the ends that should guide public action, to point the way forward to the realization and improvement of progress by managing and making it rational. In other words, development—by way of the positive approach to knowledge—was first invented in the early nineteenth century as a way to ameliorate the chaos apparently brought on by progress itself.

As we will see, the post-Enlightenment meaning of development as state practice rather than immanent process would come to play a pivotal role in the emergence of scientific colonialism and the rise of the expert in the first half of the twentieth century. In the interim, however, the more dominant strand of eighteenth-century thought, which assumed improvement to be a naturally occurring trend, continued to buttress the liberal view of progress, in Britain and to a lesser extent in the empire. The strong influence of laissez-faire doctrine weighed against the realization of state-aided schemes for the amelioration of poverty in Britain; with the return of prosperity in the late 1840s, the debate subsided and such proposals were temporarily marginalized. Although the idea of intentional development of resources by the state in response to the problem of surplus population was first voiced in early industrial Europe, it would not be put forth again until the end of the nineteenth century by Joseph Chamberlain, and it would not be put into practice on any significant scale until the mid-twentieth century. But by then, as I argue in later chapters, the problem was not surplus population *in Britain,* but in the territories and dependencies of the colonial empire overseas.

The narrative of universal progress would find its champions as well as its detractors in other theaters of the British imperial world in the nineteenth century. Nowhere was this more the case than in India, where the new centralized state system constructed out of the English East India Company's military expansion between 1770 and 1820 gave birth to a new kind of administrative power, one in which the idea of improvement formed a critical part of the governing principles that sustained imperial rule over the subcontinent. In many ways, the legitimacy of the Company and later British Raj rested on the claim that, through the development especially steam, would elevate this ancient civilization into "imperial nationhood."[36] His modernizing program was given force by the 1835 Resolution on Education, which committed the British to the promotion of European literature and science in Indian higher education. The turn toward English-language instruction was further entrenched by Charles Wood's *Educational Dispatch* of 1854, which celebrated "the diffusion of the improved arts, science, philosophy and literature of Europe; in short of European knowledge" among the indigenous population of India.[37] The spread of Western learning, the authors of the *Dispatch* predicted, would "teach the natives of India the marvellous results of the employment of labour and capital, rouse them to emulate us in the development of the vast resources of their country, guide them in their efforts and gradually, but certainly, confer upon them all the advantages which accompany the healthy increase of wealth and commerce; and, at the same time, secure to us a larger and more certain supply of many articles necessary for our manufacture and extensively consumed by all classes of our population, as well as an almost inexhaustible demand for the produce of British labour."[38] The reform program reached its zenith during the tenure of Marquess Dalhousie (1848–56), who set up a separate Public Works Department in 1854 to facilitate a major infrastructure-building campaign under the motto that railways, telegraphs, and a uniform postal system were the "three great engines of social improvement" in India.[39]

The liberal impulse to reform, however, was not so easily translated into practice. The creation of a clearly defined land revenue settlement based on the net produce method and girded by a public domain of law proved impossible to implement. The immense diversity of land usages, based on enormous variations in village, household, and caste relations, required a detailed survey of existing arrangements. Information on crops, soils, stock, and agricultural castes as well as previous history had to be collected before a fair assessment could be made for each case. In the end, these detailed accounts proved unworkable and officers were often obligated to rely on customary practices regarding what was payable by individual cultivators.[40] What is more, the early British Raj faced the pressing military imperative of establishing its rule over the vast areas of its newly founded empire, while at the same time organizing the production and marketing of new, high-value crops.[41] Both of these problems were central to the state's demand for increasing revenue yields from the land, as its major source of taxation, and necessitated close political management of the economy. This left administrators little option but to reinforce the power of local authorities such as the village headmen, small zamindars, and various village communal structures, which they recognized as the most effective means of extracting higher revenues and maintaining social stability at the local level. As a result, as David Washbrook has argued, the attempt to create an effective system of individual property rights was counterbalanced by the recognition of customs and norms of local agrarian communities which entangled property relations in a layer of obligations that directly interfered with the rights of individuals to possess, use, and accumulate property in land.[42]

By midcentury, however, a market in land had begun to emerge. Incidences of land litigation and cases concerning disputes over landed property increased sharply. Important socioeconomic changes lay behind this shift. Beginning in the 1840s, the agrarian economy showed signs of improvement, brought on by the demand from Europe for sugar and China for cotton, while at the same time population throughout India was on the rise, leading in some areas to land scarcity.[43] Land was becoming increasingly an object of competition, which elevated land values and led to the emergence of effective property rights irrespective of state apparatuses. In this context the ryotwari system, radicalized in the hands of a new generation of Utilitarian-inspired settlement officers, played an effective role in breaking down corporate and status-based forms of landholding by clearly defining individual proprietary title to land and by negating the role of traditional intermediary elites. The conferring of proprietary title on the cultivator brought with it the right to freely let or sell one's holding, but it also left proprietors responsible for the repayment of any debts. Peasant indebtedness escalated in the context of excessive land revenue demands by the state, itself a product of Utilitarian logic. Following upon indebtedness came the transfer of land through forced sales, in which moneylenders took title but retained the peasant-cultivator as tenant. By the 1850s, such transfer of land was becoming widespread.[44] Moreover, the superseding of intermediary elites by way of the revenue settlements led to a growing resentment among these segments over their loss of status and economic position. The extent of their resentment was revealed during the Rebellion of 1857, in which many of these intermediary classes were instrumental, especially in the newly settled area of Awadh where the *talukdar*'s position was more firmly entrenched than elsewhere. What is more, many of the village cultivators of Awadh supported the dispossessed talukdars and joined the revolt. The irony of these events surprised colonial officials and threw doubt on fifty years of agrarian

Maine argued that codification required the adaptation of existing laws and customs, which dictated a thorough knowledge and appreciation of the indigenous institutions of Indian society. Much of the tentativeness and conservatism of the post-Rebellion generation of Indian civil administrators can be ascribed to the kind of mindset these lessons instilled.[52]

Maine's prophylactic philosophy paralleled the wider intellectual currents and ambivalences of late Victorian and Edwardian Britain. With the rapid acquisition of extensive territories in Africa and Asia in the late nineteenth century, Britain became the center of a vast and growing tropical empire. The focus of imperial ideology, as a consequence, began to shift away from the colonies of settlement toward the non-European world. Much of what was termed the "dependent" empire—from India to the West Indies to sub-Saharan Africa—was instilled with tropical otherness. The "tropics," as David Arnold notes, were defined not only as being culturally and environmentally distinct from Europe but also as possessing a common set of attributes shared by all their constituent parts. The landscapes, peoples, and politics of the tropical world were thought of as mentally and spatially divergent from the supposed normality of the northern temperate zone.[53] The backwardness of Indian agriculture, the decline of West Indian plantation society, and the perceived lack of material culture in sub-Saharan Africa were all blamed, in part, on tropical climatic conditions and influences, which it was believed had produced apathy and poverty among indigenes.

As Richard Grove notes, the exotic nature of the tropics has long been a powerful cultural trope in Western thought, stretching back to early European ventures to the New World and Asia and the increasingly frequent contact they occasioned with the tropical and oceanic islands of the South Atlantic and later the Indian Ocean and Pacific.[54] The representation of the tropics as an earthly paradise of natural exuberance and fertile abundance would continue to shape the imaginations and perceptions of Europeans until well into the twentieth century. But the fecundity of the tropical environment could also elicit fears about the dangers and distinctiveness of these new landscapes. Europeans quickly became aware of the alien and often hazardous climatic and biotic features of these newly colonized lands, especially the distinctiveness and intensity of disease, which in places such as the west coast of Africa killed off somewhere between 25 and 75 percent of all new arrivals within a year during the eighteenth century.[55] The high rate of white mortality raised fundamental questions about whether it was possible for Europeans to successfully adapt and establish colonies of settlement in the tropics.

By the early nineteenth century, it was increasingly argued that whites were incapable of prolonged physical work and their offspring vulnerable to moral and physical degeneration when exposed to the harsh atmospheric and pathogenic conditions of the tropics.[56] The debilitating climate was also held responsible for the supposed moral and physical weakness of local inhabitants. The natural exuberance and greater heat produced indolent, overly passionate peoples who were incapable of advancing toward a higher form of civilization.[57] The growing alienation of Europeans from tropical environments and peoples also contributed as the century wore on to more biologically rooted explanations for the failure of white acclimatization and native indolence. From the 1820s, West Indian and Anglo-Indian medical texts began to attribute the different susceptibilities to disease of whites and blacks to fundamental and relatively fixed physiological dissimilarities, which, in turn, drew medical attention to the importance of racial differences in the development of pathological anatomy.[58]

The increasing stress medical practitioners placed on apparent divergences in racial immunity paralleled similar transitions in the concept of race. The second half of the nineteenth century saw a hardening of racial categories and increasing distinction made between different social classes and groups both in Britain and the colonies.[59] New intellectual trends were worked together into a renewed imperial creed designed to reaffirm the Victorian sense of progress and superiority in the face of colonial challenges and the signs of growing social inequality and poverty in Britain itself. One of the most important shifts in the Victorian world view was the replacing of Biblical chronology with a new geological sense of time, in which the earth was now thought to be millions of, rather than several thousand, years old. This extended periodization of time was projected onto the axis of geographical distance, creating what Anne McClintock describes as "anachronistic space."[60] Imperial progress across the geographical space of empire was refigured as a historical journey backward in time to an atavistic moment in prehistory. Contemporary primitive societies could be understood as preserved "survivals" that were roughly analogous to the early reaches of modern European society, and could thus help shed light on the origins and development of "progressive" nations.

Explicit in this type of writing were ideas of social evolutionary progress and scientific racism. Particularly after the publication of Charles Darwin's *The Origin of Species* in 1859, it became popular to see the whole of mankind as members of a single family tree in which different peoples where, despite natural riches and abundance, progress seemed to be delayed. Here, European labor and settlement were ill-suited and could not hope to replace the "native races."[67] The essential prerequisite for the development of these regions was for white men to induce others more accustomed to the tropical environment to do their labor for them. Dilke was nevertheless enthusiastic about the potential of the new possessions of tropical Africa, which he believed contained a treasure house of unimaginable wealth waiting to be had. He singled out the highlands and "healthy plains" of southern and East Africa as the most favorable parts of the continent for European enterprise.

Dilke's theme of separating out and injecting renewed enthusiasm for Britain's tropical empire was echoed by many other prominent members of the colonial lobby, who worked incessantly in the 1880s and 1890s to ignite interest in the study and development of the tropics and to draw the public's attention to the opportunities of the African continent in particular.[68] The fertile lake districts of East Africa and rich interiors of the western and equatorial zones were waiting to be opened up, they said, by the pioneering spirit and endeavor of the Anglo-Saxon race. The new imperial discourse was popularized through the periodicals, meetings, and dinners of various scientific and imperial institutions. At an 1896 meeting of the Royal Colonial Institute, Sir George Baden-Powell struck a note of unbridled approval as he stood before an audience of fellow enthusiasts and surveyed the recent shift in national sentiments.

> Wise men are all agreed that tropical Africa, practically untouched as yet, contains great stores of wealth for the human race, great supplies of those mineral, vegetable, and animal products of which industrial nations stand in such need . . . Our nation has taken over great areas of Africa where the noblest instincts of our race, the highest work with which our people are concerned, and the best business interests of our Empire, are certain to find opportunities for profitable and successful exercise . . . We may rest confident that in tropical Africa we shall be able to reap a welcome and new harvest of results conducing alike to the prosperity of our own nation and of the natives, and leading to the setting up of civilization and the extension of the blessings of Christianity over all that hitherto neglected area of the world.[69]

He went on to trumpet imperial expansion and development as an ideal palliative for Britain's "surplus population," which, in the shape of mass rural-urban migration and urban unemployment, had emerged as a problem of growing concern for late Victorian and Edwardian political pressure groups and social reformers.[70] He broke Africa into two great geodetic regions: there were the ports, coastal settlements, and river estuaries of the lowlands, where life was essentially tropical, and then there were the more "temperate-like" highlands. In the lowlands, where the "native races" were densely populated and the climate prohibitive, Europeans were unable to settle permanently but there would be plenty of openings for a white managerial class providing the "brains" and "spirit" for development as these areas became more and more occupied. Even more promising were the highlands, where the healthier climate was conducive to considerable white emigration. He predicted that "flourishing white communities will soon astonish the world in these highlands of Southern tropical Africa."[71] Some were even more optimistic, suggesting that sanitary science would soon enable whites to colonize and inhabit the wet tropics as well.[72]

The lure of the lush tropics was repeated like a mantra over the course of the late nineteenth and early twentieth centuries. One of the more stirring appeals came from the social evolutionist writer Benjamin Kidd, whose hugely popular book *The Control of the Tropics* in many ways set the tone for the age. "The great rivalry of the future is already upon us," Kidd explained in 1898. "It is for the inheritance of the tropics, not indeed for possession in the ordinary sense of the word, for that is an idea beyond which the advanced peoples of the world have moved, but for the control of these regions according to certain standards." For Kidd the standards were clear: "In the immense regions concerned there are embraced some of the richest territories on the earth's surface. But they are territories as yet, for the greater part, practically undeveloped . . . With the filling up of the temperate regions and the continued development of industrialism throughout the civilized world the rivalry and struggle for the trade of the tropics will, beyond doubt, be the permanent underlying fact in the foreign relations of the Western nations in the twentieth century . . . It is not even to be expected that existing nations will, in the future, continue to acknowledge any rights in the tropics which are not based both on the intention and the ability to develop these regions."[73] A decade later, Sir Charles Bruce was saying much the same thing. Like Kidd, Bruce saw the control and development of these "tropical estates" by the rival nations of Europe as the great challenge of his time.[74]

As these examples illustrate, many publicists and politicians eyed the possibilities of tropical development with renewed confidence and power of the state. Dernburg also establish the Hamburg Colonial Institute in 1908 to provide colonial recruits with scientific training beyond linguistic studies.

There is no doubt that British statesmen and observers were aware of these trends among the other leading imperial powers, especially Germany, and that this contributed to the formulation of similar doctrines in Britain.[80] The crucial juncture came with the appointment of Joseph Chamberlain as colonial secretary in 1895. Chamberlain, more than anyone else in Britain, was responsible for bringing the ideological currents of the new imperial movement together with a novel vision of the role and power of the state, one in which the application of science and technology was central. Chamberlain's doctrine of empire negated the passive conception of the state championed by Victorian liberalism, offering instead an active and interventionist policy aimed at strengthening British industrial and national competitiveness and efficiency. The salvation of British industry and trade, he suggested, lay in harnessing the untapped possibilities of the empire, with its potentially lucrative markets and vast natural resources and raw materials. Opening up the colonies, he believed, would boost industrial production and help avert the problems of surplus population and labor unrest at home by stimulating employment and the creation of a well-paid, more productive workforce. Colonial development abroad was thus linked to social reform at home in an effort to unite capital and labor in the cause of empire. He tried to push government beyond the limits of free-trade orthodoxy by calling for various tariff adjustments and preferences in favor of British and empire interests. He and his supporters envisioned an imperial Zollverein wherein Britain would be drawn into closer union with the white settler colonies, surrounded by high tariff walls and collectively exploiting the vast resources of the undeveloped regions of the empire. Ultimately, they looked forward to a strong and autarkic imperial federation that would be able to hold its own against the other great powers of the twentieth century.[81]

Chamberlain's years as colonial secretary marked the effective beginning of the Colonial Office's engagement with science and technical expertise to solve the hitherto intractable problems of tropical development.[82] When Chamberlain took up his duties in 1895, the Colonial Office was structured around five main geographical departments, each responsible for a particular region or group of related colonies, and a general department, which dealt with office procedures and matters which affected the colonies as a whole.[83] All upper-division staff were what might be called administrative generalists, drawn from the English public school and Oxbridge tradition, with academic backgrounds primarily in the liberal arts and classics.[84] There was no medical, agricultural, technical, or scientific staff employed in the Office itself. With the territorial expansion of the late nineteenth century, however, there was an increasing demand for technical advice and officers qualified for technical posts. The CO looked to the Crown Agents, among other auxiliary organizations, to provide this service, especially in engineering, since the Office itself provided no training.[85] From 1896, technical advice was also obtained from the Scientific and Technical Department of the Imperial Institute, which was set up to create new openings in trade and promote agricultural and industrial development.[86] In addition, the CO had a long working relationship with the Royal Botanic Gardens at Kew.[87] The services of Kew were at the disposal of both the self-governing and the Crown Colonies on a vast range of agricultural and botanical matters. Chamberlain built upon and extended this bureaucratic tradition, recognizing that the problems surrounding the administration and development of the Crown Colonies could no longer be adequately addressed through the old nineteenth-century geographic organization of the CO. As the next chapter details, his imperial estates program enlisted the first permanent scientific experts and advisory committees as part of the institutional structure of the Office, especially in the areas of tropical medical and agricultural research and training.

But the New Imperialism, as Drayton has rightfully argued, must also be seen as an extension of a wider "revolution in government" which gave late Victorian statesmen new confidence in the capacity of the state to manage and direct social and economic crises in the name of the common good, both at home and overseas.[88] The recruitment of expertise and the expansion of bureaucratic power became important components in the policy initiatives of the late Victorian and Edwardian periods.[89] Not coincidentally, a pivotal transformation in the political, economic, and social status of science in Britain took place around this time, sparked by concerns for the nation's economic backwardness in comparison with newer industrialized nations, especially Germany.[90] A science lobby emerged, centered on the British Science Guild, which drew a close connection between science, national economic strength, and political power in a competitive international system.[91] Seeing Germany as the model, many social reformers in Britain at the turn of the century demanded dramatic reforms in state machinery in the name of greater efficiency.[92] For them,

Congo Free State exposed the glaring incongruities of fin-de-siècle imperial rhetoric.

Out of this antipathy evolved an alternative doctrine of development, first spread by Mary Kingsley, the itinerate traveler and ethnologist, who after two journeys to West Africa in search of "fish and fetish" emerged in the late 1890s as one of the leading authorities on the continent.[101] She was also one of the most outspoken, objecting to what she called the "missionary party's" caricature of natives as helpless children and savages, and faulting them for trying to turn Africans into "black Englishmen."[102] From her own philosophy of polygenesis and separate development of the races and sexes, Kingsley argued that Africans were an entirely different species who should follow their own separate cultural path of development.[103] She was also highly critical of Chamberlain's brand of expansionary imperialism, which, as she saw it, was responsible for the shift toward greater formal control in West Africa through the establishment of the Crown Colony system. This new interventionist policy, Kingsley argued, was disrupting trade and undermining African legal and cultural systems, leading to rebellions such as the Hut Tax War.

To counter the Colonial Office's schemes, Kingsley put forth an "Alternative Plan" in which a grand council made up of members of the British Chambers of Commerce would form the central decision-making body. The council would appoint a governor general of West Africa, whose duties would include regular six-month visits to the territories under his control.[104] Under the grand council would be two subcouncils; one made up of medical and legal experts and the other of African chiefs. In this way, control would be put in the hands of "experts" rather than "clerks and amateurs." Her efforts to form a traders-in-government scheme were premised on the belief that those who ran Africa must be professionals who had an intimate knowledge of the place, its people, their laws and customs, and not by the ignorant "sons of gentlemen" to be found among the permanent officials at the CO. West African traders were experts in colonial affairs by virtue of their position and practical experience. Their knowledge of and interest in local affairs made them the least likely to break up, as she put it, the "true Negro culture state." She also advocated the use of anthropology as a tool of imperialism, especially investigative fieldwork into native legal and social institutions, arguing that those who would govern Africans must first know them. In many ways, then, Kingsley attempted to bring the interests of the Liverpool merchants in line with the new knowledge of science, especially that of anthropology and ethnology, in a bid to forge an alternative and more "enlightened" colonial policy—one based on an understanding of African culture that would leave its social organization intact.[105] She wanted a synthesis between "old hands" and new methods, between the "man who knew his natives" and the "man who knew his science."[106]

In June 1900, Kingsley died of typhoid fever while nursing Boer prisoners of war in Simonstown, South Africa. She was only thirty-seven years of age. Her agitation against CO "amateurism" and her ideas about the study of contemporary African cultures as part of colonial administration did not die with her, however. In the years before the First World War, "Kingsleyism" spread among Liberal critics and helped define the debate on imperialism that erupted in the wake of the Anglo-Boer War.[107] Her untimely death prompted a close friend, Alice Stopford Green, to take up an earlier idea of Kingsley's of forming an "African Society" that would attract both political and scientific authority. In 1900 the African Society was founded in her memory as a forum for the exchange of knowledge among traders, academics, and officials.[108] The purpose was both to investigate the customs, traditions, and languages of Africans and to facilitate the commercial development of the continent in a manner best fitted to secure the welfare of its inhabitants.[109] Its early membership reflected Kingsley's vision of a synthesis of interests, including as it did West African traders and merchants, prominent Liberal and Radical politicians, and some of the most famous scientists and academics of the day—most especially, of course, anthropologists.[110]

African Society members and others carried forward Kingsley's campaign for a more enlightened colonial policy. More than anyone, it was her protégé, the Liverpool journalist E. D. Morel, who was responsible for transforming her ideas into the new imperial doctrine of "development along native lines."[111] Perhaps most receptive to the "native lines" doctrine were the agents of British merchant capital, such as the Liverpool and Manchester Chambers of Commerce and the British Cotton Growing Association. The "Third Party," as Morel coined that section of the Liverpool merchant community headed by John Holt, advanced the view that the rightful place of the African was as a peasant cultivator and proprietor with security of tenure, producing goods for trade under British rule.[112] Commerce, in conjunction with peasant commodity production and freed from concessionary monopolies and overburdening administrations, was the soundest policy for tropical development.[113] Such a policy would guard against mismanagement and reckless or ill-conceived schemes by argued, the early phase of colonial rule in British tropical Africa, stretching from the beginnings of effective occupation in the 1890s until the outbreak of the First World War, was marked by bold experiment and intervention as the new regimes sought to transform local societies based on the widespread introduction of private property and wage labor relations.[122] Despite intentions, however, the ability of the colonial state to carry through on its proclaimed initiatives was severely constrained. Efforts to secure an adequate supply of cheap wage labor proved insurmountable in the face of a relative scarcity of population and an abundance of land, sparking considerable indigenous resistance and public debate. The colonial state possessed neither the military clout nor the economic resources to impose such changes by force. The inadequacies of colonial power soon forced a move away from ambitious development plans and visions of universal progress based on the imposition of capitalist relations of production. Colonial rule increasingly came to depend on alliances with traditional elites and local chiefs in order to secure political order, collect taxes, guarantee the supply of commodities for the market, and recruit an indigenous labor force. The shift was perhaps most apparent in West Africa, where, in the years just prior to World War One, a complete turnaround took place, toward a policy of peasant commodity production and the upholding wherever possible of all existing native customs with regard to the use and occupation of land.[123] Most crucial of all was the rejection of private property in land as the basis of progress, which would have a determining effect on the course of action not just in West Africa, but throughout the colonies right down to the final years of empire and beyond.[124] From this perspective, as Frederick Cooper observes, "the much celebrated policy of 'indirect rule' in British Africa . . . represented an attempt to make retreat sound like policy."[125] As in India earlier, the colonial state undertook to block the very process of capitalist development and social transformation which it had helped to set in motion in the first place, and yet, in the end, ironically, it was powerless to keep intact even this surrogate solution.[126] The attempt to order the chaos of development ran up against the demands of new social forces and relations of production, most notably the emergence of an African working class and petite bourgeoisie, which by the 1930s were threatening to burst the provisional boundaries of colonial rule itself.[127]

Although Chamberlain's ambitious imperial initiatives met with considerable resistance both from Liberal critics in Britain and from local indigenous societies in the colonies, his linking of the development imperative with the growing importance of trained scientific knowledge and state planning would continue to resonate throughout the Edwardian period and beyond. His importance lies not so much in the actual schemes he helped initiate, as in instilling a new breadth of vision among those concerned with imperial and colonial affairs, and in fixing the idea of development onto the British government's agenda.[128] Although his opponents disagreed with many of the means and aims of his "forward" policy, they accepted its underlying premises: that the peoples of the tropical regions of the world were incapable of running their own affairs; that responsibility therefore had fallen to the British to act as imperial trustees; and that this implied a move toward greater state intervention and control. More pertinently, as we will see in subsequent chapters, the use of technical experts and advisory bodies in planning and implementing a more systematic and scientific approach to colonial rule provided a shared and accepted practice. Even if the more intrusive and costly aspects of Chamberlainite development doctrine proved beyond the prevailing orthodoxies of the day, the imperial government could still demonstrate that it was doing something about the problem of tropical development by enlisting the services of technical advisers and applying indirectly the benefits of science and technology. In many senses, then, science became a seductive, if contradictory and contested, medium through which the political implications and debates surrounding development in Africa and elsewhere in the twentieth century were filtered, neutralized, and sanitized, and through which various actors both at the center and on the periphery attempted to influence and modify colonial policy.

Kew, the Imperial Institute, and the Institutionalization of Tropical Agricultural Science in the West Indies and West Africa, 1895–1914

Chamberlain's image of a vast, undeveloped "imperial estate" was more than mere metaphor. It spoke to the heightened attention given to agriculture in the tropics at the time. The spectacular collapse of coffee planting and the equally impressive rise of the tea industry in Ceylon; the economic decline of the West Indian sugar industry, once the prize of the British Empire; the sudden rise of rubber planting in Ceylon, Malaya, and other tropical countries as well as the success of cocoa in West Africa; the depression in cotton followed by the formation of the British Cotton Growing Association and the extension of this crop to the West Indies and Africa—all of this generated new interest in and awareness of both the possibilities and the challenges of tropical agriculture. "Effective occupation" meant more than laying down hundreds of miles of railway track and pushing feeder roads through near-impenetrable rainforest. It implied—quite literally, as Richard Drayton reminds us—an ecological regime premised on the fundamental principles of sedentary agriculture and husbandry, private property, and, above all, the transformation of these territories into thriving producers of agricultural commodities and other natural resources for the global market.[2] To make good their claim to the tropics, Britain and the other European powers had to develop and improve them, which mandated above all the introduction and extension of export agriculture.

Initially, plantation agriculture remained the preferred model of colonial development, with the trend toward large-scale enterprises appearing all over the world: in Ceylon, Java, Sumatra, Malaya, the West Indies, South America, and parts of tropical Africa.[3] That said, the mixed economic fortunes of the late nineteenth century drew attention to the inherent structural weaknesses of the old tropical plantation system. The prosperity of such colonies rested on an unstable footing: they were highly specialized, open economies that were acutely vulnerable to the unpredictable fluctuations of the world market; they depended on regular supplies of cheap and often highly exploited or indentured labor that was prone to endemic resistance; and they employed uniform and intensive monocultural systems that placed heavy ecological stress on soils and forests, while crops were highly susceptible to the rapid spread of diseases and insect pests. Sooner or later, overproduction, attacks from disease, and competition from other countries would bring about depression and, sometimes, total collapse. The dramatic ruin of coffee planting in Ceylon and the steady decline of the old sugar colonies in the British West Indies were perhaps the most stunning examples of these lessons, and led many to question the wisdom of overspecialization and to look to alternative modes of agricultural production. The vagaries of the planting industry might be offset, it was felt, by encouraging a policy of diversification into new agricultural products, and by promoting as many forms of agriculture as possible, from the largest capitalist organizations down to the smallest and simplest forms of village cultivation. More than this, the 1880s and 1890s marked the beginning of concerted efforts to promote agricultural science as the means of resolving the hitherto intractable problems that haunted the plantation system, as well as making native farming methods and practices more efficient. The rise of experimental agriculture and rational management offered hope that decline was not inevitable if agricultural industry could be run along scientific lines.

Since the eighteenth century, the British Empire had been well served by the collecting and acclimatizing of new plant introductions at the Royal Botanic Gardens at Kew and its network of satellite gardens throughout the world.[4] Under Sir Joseph Banks and subsequent directors, Kew became a great botanical exchange house for the empire, transmitting economic crops and horticultural plants from one hemisphere or continent to another. The zenith of Kew's influence came in the second half of the nineteenth century with its efforts to establish cinchona, rubber, and tea cultivation in South and Southeast Asia. Under the guidance of William Thiselton-Dyer, first as assistant director in 1875 and then as director a decade later, Kew's association with the Colonial Office grew closer, and Kew's focus became less concerned with herbarium and taxonomic work and more with the application of botanical knowledge to practical agriculture. Its system of botanic gardens was substantially expanded and reorganized in the 1880s and 1890s with the creation of a second tier of smaller botanical stations in the West Indies and West Africa that were intended to act as centers of experimental agriculture and extension.[5]

Kew's more active role as a center for scientific and economic advice for empire was, in many ways, a direct response to important changes transforming botany and agricultural science in the late nineteenth century. On the Continent, new knowledge of the principles of chemistry and physics was driving a revolution in experimental plant physiology, pioneered by the work of Julius von Sachs, Wilhelm Pfeffer, Heinrich Anton de Bary, and others, who inspired a whole generation of younger ameliorative measure in which peasant cultivation and estate agriculture would exist side by side to the mutual advantage of both.[13] For the first time, the problem of relatively rapid population growth, and the consequent effects of landlessness and rural discontent, motivated a critical change in Colonial Office attitudes and policy toward native welfare. As we will see in subsequent chapters, it would not be the last. Equally novel was the commission's emphasis on the importance of scientific knowledge and research in aiding both estate and peasant cultivation. Kew's assistant director, Daniel Morris, who was enlisted as the commission's "expert adviser in botanical and agricultural questions," submitted a separate report on the islands' agricultural resources in which he recommended that sugar cultivation be confined to the best lands only, aided by scientific research and mechanical innovation, with the aim of producing the highest quality cane at the lowest cost.[14] Measures were also needed to improve peasant cultivation techniques and to establish extension services and instructional training among the small farmers and peasantry. Such assistance, Morris suggested, was best given by the state through the creation of a special department of economic botany that should be paid for by the Imperial Exchequer. The new imperial department would coordinate and work through the existing regional network of botanic stations to create an effective research and extension service for both large planters and smallholders.

Through Chamberlain's active endorsement of the commission's findings, imperial grants-in-aid were given to establish central sugar factories and improve the region's transportation network. Morris's proposal for an Imperial Department of Agriculture in the West Indies was also approved. The new department began operations in 1898 with its headquarters in Barbados and with Morris recruited as its director and the region's imperial commissioner of agriculture. Under his direction the new imperial department was staffed with an impressive team of experts, and the old botanic gardens and local experimental stations were brought under the fold of the new central organization. Most laboratory work was moved to Bridgetown, and the staff's research findings were published in the department's own scientific journal, the *West Indian Bulletin.* The fact that the department was located in Barbados, the most important sugar-producing island in the British West Indies, and that it needed to garner the patronage and support of the islands' elites in order to justify its continuing survival, meant that from the beginning the interests and outlook of the plantation sector thoroughly permeated the agenda. Morris actively courted the West Indian planting community, holding conferences and promoting the utility of the department's work for the industry. He and his team paid much less attention to raising the standard of peasant cultivation and diversifying production through the development of alternative export crops. Whatever may have been the original intention of the commission, in practice most of the new department's activity revolved around providing useful scientific advice for the West Indian plantation industry, and, above all, acting as the leading center for sugar cane breeding research. In the West Indies, in any case, the marriage of the new botany with tropical agriculture—perhaps the most tangible result of Chamberlain's constructive imperialist vision—proved to be an unmitigated success for colonial policy, but the benefits of the alliance continued to be monopolized by those who dominated the plantation sector. Despite the trepidations of officials in London, estate agriculture remained the preferred model of development in the region until well into the twentieth century.

The founding of the Imperial Department of Agriculture in the West Indies marked the high point of Kew's special relationship with the Colonial Office in the last decades of the nineteenth century. Perhaps in recognition of his service in bringing the knowledge and expertise of Kew to bear on the problem of colonial development, Chamberlain made Thiselton-Dyer the CO's official botanic adviser in 1902. In retrospect, the appointment was largely anticlimactic. With Chamberlain's resignation the following year and Thiselton-Dyer's retirement shortly thereafter, the importance of Kew's scientific empire of botanical gardens and experimental stations rapidly faded. Signs of a rift between Kew and the CO could already be seen in the late 1890s over the question of establishing botanic gardens in West Africa. Many local colonial officials in the region expressed dissatisfaction with Kew's men in the field. When the curator of the Kofu botanical station in Gambia left in 1895 to take up a similar post in Sierra Leone, for example, the governor decided not to fill the vacancy, suggesting that the colony was not ready for a full botanic station under the supervision of a skilled officer.[15] The governor of Lagos, William MacGregor, also believed the colony could not afford to maintain both a model farm, to demonstrate new methods to native farmers, and a botanic station, expressing his preference for the former.[16] Similarly, the high commissioner for Southern Nigeria, Ralph Moor, noted that the status of the botanic station at Old Calabar, which he had not been in favor of establishing in the first place, had become a vexed and controversial question, and recommended it be replaced by a forestry department and, in time, an expert agriculturist capable of determining which plants and soils were likely to give the best economic return.[17] Thiselton-Dyer fired back, defending Kew's reputation and reminding CO officials that a great deal of money and effort had already been put into the establishment of these botanic stations. It would be a matter of great regret to allow them to lapse into disuse—or to be placed under the authority of a forestry officer, given that horticulture and forestry were distinct subjects with different aims.[18] Despite his objections, officials in Lagos and Southern Nigeria moved ahead with their plans, creating new government forestry departments that combined botanical and forestry operations; in Northern Nigeria, despite the lack of timber tracts, a similar scheme was introduced by Frederick Lugard in 1902.[19] It must have been with some frustration and disappointment over what appeared to be mounting official indifference to the achievements of the botanical garden system he had worked so ardently to create that Thiselton-Dyer made the decision to retire as Kew's longtime director in 1905.

But the fading of Kew's brilliance in the last years of Thiselton-Dyer's tenure should not obscure its lasting influence on the organization of tropical agricultural science and its integration into colonial government. The success of the Imperial Department of Agriculture (IDA)—arguably Thiselton-Dyer's and Morris's most innovative achievement—stood as an exemplary model for other colonial administrations to emulate.[20] In 1899, for example, the Jamaican government moved forward with plans for bringing together its various botanical and agricultural operations under a single managing Board of Agriculture, and the following year it hired an agricultural chemist, H. H. Cousins, to head the Central Experimental Station at Hope. In 1908, the board's functions were taken over by a newly amalgamated Department of Agriculture with Cousins promoted as the first director.[21] The same year, perhaps as a means of asserting its independence from imperial control, the Barbados legislature set up its own local Department of Agriculture, and soon other, smaller colonies followed suit. Similar institutional changes were afoot in the Eastern dependencies and West Africa in the early twentieth century. The traditional acclimatization and horticultural work of Peradeniya in Ceylon, for example, was gradually eclipsed in the late nineteenth century by demands for the application of botanical knowledge to practical agriculture, which eventually led to the formation of a new Department of Agriculture and School of Agriculture in 1913.[22] A similar course of action was taken on the sugar island of Mauritius, where the influence of the IDA model was particularly strong.[23] In West Africa, the Gold Coast took the lead, establishing a Botanical and Agricultural Department in 1904 under the direction of W. H. Johnson, the former head curator of Aburi.[24] Meanwhile, in the Nigerian Protectorates it was increasingly recognized that the future welfare of the territories rested on the development of their agricultural resources and that this required rethinking the initial decision to amalgamate botanical and forestry operations under the direction of a forestry department. By 1910, both Northern and Southern Nigeria had resolved to obtain expert agriculturists to head and oversee the creation of separate agricultural departments.[25] Thus, by the onset of the First World War most British colonial territories in Africa as well as in the West Indies and Asia had placed agricultural research and resource development under some form of departmental structure modeled on the example of the IDA. Nor did the influence of the IDA end there, for many of the young experts enlisted and groomed in the West Indies by Morris, and, later, Sir Francis Watts, would go on to become rising stars of British colonial science, working throughout the empire from India to Ceylon, Malaya, and tropical Africa and later serving as some of the most influential advisers to the Colonial Office in the 1930s and 1940s.[26]

But if Kew's authority continued to live on indirectly through the institutional models and personnel that helped shape the new colonial agricultural departments, an obvious shift in metropolitan policy and advisory arrangements nonetheless took place in the first decade of the twentieth century. With Thiselton-Dyer's resignation as director of Kew in 1905, never again would the institution enjoy the kind of access to and influence in government circles that it had in the last quarter of the nineteenth century under his administration. Kew's sudden fall from grace was in part due to the changing political fortunes of Chamberlain and the Tory-dominated Unionist Party, which went down to defeat in the 1905 national election. The new Liberal majority government that swept to power was highly critical of Chamberlain's colonial policies and may have regarded Kew as too closely aligned with his agenda and benefaction. Over the next decade a series of colonial secretaries—Lord Elgin, Lord Crewe, Lewis Harcourt, W. H. Long—sought to carve out their own brand of Liberal "social imperialism." This did not mean abandoning Chamberlain's imperial rhetoric, however: the colonial undersecretaries of state for the new Liberal government, Winston Churchill and his successor, Colonel J. B. Seeley, continued to allude to the image of a "great estate" in their writings and speeches, and to reaffirm the CO's commitment to the systematic and scientific development of the empire's resources.[27] The new colonial secretary, Lord Elgin, following the parting advice of Thiselton-Dyer appointed a permanent commissioner of agriculture for West Africa, similar to the position held by Morris in the West Indies.[28] Elgin, however, did not look to Kew to help find a suitable candidate but instead offered the job to Gerald C. Dudgeon, a former planter from India who had studied tropical agriculture in Egypt and Ceylon. Even more significantly, the new post was based at the Imperial Institute in South Kensington and not affiliated with Kew in any way. It is clear from official correspondence that, from 1905 until the First World War, it was the director of the Imperial Institute, Wyndham Dunstan, who served as the CO's chief source of expert advice on all agricultural and natural resource issues and problems.[29]

This may have been partly symbolic, the new regime's way of signaling that things were going to be done differently. But by the first decade of the twentieth century, as we have seen, many colonial government officials were questioning the competency of curators and botanic officers trained and sent out by Kew and casting doubt on the usefulness of the old botanic stations, which rightly or wrongly were perceived as expensive gardens raising ornamental and exotic plants of doubtful commercial value.[30] At the same time, with the retirement of Thiselton-Dyer and the departure of Morris for the West Indies, Kew lost the vital leadership that had made the institution so prominent in the nineteenth century. Into this vacuum Dunstan entered, promoting the Imperial Institute as the "Kew of Chemistry" and its natural successor as the CO's de facto department of economic botany and development. As he explained in his presidential address to the Chemistry and Agricultural Section of the British Association for the Advancement of Science in 1906, "The days when a botanical garden served the purpose of an entire scientific establishment in a colony have passed away and we now require, in order that a proper return should be obtained, and the natives assisted in their agricultural practice, a scientific department with a proper complement of specially trained officers."[31] The new "scientific" departments of agriculture in the colonies would be assisted by the Imperial Institute in London, which Dunstan hoped would emerge as the central organization for colonial agronomy. The facilities and laboratories of the Institute's Scientific and Technical Department, he suggested, were well equipped for dealing with the pressing problems of tropical development, while its small nucleus of experts might play a vital role collecting samples of potentially valuable economic products to be sent back to South Kensington for detailed examination.[32]

Initially, officials and staff at the CO seemed won over by Dunstan's promise to deliver the practical goods, and consulted him on a number of critical administrative questions.[33] When proposals were aired for the merging of the Institute with the new Imperial College, they were met by protests from the colonial secretary, Lord Elgin, who began negotiations to have the Institute's work placed more closely under the CO's direction. From the middle of 1906 the CO took an active interest in the Institute's operations. This new working relationship was perhaps most clearly seen by the appointment of Gerald Dudgeon as superintendent of agriculture for the British West African Colonies and Protectorates in 1906. Dudgeon acted as a kind of traveling commissioner, touring each of the West African colonies and protectorates, inspecting and assessing their current and future potential for agricultural development, and sending plant and soil samples off to the Institute for examination. At the end of each tour he submitted a report to the CO on the agricultural and forest products of each territory. Dunstan, in turn, reviewed the reports and provided the CO with assessments of the findings and recommendations. Dudgeon made his first tour of the region in 1906, visiting Sierra Leone, the Gambia, the Gold Coast, and Southern and Northern Nigeria.[34] Echoing Dunstan's sentiments, Dudgeon portrayed the region's botanic stations as expensive and unnecessary relics of the past, suggesting instead that the colonies' embryonic agricultural staff and limited resources should be directed to rousing indigenous cultivators to the possibilities of cash crop farming. The new departments of agriculture needed, therefore, to steer their efforts primarily to the improvement of indigenous methods of planting and cultivation through what would later be termed agricultural extension work: demonstration plots, traveling instructors, and agricultural education in local schools.

As Dudgeon's recommendations suggest, the eclipse of Kew by the Imperial Institute reflected more than a partisan change in advisership; it mirrored the shifting contours of colonial development debate. As noted in chapter one, opposition to Chamberlain's "forward" view gained currency among many Radicals and prominent Liberals, as well as many in the merchant community, in the face of indigenous rebellions and scandals in Africa at the turn of the century. The new Liberal ministry was receptive to the views put forth by Chamberlain's critics, like E. D. Morel, who advocated an alternative colonial strategy, one that "would develop stem in part from the opinions expressed by local colonial government officials in West Africa, who came to seriously question the value of the Institute's scientific services. Given its remote location in London and the problem of communication that this entailed, many of the new colonial agricultural departments found it easier to carry out the kind of simple analytical investigations performed by the Institute's Scientific and Technical Department themselves. More crucially, local authorities were having doubts about the advice and competency of the Institute's visiting experts like Dudgeon. For Dudgeon, the fertility and natural abundance of the region was axiomatic; what was needed was a change in government priorities and African character. African farmers, in his view, were inherently conservative, stubbornly clinging to their old customs and apathetic to new ideas and practices. They would have to be induced to adopt a deeper form of cultivation using bullock-pulled plows, rather than heaping or lightly scratching the soil with short hoes, and a more continuous system of rotational farming, rather than replenishing the soil through what he described as the "obviously wasteful" practice of shifting cultivation. Though most government authorities agreed that the adoption of such practices were desirable improvements, they doubted whether local people would ever voluntarily give up their customary practices in favor of the new methods. As one traveling commissioner in the Gambia remarked, "I have repeatedly pointed out to them the great advantages of a plough, but their reply is always the same . . . what was good enough for our grandfathers is good enough for us." Another, who at least had some sense of the difficulties involved, noted "that their fields are, in most places, made in bush, and that the stumps of the trees and bushes would hinder the use of the plough, and that it would take too long to remove these stumps."[41]

By the time he had completed his second regional tour in 1907, questions about Dudgeon's status as a tropical agricultural expert were beginning to be raised. In the Gold Coast, for example, the acting director of agriculture, A. E. Evans, disagreed with Dudgeon and Dunstan's view that the old BCGA plantation at Labolabo was an unsatisfactory site for cotton experimentation. As Evans ironically noted, the area around Labolabo was in fact *the* center of the cotton industry in the colony and was the only district where cotton was likely to be grown, since it was the only district where more profitable crops like cocoa, rubber, and palm could not be grown.[42] Perhaps the most damning critique came from Northern Nigeria, where Dudgeon and Dunstan had expressed such high hopes for the development of a new cotton industry. The governor, Sir Henry Hesketh Bell, submitted these biting observations after touring the area marked out by them for cotton cultivation:

> The representations which have been made in the past concerning the actual production of cotton and the enormous possibilities of the future appear to have been characterized by a great deal of optimism . . . Those who have written and reported concerning the cotton-growing possibilities of Northern Nigeria cannot, I think, have had much experience of the same subject in other tropical parts of the world. They have spoken enthusiastically about the fertility of the soil and the suitability of the natural conditions generally, but I am inclined to believe that their enthusiasm on reaching the Hausa States was considerably engendered by having, on the way up, traversed hundreds of miles of very poor, barren, and almost unpopulated country . . . I can only say that the soil of most of that part of Northern Nigeria through which I have been traveling during the past two and a half months is some of the poorest that I have ever seen under cultivation anywhere . . . the successful cultivation of any valuable tropical products such as cocoa, rubber, coffee, tobacco, etc., north of the Niger and Benue, appears to me hopeless . . . [and] the agricultural outlook in Northern Nigeria certainly does not appear to me a very promising one.[43]

Indeed, by the time Dudgeon had completed his final tour in 1910, Hesketh Bell had come to the conclusion "that very little can be expected from such 'flying visits' . . . we are more in need of actual experiments and instruction in agriculture than of reports and suggestions."[44] Not long afterward, Dudgeon's health deteriorated and he was invalided and subsequently resigned his appointment as superintendent. When the colonial secretary, Lewis Harcourt, queried the West African governors on whether a new officer should be appointed in replacement, all but one responded negatively, arguing that the continuation of the post would be superfluous and impractical. The BCGA protested bitterly. Arthur Hutton, the chairman of the BCGA, reminded the CO of Dudgeon's special connection with the association, stressing that it was imperative for the BCGA to have someone to confer with and to help foster cooperation between the association and the new departments of agriculture.[45] But, in view of the strong opposition expressed by the West African governors, Harcourt decided not to push the issue.

which meant not only large infrastructure and public works projects and the laying out of new agricultural industries but also a coordinated and continuous policy of administration. Development would be possible only when significant numbers of Europeans, both officials and merchants, could live and work in the tropics.

Chamberlain's aim coincided with the emergence of the new germ theory of disease, which helped place tropical medicine on an entirely different footing in the late nineteenth century as a new, metropolitan specialism.[50] Prior to the 1890s, the treatment of and measures to prevent tropical diseases were largely determined by the widespread belief in the miasmatic theory of infection, which argued that diseases arose spontaneously from within the body or from unhealthy environments.[51] In this view, unsanitary and climatic conditions such as filth and garbage, high temperatures and mists, as well as certain geographical locations, were seen as the primary causes of disease, and basic public health measures such as swamp drainage and town sanitation were the prescription. Beginning with the work of Louis Pasteur in the 1860s, however, germ theory paved the way for dramatic discoveries in biological and medical research that allowed for the increasing specialization of medical knowledge in Europe. Important research by Alphonse Laveran, Patrick Manson, Ronald Ross, and others in the 1880s and 1890s laid the groundwork for the development of tropical medicine as a distinct specialization concerned primarily with vector-born parasitic diseases.[52] These discoveries generated confidence that the new biomedical knowledge of the etiology of tropical diseases would result in better control measures, especially personal prophylactic treatments, which would help protect Europeans and permit them to tap the wealth of hitherto impenetrable regions.[53] The health and welfare of native populations were, at best, ancillary issues in these deliberations. The goal was clearly to make the tropics more hospitable for European administrators, traders, and experts whose agency was needed to develop the imperial estate.[54]

In 1897, largely through the intervention of Chamberlain's private secretary, Herbert Read, Britain's leading authority on tropical diseases, Patrick Manson, was obtained as medical adviser to the Colonial Office.[55] With the appointment of Manson, the new knowledge of the germ theory and malaria transmission overlapped with the practical necessities of colonial policy to produce an important paradigmatic shift in the study of tropical disease. Prior to the late 1890s, the CO had done very little to combat the problem of the high mortality rate of Europeans in the tropics, other than contracting the services of outside physicians to examine candidates for colonial appointments and to treat officials on leave.[56] Manson, with the enthusiastic support and administrative clout of Read behind him, transformed his position into that of an influential aide directly advising the secretary of state on colonial medical policy.[57] One of his first actions was to push for the creation of a special center to develop research and training in tropical medicine. In fact, in his inaugural address at St. George's Hospital in 1897, he called for a course of lectures in tropical medicine to be taught at all of the country's main medical schools, and for a new program for the training of colonial medical officers (CMOs) to be opened at the Seamen's Hospital at the Albert Dock.[58] Read, in turn, drafted an important memorandum on the need for such a program, noting that Britain's increasing involvement in Africa, along with the new initiatives in railway and harbor construction, had created a recruitment crisis in the colonial service, and that this was especially true of medical posts, where openings were numerous. It was advanced that prospective CMOs should undertake a three-month intensive course on the essentials of tropical medicine.[59] By the end of 1898, Manson and Read's initiative had led to a proposal for the establishment of a school of tropical medicine at the Albert Dock branch. Despite opposition from the Dreadnought Hospital, the Royal College of Physicians, and the Indian Medical Service Hospital School at Netley, the plan went ahead, with the new London School of Tropical Medicine opening in October 1899.[60]

In July 1898, Chamberlain wrote to Lord Lister, requesting that the Royal Society establish a Malaria Committee, which in turn was asked to organize an expedition in Central Africa to investigate malaria and blackwater fever.[61] Then, in 1901, Chamberlain asked the British government, the government of India, and the Crown Colonies to contribute to a central fund for the study of tropical diseases. To coordinate the collection and allocation of these funds, the Office established an Advisory Board for the Tropical Diseases Research Fund (TDRF) in 1904.[62] Most of the funds raised went to the London School to pay for lecturer's salaries. In addition, another committee, chaired by the parliamentary undersecretary, Lord Onslow, was formed to examine various practical measures for the prevention of malaria. The committees set up to study tropical diseases were a new departure for the CO in that they were composed of men noted for their expertise rather than their social position or political connections, and they made crucial decisions concerning the allocation of funds and the priorities of research.[63]

Six months before the London School even opened its doors in October 1899, a group representing local commercial and merchant interests, headed by the president of the Elder Dempster Shipping Company, Sir Alfred Lewis Jones, established another School of Tropical Medicine in the "second city of empire," Liverpool.[64] The first dean of the new school was the then Holt Professor of Pathology at the University College of Liverpool, Rubert Boyce, and the key lecturer was none other than Ronald Ross, who, disgruntled with the Indian Medical Service and tired of his years in the tropics, took up a Lectureship in Tropical Medicine in April 1899. The formation of the Liverpool School was, as Helen Power notes, the "unlooked for result" of the Colonial Office's own propaganda. Initially, the CO took a lukewarm position toward the rival school, failing to recognize it as on par with London until 1900.[65] Even after this, it continued to rank second, receiving much less financial assistance from the TDRF than London. The CO's response to the Liverpool School was also partly ideological, in so far as the latter expressed a countercurrent of criticism against the new official stand on tropical medicine, one that drew its inspiration from the work of Ross and his supporters, with clear interest in the environmental implications of the new tropical medical research. The Liverpool School's agenda, at least initially, overlapped in important ways with that of its commercial patrons in the Liverpool merchant community, and found support from prominent Radicals, including E. D. Morel, who were eager to invoke scientific opinion in their attack on Chamberlainite development doctrine. The Liverpool School's close association with the city's commercial community and its lack of resources also made for a much smaller organization, closely concerned with medicine and public health in the tropics.[66] In contrast to the school in London, whose primary aim was to train doctors for the West African Medical Staff (and later the Colonial Medical Service), Liverpool eked out an alternative existence, often acting, as Power describes it, as "a sanitary hit squad who would respond to reported outbreaks of disease and rush to various locations to offer their advice."[67] In all, a total of thirty-two expeditions were sent abroad by the school between 1899 and 1914.

From the start, with the school's first expedition to Freetown, Sierra Leone, in August 1899, Ross's ideas about mosquito reduction were received with controversy and scepticism.[68] Ross's research, both in India and Freetown, indicated that only certain species of *Anopheles* mosquitoes, which appeared to breed in comparatively small and isolated pools of stagnant water, could act as intermediary hosts of human malaria. These observations hinted at the possibility of easily and inexpensively eradicating them in some localities, and thus disrupting the transmission of the malarial parasite to its human hosts. Ideally, the most effective preventive measure would be the obliteration of breeding pools of *Anopheles* by drainage, but, failing this, the expedition report recommended oiling pools with kerosene. Admittedly, such operations would be considerably less effective in localities where the land was level or waterlogged or where insects bred in places that could not be easily found, but given the greater precision in attacking pools rather than draining away whole areas infected by malaria, Ross was optimistic that real, cost-effective improvements could be made in the principal centers of population and civilization.[69]

Ross's optimism, however, was not shared by all. Accompanying the official report was a memorandum by the town's health officer, Dr. William Prout, who felt that infection by mosquitoes was only one of the means by which malaria could be contracted. And while he agreed to and had long recommended drainage as the most effective prevention of malaria, he regarded Ross's new methods as mere tinkering.[70] Prout was prepared to accept the new mosquito theory, but he felt it was probable that certain species might transmit the disease through soil or water directly, rather than by biting. As Mary Sutphen remarks about the acceptance of germ theories of plague, the new knowledge was "readily accepted, in part because doctors, colonial officials, and the lay public could easily transpose them onto an older theoretical framework, finding germs in the same places where they had found the causes of other diseases in the past."[71] Significantly, Prout based his challenge not so much on the authority of science as on his long experience of the West Coast. He was the "man on the spot"—and the governor and the CO tended to agree.

The Colonial Office did, however, send Drs. John W. W. Stephens and Samuel R. Christophers of the Royal Society's Malaria Commission to West Africa in late 1899.[72] In Freetown, they concluded that Ross had seriously underestimated the number of breeding places, which were not simply confined to a few small, permanent pools, but also included runoff springs, pools that formed in the stream beds that ran through town, and various other bodies of waste water. Part of the problem stemmed from the fact that Ross had made his observations during the rainy season, when regular downpours acted as a scouring mechanism limiting the number of isolated pools, whereas in the dry season the breeding puddles followed a completely different pattern.[73] Their findings implied that it was practically impossible to effectively eradicate the mosquito population of

Freetown. As part of their investigation, however, Stephens and Christophers claimed to have made a great epidemiological discovery. Confirming the findings of the German bacteriologist Robert Koch, they observed that many African children harbored the malaria parasite, but that, once they had survived past the age of five, they rarely showed any outward signs of the disease and lived comparatively healthy lives. From this they concluded that while the adult population had built up a natural or partial immunity to the disease over time, among African children it was endemic. The children acted, as it were, as a natural reservoir of infection.[74]

Stephens and Christophers also—somewhat dubiously—observed that while "the centre of Freetown, the capital of Sierra Leone, was completely free from Anophelines . . . a quarter to half-a-mile away it was very different . . . in the outskirts we come to the streams, with their countless larvae, in the back-waters and pools, and to the native huts, infested with Anophelines, having therein a plentiful supply of blood. Why then, should they fly abroad and visit the centre of town? In fact they did not, and the result is this important condition, that in the centre of Freetown, a quarter to half a mile from deadly centres of infection, you have a segregated protected area, where the danger of contracting malaria is extremely remote.[75] From these observations they concluded that it was possible to prevent malaria among the European population by separating them from infected Africans, especially children. In practical terms, this meant the segregation of European residential areas from the African quarters of town by a buffer zone of unoccupied land that was at least a quarter of a mile wide, and a ban of all Africans from European areas after nightfall, when, it was said, ninety percent of all mosquito bites occurred.[76]

Offended by the Royal Society's dismissal of his mosquito-reduction theory, Ross set about raising his own Tropical Sanitation Fund to go to Freetown and prove once and for all that his methods were effective. James Coates, a Glasgow businessman, pledged £2,000 with no strings attached for a year's trial to test Ross's plan, and in July 1901 a second expedition arrived in Freetown under the command of Ross and Dr. Logan Taylor.[77] The recent discovery that the genus *Stegomyia* were carriers of yellow fever, as well as the fact that *Culex* mosquitoes were known to carry the *Filaria* worm, prompted Ross to extend the expedition's activity by organizing local recruits into two gangs: the *Culex* gang undertook a massive cleanup of the town, gathering all the refuse and water vessels from the houses and compounds, while the *Anopheles* gang was responsible for emulate. There, a system of sanitary commissioners, reports, and regulations for each presidency ensured that "stagnation in sanitary matters" was not allowed. After much wrangling, however, Chamberlain rejected the plan on the grounds that Ross, who the three Chambers of Commerce had nominated to be the commission's scientific expert, was not a sanitary engineer, and was therefore not qualified to report on the costs of reforms or the means of meeting them.[93]

The problem of sanitation in West Africa was not allowed to rest with Chamberlain's rebuttal. There is evidence that, throughout 1904 and 1905, Ross and William MacGregor kept the idea of a Sanitary Commission alive.[94] With the coming to power of the new Liberal government in December 1905, relations between the Liverpool School and the Colonial Office improved, which may have encouraged the former to renew its efforts for sanitary reform. In 1907 the undersecretary of state for the colonies, Winston Churchill, convinced the Treasury to authorize an increase in the annual grant to the Liverpool School from £500 to £1,000, and to double its own contribution to the TDRF to £1,000.[95] It seems also that Ross was on closer terms with Liberal MPs like Colonel John B. Seeley, Churchill's successor, who managed to obtain a seat for Ross on the CO's Advisory Committee of the TDRF in 1908.[96] In 1906 another deputation was sent to the Colonial Office, but once again the proposal was "put by."

With the spread of bubonic plague to the Gold Coast in January 1908, however, alarms over unsanitary conditions were once again heard. On the advice of Manson, the CO invited Professor William Simpson to investigate the outbreak.[97] Quite apart from the antiplague measures, Simpson was also instructed to visit some of the major towns and centers in Sierra Leone, Gold Coast, and Southern Nigeria to investigate their sanitary conditions and report on the effectiveness of the existing medical and sanitary services.[98] In his report, Simpson noted that the practical application of Manson's and Ross's findings had not progressed in the region to any great extent. It was necessary to address such unsanitary conditions in the towns as poor surface drainage and water supplies, unnecessary overcrowding, poorly constructed and ventilated housing, numerous breeding pools, and rank vegetation close to houses. Conditions that affected the local population would in turn affect Europeans unless the two could be separated. He also suggested that the health of Africans no less than Europeans was important for the development of the colonies, and that systematic and continuous preventive measures in the interest of public health would have to be undertaken to protect the filling in, channeling, draining, sweeping, and oiling all the pools and puddles.[78] Operations wound up by the autumn of 1902, and the results were far from reassuring. Critics charged that the experiment had done nothing to reduce the occurrence of malarial fever. Ross and Taylor fired back, accusing Freetown authorities of moving too slowly in grading the streets and refusing to follow up once the expedition had left.[79] The Colonial Office, for its part, failed to mention mosquito extermination in its official pamphlet on the prevention of malaria, concluding instead that personal protection through the use of mosquito curtains, gauzes, and nets, along with residential segregation, were the best practical ways to protect the health of Europeans in the tropics.[80]

Ross did, however, find crucial support from one important West African official: Sir William MacGregor, the governor of Lagos, who maintained, contrary to Koch's theory of acquired immunity, that African adults also suffered from malaria.[81] MacGregor was also highly critical of the Royal Society's recommendation of residential segregation, noting that "From the administrative point of view it is an unacceptable doctrine . . . The policy followed in Lagos in this as in other matters is to take the native along with the European on the way leading to improvement. Here they cannot live apart nor work apart, and they should not try to do so. Separation would mean that little, or at least less, would be done for the native, and the admitted source of infection would remain perennial. To simply protect the European from fever here would never make Lagos the great commercial port that it should become."[82] In Lagos, MacGregor's principal medical officer, Henry Strachan, devised a multi-pronged approach to attack the disease at its source among the African population—distributing free quinine to all the inhabitants of the Lagos area, teaching elementary sanitation in government-aided schools, and disseminating general sanitary information through the Lagos Board of Health and through village lecture tours carried out by African medical officers using the local vernaculars.[83]

Support was also forthcoming from other influential sources, including Professor William J. R. Simpson, chair of hygiene at King's College, London, who from 1904 onward became a staunch advocate of Ross's measures for mosquito reduction and sanitary reform. Simpson had been among the pioneers who, together with Patrick Manson and James Cantlie, helped establish the London School of Tropical Medicine, where he lectured on tropical hygiene for nearly twenty-five years. Simpson, moreover, had spent ten years in India as the first medical officer of health for

Calcutta, where he gained a controversial reputation as a leading advocate of sanitary reform and germ theories of disease.[84] It was largely through Simpson, whose expertise as a special consultant was called upon by the CO no less than eight times between 1900 and 1924, that the new knowledge of tropical hygiene was widely diffused throughout the empire, from Hong Kong to Singapore, to Calcutta, to sub-Saharan and Southern Africa.[85] At every turn, Simpson pushed for the creation of state-directed sanitary services in the tropical colonies.[86] He and Ross naturally gravitated to one another, becoming close friends and allies.[87] In many ways, in fact, it was Simpson who carried on the fight for sanitary reform which the increasingly embittered and resigned Ross had begun. It was Simpson, with his unrivalled reputation and experience as "the most distinguished tropical sanitarian in the British Empire," who became the Colonial Office's principal sanitary adviser in the early twentieth century.[88]

At the 1907 meeting of the British Medical Association, Simpson suggested that the progress which had been made since Ross's discovery had rendered the prevention of malaria a practicable reality, provided that the necessary machinery and organization for such an undertaking were made available.[89] He pointed to the success achieved by Dr. Malcolm Watson, the district surgeon of Klang in the Federated Malay States, who with generous government backing embarked in 1901 on a bush-clearing and land-draining campaign aimed at freeing the town and neighboring Port Swettenham of a recent and serious epidemic of malaria, with the result that by 1903 cases of fever had dropped to a tenth of their former number.[90] Similar action was needed elsewhere, and, to this end, Simpson proposed the formation of a Colonial Sanitary Service, which would be supervised by a sanitary commissioner who would be responsible for preparing annual reports. This organization should be supplemented by a Colonial Office advisory body, which would read the commissioner's reports and advise the government on health matters in the colonies.[91] In proposing such an organization, Simpson was picking up where Ross and the Liverpool School had left off after their first expedition to West Africa. In March 1901 a delegation of the Liverpool, Manchester, and London Chambers of Commerce and the Liverpool School met with Chamberlain to propose the establishment of a central sanitary authority for all of British West Africa.[92] Under the proposed plan, a sanitary commissioner was to be appointed who would, with the assistance of a sanitary engineer, travel from port to port to examine the sanitary conditions of each, and then report directly back to the CO. India was cited as the model to inhabitants, colonizer and colonized alike, against the ravages of disease. In view of this, Simpson recommended the formation of a special sanitary organization for the West African colonies, emphasizing the need for a special health department in each colony whose members would be specially trained and whose whole time would be devoted to public health duties. While the health branch would remain part of the West African Medical Service, its functions should be separate and distinct. The major towns and centers would have special medical officers of health (MOHs); in all the other stations it would be the task of the regular medical officers to carry out health duties under the guidance of the health department. Simpson wanted the health department in each colony to be under the command of a sanitary commissioner who would have complete executive control over the health organization, although he would remain subordinate to the principal medical officer (PMO). Finally, in order to coordinate the work of the whole West African service and to act as a link with the Colonial Office, he recommended the appointment of a regional inspector-general.

The influence of Simpson's *Report on Sanitary Matters in Various West African Colonies,* published in June 1909, was more extensive than might have been expected, for his investigation coincided with a general reassessment of the conditions and organization of the West African medical services. Despite the advent of new training facilities at Liverpool and London and the amalgamation of the West African Medical Staff (WAMS) into a unified service in 1901, there continued to be bottlenecks and shortages of trained CMOs, as well as frequent complaints from the field of low and unfair salary scales, unreasonable workloads, and a lack of opportunities for promotion.[99] Throughout the early 1900s, there remained no standard policy concerning promotion, salary scales, and pensions. Thus, in 1908, a Departmental Committee on the West African Medical Staff was appointed to inquire into recruitment and conditions. As part of its terms of reference it was asked to consider whether, as Simpson proposed, distinctions should be made between the medical and sanitary duties of staff, and, more importantly, whether a separate branch of the WAMS should be established to deal with sanitation.[100] In the event, the committee decided against a sharp division of responsibility.[101] The problem of sanitary administration, it suggested, could best be solved by the appointment of a limited number of senior officers of staff who would devote themselves entirely to sanitary matters. It was seen as imperative that a unified service be maintained, with sanitary officers (SOs) included as members of the medical department under the control of the principal medical officer.[102] The committee also rejected for the moment the idea of a director-general of the West African Medical Staff, proposing instead the preliminary step of holding annual conferences of principal medical and senior sanitary officers for each colony.

Perhaps the most crucial of the committee's recommendations, in hindsight, was for the creation of a central advisory committee on medical and sanitary questions for all of tropical Africa. It felt such a body would enable the Colonial Office to maintain closer relations with the medical profession and science, thus making it easier to obtain recruits for the Colonial Medical Service by forming a link between the CO and other agencies, such as the Schools of Tropical Medicine and the University Appointment Boards. It would also advise on the selection of candidates for the service and the filling of vacancies by promotion in the senior ranks of staff. More importantly, it would comprise a central body capable of offering expert advice on questions connected with tropical medicine and hygiene, which, the committee noted, "[a]t the present time, the Colonial Office appears somewhat at a loss for."[103] Acting on the committee's findings, the colonial secretary, Lord Crewe, set up the Advisory Medical and Sanitary Committee for Tropical Africa (AMSCTA) in November 1909, with Sir Patrick Manson serving as chairman.[104]

The creation of a colonial sanitary branch and the advent of the AMSCTA were the culmination of nearly a decade of drawn-out debate. But their appearance did not signal the triumph of sanitary and public health reform in the African colonies. For many local administrators the new sanitary order was anything but a welcome adjunct to the main task of securing power and control in the newly annexed tropical colonies and protectorates. The gap between the rhetoric of reform and reality remained very wide indeed. Colonial medical and sanitary policy continued to be riddled with personal tensions, structural limitations, and ideological ambiguities, which often reflected competing agendas and visions of how the colonies should be managed and developed. The unwillingness of colonial authorities to move forward with plans to clean up the tropical slums of Africa and extend health care systems was clearly dictated by the constraints of a penny-pinching empire, which prided itself on colonial financial self-sufficiency with minimal cost to British taxpayers. The lackluster performance of colonial medical practice was more than a question of scarce resources, however. The sweeping sanitary reforms and schemes of metropolitan experts and medical health officers were held in check by local administrators, who feared such invasive measures might provoke popular discontent or alienate Native Authorities and local rulers, whose cooperation was essential to the state's legitimacy and power.

Nothing illustrates this point better than the policy of residential "health segregation," which, as noted earlier, was first proposed by Stephens and Christophers at the turn of the century. The policy was officially endorsed by the AMSCTA in 1910, and over the next decade, largely at the urging of Professor Simpson, the committee spent more time on the issue of urban segregation and town planning than any other. In West Africa, for example, Simpson regarded separate European residential and business districts as the first important step toward an effective sanitary order. It was, according to Simpson, the overcrowded and deplorable sanitary condition of the "native towns" that was at the root of such epidemic diseases as the plague, which in due course had a detrimental effect on the health of Europeans. Beyond separation, however, the next and more essential move was to control infectious diseases among the African population[105] Simpson would later draw similar conclusions in East Africa, where the opening up of the Protectorate had given rise to the spread of epidemic diseases such as sleeping sickness and cerebrospinal fever, leading to the even more serious problem of depopulation.[106] There could be improvement only through the creation of a special health service with full and effective powers to direct sanitary administration. In the segregated European areas, a significant measure of protection against malaria and other mosquito-borne diseases could be achieved through mosquito-proofing houses and individual precautionary measures such as bed netting and quinine. In the African towns, however, things would have to be done differently, as the generally unsanitary state of housing there would never be remedied by leaving it to "the primitive ideas of a primitive people."[107] Sanitary authorities required the power to do away with the congested areas that characterized every "native town:" pulling down unsanitary houses, demolishing and remodeling overcrowded sections of town, and enforcing good building laws. In West Africa, Simpson sought to gain control over the coastal ports and the new and fast-growing trading and mining centers of the interior, alarmed as he was at how they were chaotically bringing together all "sorts and conditions" of people with the potential for introducing infections and epidemic diseases. He was particularly critical of new towns like Sekondi on the Gold Coast, which was connected to the interior by a railway that ran through such important gold mining centers as Tarkwa and Obuassi before reaching the inland terminus of Kumasi, the capital of Ashanti. If an epidemic broke out in a place such as Sekondi, it could easily spread along the transportation and communication networks of the colony, crippling production and trade throughout the Gold Coast. The impulse to civilize was no doubt strong in Simpson, but it was the economic imperatives of colonial development that were stressed in his reports.

Simpson's warnings proved prophetic when an outbreak of yellow fever swept through Sekondi and much of the rest of British West Africa in 1910. Sir Rubert Boyce, dean of the Liverpool School of Tropical Medicine, was sent to investigate the epidemic. His investigation was controversial, marking a turning point in the study of yellow fever in West Africa. Based on an extensive *Stegomyia* survey and an examination of the hospital casebooks and annual medical and sanitary reports for the various colonies, Boyce determined that there had been four unconnected outbreaks of yellow fever in 1910, and frequent previous outbreaks along the coast. Noting that the most common mosquito of the coast towns was the *Stegomyia calopus,* Boyce conjectured that yellow fever had often been mistaken in the past for a variety of malarial-type fevers by medical officers who betrayed, or so he had found, ignorance of the disease, its prevalence, and its treatment. According to Boyce, the evidence clearly pointed to an almost continuous occurrence of yellow fever dating back to at least the 1890s. Boyce held that the disease was, similar to malaria, endemic among the "dense native population," surviving in a mild but chronic form which became manifest only when there were sufficient numbers of newly arrived "non-immunes" to cause an epidemic.[108]

Not everyone found Sir Rubert's logic convincing. His report in fact sparked heated debate among other members of the AMSCTA, and, even after protracted discussions throughout the autumn of 1910, the committee was unable to come to an agreement and decided to abstain from pronouncing on the question of endemicity until further expert examination could be carried out. It did, however, endorse the measures advocated by Boyce for the prevention of yellow fever, attaching great importance to European residential segregation, as well as to the destruction of *Stegomyia* and the isolation of infected persons.[109] At Manson's request, the new colonial secretary, Lewis Harcourt, sent a circular to all West African governors, drawing their attention to Boyce's opinion that yellow fever was endemic among the indigenous peoples of the coast and asking all medical officers to report any suspicious fever cases.[110] In his despatch, Harcourt was unequivocal: "Emphasis has often been laid upon the desirability of ensuring separation of the dwelling places, and more especially the sleeping places, of Europeans in West Africa from those of natives; and there can be no doubt that one of the most important deductions from the history of the recent outbreak is the enforcement of this policy."[111]

The reaction to the CO's position on segregation in West Africa was generally one of cautious acceptance, but without any firm commitment in concrete terms. In the Gold Coast, where the yellow fever epidemic had hit hardest, the incoming governor, J. J. Thorburn, was much impressed by the senior sanitary officer's opinion that yellow fever was in all probability endemic to the West Coast. So much so, that he recommended a substantial increase in expenditure for preventive sanitary measures.[112] But as to the proposed creation of a segregated European residential area in Sekondi, he chose not to take any definite action until the railway terminus had been completed and a tangible plan for the utilization of the area had been approved.[113] In the Gambia, Governor George Denton felt that the proposed construction of a cantonment at Cape Saint Mary for European officers, and the laying of a tramline connecting it to Bathurst some seven miles away, was both too costly and unwarranted given the declining rate of sickness among Europeans in the town.[114] His senior medical officer, Dr. T. Hood, stressed that, given the present shortage of sanitary staff and money, it would be infeasible to carry out even the most routine antimosquito measures, let alone large segregation projects beyond the principal towns of Bathurst and McCarthy Island.

In Southern Nigeria, the principal medical officer, Henry Strachan, who as noted earlier was a longtime opponent of segregation, persuaded Governor Walter Egerton that Boyce's theory on the endemicity of yellow fever was inconclusive. In a memorandum on Boyce's report, Strachan agreed that yellow fever was endemic in some parts of West Africa, but he refused to accept that it was endemic everywhere, and went so far as to say that the great majority of the African population of Lagos could not be immune since there had been no recognized epidemic of the disease for quite some time.[115] Strachan's rejection of the theory was motivated in part by his fear that it would divert attention away from what he considered to be the most pressing health problem: deaths from malaria among African infants and children. It was also an implicit rejection of the theory's main conclusion, that European segregation was the only practical prophylactic measure against infectious disease in West Africa. Strachan urged authorities instead to continue to attack the mosquito through extermination campaigns and the reclamation of swamplands.

Even in Northern Nigeria, where the new senior sanitary officer, Dr. Cameron Blair, was a strong advocate of the new policy, key officials such as Charles Lewis Temple and H. S. Goldsmith cautioned against too rigid an enforcement of the principle. Describing it as "a question of such vital importance that . . . it ought to be finally settled now," Dr. Blair pressed for a new General Standing Order to be drawn up making residential segregation compulsory. In the name of fairness, he insisted that the same principle apply to the "African non-native"; that is, African traders from the coast, whom the British viewed as aliens among the predominantly Muslim population of the north.[116] Acting Governor Charles Temple supported the idea of European segregation and foresaw no problem in enforcing such a principle in any new stations, but he disapproved of any drastic alteration in the established stations, and he refused to accept any regulation that would apply the same criterion to the "African non-native," citing the serious discontent among the coastal traders and the disruption to commercial development that such a measure would cause.[117] The governor, Sir Hesketh Bell, agreed that in the interests of general health each race should live in separate quarters, but he considered such measures impracticable.[118] He was not prepared to enforce by legislation the rigorous separation of European residences from those of natives, nor to enact any laws that would prevent African merchants and artisans from occupying sites in the commercial quarter of town. On the issue of compulsion, he thought his senior sanitary officer was impelled by his enthusiasm for reform to advise a "counsel of perfection."[119] So, too, it seems, did the AMSCTA, which, after weighing matters, was not prepared to go against the opinions of Temple and Bell.[120]

In fact, the only attempt to make European segregation compulsory by law was that of Sir Frederick Lugard when he became governor of an amalgamated Nigeria. In the event, however, his 1917 Townships Ordinance was never really implemented, largely because subsequent governors rejected it as one of the excesses of Lugardian Indirect Rule. His immediate successor, Sir Hugh Clifford, argued that it accorded special status to the townships, which, to his mind, along with the exclusion of immigrants from the jurisdiction of the Native Authorities, was a protectionist policy intended to slow the rate of progress.[121] This, he felt, would act as a barrier to the growth of trade and commercial enterprise in the Northern Provinces by restricting the freedom and intercommunication of different groups.[122] Clifford became the toughest critic of health segregation since Governor MacGregor of Lagos. While not rejecting the principle outright, he refused to permit any plan requiring the forcible removal of Africans from their places of residence. As governor of the Gold Coast, he declined to carry out the recommendations of Boyce and the AMSCTA to remove the village of Essekado at Sekondi in order to make way for the unfettered expansion of the European residential area.[123] He was also highly critical of the disproportionate amount of resources and attention spent on European health in the tropics. In an important confidential despatch of 1916, Clifford explained that

> the energies and activities of the Medical Department in [the Gold Coast] and in its Dependencies are today too exclusively preoccupied by the care of Europeans . . . that the benefits of modern medical science are not rendered available, in existing circumstances, to the bulk of the indigenous population, who are also the vast majority of the taxpayers; and that the Principal Medical Officer and his advisers should be invited to devote to this aspect of the West African medical question far closer attention than has hitherto been given to it, with a view to the preparation of a scheme, for adoption at some early future date . . . It should, I think, be our aim ultimately to establish a complete system of rural hospitals, etc., similar to that which today exists in Ceylon.[124]

If the British intended to live up to their imperial responsibilities, he went on to suggest, then it was time that they thought about ways of providing medical facilities and resources to as many Africans as possible. He wanted to see the training of a large African support staff in order to extend medical and sanitary services into the countryside through the creation of rural dispensaries and health clinics.

Thus, throughout the West African colonies and protectorates, the official policy of "health segregation" promulgated by metropolitan experts like Simpson and Boyce was received with caution, and, in some cases, outright disregard. Meanwhile, the CO's support for the "segregation principle" continued to be expressed in memoranda issued to various colonies by the Advisory Medical and Sanitary Committee until the early 1920s.[125] Despite ongoing official pronouncements of the sort, however, local commitment to the policy remained equivocal. Such evasion by local colonial administrators was not arbitrary. It was dictated by local realities, including the stringency of colonial financial systems and resistance from local communities, most notably that of European and African merchants who refused to cooperate with what they perceived as a draconian vision of life in the tropics. It was simply bad business for European merchants to live apart from the African quarters of town. Most of the large merchant trading companies required their local agents to sleep on the premises of their trading depots, which were generally located in or near bazaars, in order to be close to African middlemen bringing goods and products down from the interior and to guard against possible thefts and property damage.[126] Moreover, most of the older West African towns and ports contained sizeable communities of African merchants and professionals, who were vehemently opposed to schemes apparently designed solely for the benefit of whites, and who rightly feared that "health segregation" was part of a larger trend that would see the setting up of color bars in the professional and civil services, including the West African Medical Staff.[127] In the face of local recalcitrance, the grand designs of sanitary reformers and outside experts proved superfluous and unworkable.

The period from 1895 to 1914 marked the initial phase of the British Colonial Office's engagement with scientific assistance and expertise in the name of tropical development. Beginning with Joseph Chamberlain's imperial estates doctrine and continuing with the social-imperialist visions of succeeding Liberal ministries, officials and staff at the CO for the first time actively enlisted scientific experts and appointed technical advisers and advisory committees, in the hopes that, by way of the new botany and the germ theory of disease, they might gain the means of mastering tropical nature. The CO drew upon the external services of existing organizations like Kew Gardens and the Imperial Institute as never before, and took the lead in establishing or encouraging new ones, most notably the schools of tropical medicine. At the same time, with the assistance of such metropolitan research and training institutions, the old Victorian system of botanical gardens was gradually supplanted by a series of colonial agricultural departments, while colonial medical services were enlarged and reorganized to include new sanitary operations. A fledgling colonial agricultural service was built up, and, although its growth was sporadic and uncoordinated, by 1914 agricultural officers and scientists could be found in the employ of almost all the colonial territories. Even greater progress was made in colonial medicine: a unified West African Medical Staff was established in 1901, and a similar service for East Africa in 1915. It seemed that, with the new alliance of science and colonial government, the fabled riches of Britain's great tropical "estate" might at last be effectively developed.

Still, as this chapter reveals, success was more impressive on paper than in practice. In the case of both tropical agricultural and tropical medical science, the initial enthusiasm of Chamberlain and senior CO officials had, by 1914, become more muted and cautious. The Imperial Institute's attempt to focus and coordinate colonial agricultural policy and provide central leadership for the new colonial agricultural departments proved woefully ineffective. The newly appointed superintendent of agriculture for West Africa was unceremoniously terminated within a few short years, as were plans for a tropical agricultural advisory committee in London. The discipline of colonial medicine, and related medical and sanitary services, were accepted as more advantageous, however. Schools of tropical medicine and a permanent advisory medical and sanitary committee were established that would last in one form or another right through to the end of empire in the 1960s. Yet, even here, as we have seen, investigative reports and policy directives emanating from the committee and its expert members were often ignored and even rejected by authorities on the ground.

The aborted efforts of metropolitan institutions to centralize and shape the colonial agenda instilled some important lessons. It became clear that the complexities and difficulties of developing the tropics could not be adequately orchestrated by outside experts who made brief, periodic visits, or by a group of prominent scientists sitting in London reading reports and memoranda sent back from the colonies. Trained scientific officers and research practitioners were needed to work in the field, as part of the colonial technical services, intimately aware and knowledgeable of local conditions and problems. Only in this way, and only after a long period of experimentation and sustained effort, might the promise of employing science for tropical development finally be fulfilled.

Ultimately, however, the shortcomings of the initial phase of scientific colonialism were causally linked to the wider fissures of colonial practice in Africa and elsewhere. As noted in chapter one, the ambitious experiments and sweeping plans for intervention that characterized the early colonial state were quickly dashed in the face of limited bureaucratic power and continuing indigenous resistance. The colonial state simply did not have the military backing or the financial resources to impose its vision by will. The inadequacies of colonial power would force officials to abandon their earlier schemes for rapid agrarian transformation or new sanitary orders, and to seek greater accommodation with indigenous authorities in order to secure social stability and an adequate supply of the factors of production, especially labor. In this, the British were not alone. As recent studies have shown, the "forward-looking" policies of other leading European powers at the time, particularly the grand designs for railroad construction and public health measures initiated by Ernest Roume, the governor general of French West Africa, also met with great difficulties and local confrontations that would slow progress and prompt important changes in colonial ideology.[128] These early setbacks and reversals offered officials a premonition of the shape of things to come.

Nevertheless, the decades leading up to the First World War did set an important precedent. In the interest of developing the imperial estate, new tropical sciences were inaugurated and institutionalized. The appointment of technical advisers and advisory committees by the Colonial Office became an accepted practice. New technical departments, their services increasingly in demand, were established and, gradually if haphazardly, augmented. And, in their own right, the scientific and institutional models and patterns established before 1914, as we will see in subsequent chapters, would lay the groundwork for further waves of expansion and consolidation of tropical agricultural and medical expertise in the years between the wars.

to immediate shortages of critical intermediate inputs and chemical materials for British industry.[5] Many of the input shortages occurred in relatively new fields of industry that were based on the application of new knowledge in the physical sciences, especially chemistry. In response, the government set up an Advisory Council on Scientific Research to help increase the supply of trained research workers and to provide grants for industrial research. As the war drew to a close, the council's work was placed on a permanent footing with the creation of the Department of Scientific and Industrial Research (DSIR).[6]

At the same time, the large-scale mobilization of domestic and imperial resources in order to supply and arm Britain's military forces placed unprecedented demands on the state for greater social and economic intervention.[7] State bureaucracy was transformed by the incorporation of technical experts and consultants from business, trade unions, the professions, and academia.[8] State power was extended through the creation of new government departments, expansion of state production and services, and imposition of far-reaching controls on prices, finances, supplies, and labor.[9] In 1917 a separate Ministry of Reconstruction was established in preparation for the postwar rebuilding of the economy, and a Committee on the Machinery of Government under Lord Haldane was appointed, which recommended a more extensive use of advisory bodies and experts in the work of government.[10] This was echoed by Balfour's Committee for the Coordination of Research in Government Departments, established in 1920 to organize periodic joint conferences between all departments engaged in scientific and economic research.[11] Thus the "triumph of the expert" had its metropolitan equivalent, as the war and later the Depression spawned a new realization of the need for greater integration of government and expertise, and a growing enthusiasm in the interwar years for central planning as the basis for social and economic progress.[12]

Significantly, many of the advocates of the new government science policy were also staunch supporters of imperial development. With the resignation of Asquith as prime minister and his replacement by Lloyd George in 1916, a reconstructed and streamlined Imperial War Cabinet replaced the coalition government, bringing to the center of power and policy-making a circle of political actors fired by the ideals of imperial unity and keen to use state enterprise and financial assistance to foster the development of national industries and resources, scientific and technical research, and tariff reform.[13] The change in regime generated a good deal of support for imperial preferences and development in many quarters outside of government as well. This new turn toward empire was based on the belief that free trade had left Britain vulnerable to German control of important supplies of commodities and metals, and that in the interest of national security the empire must strive to be more autarkic. At the Imperial War Conference in 1917, for instance, resolutions were passed supporting the principles of imperial tariff preferences, empire economic development, and empire settlement.[14] It was out of this confluence of empire and science, as Michael Worboys notes, that a renewed "science for development" movement emerged in Britain in the 1920s. In the hopes that science and expertise would "unlock" the wealth of the empire's "undeveloped estates," supporters of the movement worked ardently to set the Colonial Office on a more rational and systematic footing. One of their more notable achievements, as this chapter details, was to revive the idea of a central organization in London capable of directing and guiding the colonial agricultural research and technical services.

Milner, Amery, and the "Imperial Estates" Redux

Those who dreamed of rationalizing the empire celebrated Lloyd George's appointment of Viscount Milner as colonial secretary in January 1919 as a victory in the struggle for ascendancy.[15] His appointment raised the possibility of connecting the new science policy with a renewed program of "constructive imperialism" and empire development. Once again the strategic engagement of science, development, and expertise initiated by Chamberlain was brought to the fore, as the CO became receptive to pressure from the scientific lobby for more research and expert involvement in "tropical development." Shortly after taking office, Milner gave his support to a proposal for a £100,000 grant for research, including economic inquiries and surveys, to develop the natural resources and trade of the colonies. The proposed grant was conceived by CO staff as a scaled-down version of the DSIR's £1 million grant for the promotion of research in the United Kingdom.[16] Surprisingly, the Treasury gave the scheme the green light, and the CO promptly set up a Colonial Research Committee under the chairmanship of the Oxford geographer and ardent tariff reformer, Sir Halford Mackinder. But in the wake of postwar austerity measures, the committee's grant was reduced to a paltry £2,000 per annum in 1922 and was eventually cut completely. After 1922 it did little more than rubber-stamp existing projects.[17]

of new civilizations we are able to assist in building upon the sure foundations of their knowledge."[27] Ormsby-Gore made similar pronouncements. As parliamentary undersecretary of state under Amery, he was the first CO minister to travel widely to the colonies, issuing his findings in a number of detailed reports.[28] He possessed an unbridled faith in the problem-solving potential of science, especially of research on eradicable tropical diseases, agricultural pests, and soil problems, and though he was never very explicit as to how the results of such research were to be applied, he felt certain that achieving a critical mass of exploratory knowledge of colonial conditions was the first step in overcoming the biological and ecological barriers to tropical development.[29]

The commission reserved some of its strongest criticism for the derelict condition of the East Africa Agricultural Research Station (EAARS) at Amani, Tanganyika, a research station of international reputation which had been founded by the Germans in 1902. With the transfer of the territory to Britain after the war, however, Amani's operating budget was cut back from £20,000 to £2,000 a year. As a result, most of the station's research work on soil conditions, irrigation methods, and plant physiology had to be abandoned.[30] In light of the imperial government's dismal record on colonial research, the commission felt it was essential that greater encouragement and better pay be given to scientific officers in order to attract highly trained and qualified men to the colonial technical services. It also recommended greater central coordination through the strengthening of the Colonial Research Committee, which it hoped might serve as a coordinating link and liaison body between the colonies, the universities and training colleges, and tropical research institutions in Britain, "but also formulate a definite policy and general programme for research, and, where necessary, make proposals and give advice to the Colonial Office regarding both."[31]

The publication of the East Africa Commission's report generated a great deal of interest and enthusiasm both in the British scientific press and at Whitehall for a more systematic approach to colonial problems.[32] Backed up in part by the commission's findings, Amery embarked on a major overhaul of imperial administrative machinery. Even before accepting the post of colonial secretary, Amery had made clear his intention of forming a new and entirely separate office to deal with the Dominions, which he did in 1925.[33] This left Amery's hand free to reorganize colonial administration more systematically.[34] The separating off that Amery initiated of the temperate "white dominions" from the predominantly "tropical dependencies," and his reorganization of the CO along more functional lines, marked the start of an important shift in thinking about the role of metropolitan planning and initiative in colonial development—one that would culminate with the Colonial Development and Welfare Acts of the 1940s.[35] Traditionally, the bulk of the CO's work was handled by the geographical departments, with each department operating as a self-contained unit and with very little coordination or exchange of ideas or personnel across departments. Officials in one colony or region were often completely unaware of the conditions or problems faced by authorities elsewhere. CO administrative staff, as we have seen, tended to frown upon uniform metropolitan solutions, stressing instead the principle of noninterference in local colonial administration whenever possible, and arguing that the diversity and specificity of local conditions made general, "one size fits all" policies inappropriate. Given the financial stringency that most colonies faced, however, the emphasis on local autonomy could also serve as a deterrent to local initiatives. More than this, it was becoming evident through pan-colonial forums like the 1927 Colonial Office Conference that there were many problems surrounding development and trusteeship that all or most territories shared.[36] While support for the geographic division of work remained strong among CO permanent staff, the 1920s and 1930s witnessed a dramatic increase in subject specialization, especially in the social and economic fields. One of the more tangible results of the greater coordination of colonial operations at the center was a steady expansion of the CO's existing advisory network through the appointment of a series of specialist subject advisers and advisory bodies.[37] At the same time, the General Department was substantially restructured into a division with several new subject departments. Providing support staff and serving as a liaison to coordinate the activities of the geographical departments, on the one hand, and establishing an expanding network of advisory bodies and experts, on the other, would turn out to be key tasks of the new specialist subject departments created in the interwar period.[38] As the next section explains, tropical agricultural research and expertise were, above all, the areas that colonial ministers like Amery and Ormsby-Gore considered to be most in need of support.

A Colonial Agricultural Research Service? Tropical Agricultural Science and the Origins of the Colonial Advisory Council on Agriculture and Animal Health (CAC)

World War One left many colonial administrations gutted of European officers, who either joined the war effort in Europe or enlisted in locally based regiments to fight in the colonial campaigns.[39] After the war, colonial agricultural departments as well as the other technical services were faced with serious shortages of qualified trained personnel. As conditions in Europe returned to normal and demand for colonial products increased, the need for new recruits and trained staff heightened. The chronic shortage of qualified personnel in the colonies after the war, particularly in the scientific departments, prompted Milner to appoint a set of committees in 1920 to look into the problem of recruitment and staffing in the colonial agricultural, veterinary, and medical services. All three bodies recommended expansion and salary increases on the grounds that a large influx of staff was a necessary prerequisite for further imperial consolidation.[40]

Priority was given to the expansion and reorganization of the colonial agricultural service, including research and training, since colonial development was viewed primarily in terms of the growth of tropical agriculture along scientific lines. In 1922 the Imperial Department of Agriculture in the West Indies was restructured and relocated to Trinidad to form the Imperial College of Tropical Agriculture (ICTA), with the goal of producing a professional cadre of agriculturists with sufficient knowledge and training to be stationed anywhere in the tropics.[41] Despite the creation of the ICTA, however, shortages of qualified candidates continued, prompting the head of the CO's Patronage Department, Ralph Furse, to request the formation of another committee under the chairmanship of Lord Milner to work out the details for an agricultural probationers' scheme. The Milner committee viewed its mandate in light of the debate over imperial development and cooperation, noting that future staff would need to deal with new problems such as the introduction of more stable and intensive systems of agriculture and the dissemination of knowledge of improved methods of cultivation among the indigenous and other local populations.[42] On the committee's urging, the Colonial Office introduced a colonial agricultural scholarship scheme in 1925 requiring all agricultural probationers to take two years of postgraduate training, the first year of which was spent at the Cambridge School of Agriculture studying crop husbandry with an emphasis on statistical analysis of field experiments and plant breeding, before going on to the ICTA, where they received a general introduction to the practice and economics of tropical crop production.[43] Thus, the syllabus and courses of the ICTA's postgraduate training program were designed to provide a tropical complement to the Cambridge course, furnishing tropical illustrations of fundamental principles learned at the home institution along with participation in field experimentation.[44] The ICTA, in turn, emerged as the most powerful institutional influence affecting British tropical agriculture, having admitted over fifteen hundred students by the time its doors closed in 1961, half of them with postgraduate qualifications, destined largely for careers in the colonial agricultural service.[45] The "Trinidad Mafia," as N. W. Simmonds referred to them, was enormously influential in the development of tropical agricultural research: virtually every agricultural officer posted to the British colonies from 1924 until independence was trained at Trinidad, and by the postwar years the majority of senior positions in colonial agriculture departments were occupied by ICTA grads.[46]

Beyond providing adequate training for agricultural candidates there was the broader problem of how to improve conditions in the field. In light of the findings of the East Africa Commission, Amery decided to extend the terms of the original Milner committee to examine the general efficiency of agricultural administration and research in the colonies. With the death of Milner in 1925, however, Lord Lovat took over as chair of the committee, which prepared an interim report for the Imperial Conference of 1926.[47] The report was steeped in the "science for development" ideology, arguing that the non-self-governing dependencies possessed large areas of highly fertile land which in most cases supported only a relatively small indigenous population. Given the increasing industrialization of the Dominions and India, the dependencies, if developed wisely, were likely to become the most important markets and sources of raw materials for British manufactures and industries. Science, particularly agricultural science, it was suggested, might hold the key to the whole field of development, and this required fully equipped agricultural departments with well-trained staff, supported by an organized system of research and facilities for the collection and dissemination of information.[48] To this end, the Lovat Committee felt the time had come for the provision of an efficient and well-equipped Colonial Agricultural Research Service, including the establishment of a chain of central research stations of which the Imperial College of Tropical Agriculture and the East Africa Agricultural Research Station at Amani, Tanganyika, were to be the first links. The service was to be centrally administered in London by a Colonial Agricultural Research Council with its own secretariat housed at the Imperial Institute. It also recommended the appointment of an agricultural adviser to the secretary of state, who would act as the council's chairman. The council would maintain close links with the colonial agricultural departments and central research stations, as well as with various government bodies and scientific institutions in Britain such as the Committee on Civil Research, the Empire Marketing Board, the Imperial Economic Committee, the Rothamsted Experimental Station at Harpenden and the Royal Botanic Gardens at Kew. The report also suggested that the work of the central advisory body in London might be widened to include veterinary science and forestry research and to extend to parts of the empire other than the colonies. It was thus indicative of larger ambitions at the time to create a pan-imperial scientific community which, it was hoped, might serve as a model of cooperation for constitutional and economic relations within the empire.[49] Amery and his staff agreed with the committee's recommendations, except for one crucial detail: the agricultural adviser, his staff, and the advisory council, it was felt, should be headquartered at the Colonial Office, and not the Imperial Institute, in order to bring it in closer touch with the agricultural departments overseas.[50]

It was with these aspirations in mind that Amery called the first Colonial Office Conference in May 1927. The primary object in bringing the governments together, according to Amery, was to raise awareness of the colonial empire as a separate entity and to foster closer cooperation and unity in its administrative and technical services. Most critically, Amery wanted to generate greater interest in research, and, in particular, to persuade the governments to contribute to a common pool for the financing of an agricultural research service.[51] In his opening address to the conference, Amery made his pitch.

> The importance of scientific research and organization is being recognized increasingly every year in this country and I think the general movement of thought is one in which we are certainly not being left behind in this Office or, I think in the Colonial Empire generally. There is a real consciousness of the fact that we have immense undeveloped resources which science, and science alone, can bring to rapid development . . . I am sure that the case for pooling resources sufficiently to create some sort of unified service, at any rate in the higher research grades of scientific and technical work, is one to which we ought to give the fullest and the most earnest consideration at this Conference . . . I am so convinced of the importance of this subject that I believe that it would be a pity if we separated without finding at any rate some solution of that problem which would enable us to create a more effective instrument for the scientific development of our almost unlimited resources.[52]

The conference considered a number of schemes for the reorganization of the colonial research services, including a proposal by the director of medical and sanitary services in Kenya, John Gilks, for an Imperial Science Service for the whole empire.[53] In the end, this was judged to be too "stupendous," with delegates favoring the original Lovat Committee proposal for a Colonial Agricultural Research Service instead. A further committee under the chairmanship of Lovat was formed to draw up a practical plan for submission to the various colonial governments.[54]

The Colonial Agricultural Service Organization Committee issued its report in March 1928.[55] The new committee cautioned against the creation of a separate research service without the inclusion of general agricultural officers, proposing instead a unified colonial agricultural service to be divided into "specialist" and "agricultural" wings. This they believed would help ease the tension and distance that existed between specialist research officers, who tended to be centrally located, and the more numerous general agricultural officers at the district level. The Organization Committee also endorsed plans for an empire-wide chain of central research stations.[56] Most importantly, it advocated the appointment of a chief agricultural adviser and the creation of a joint colonial advisory council with two committees, one for agriculture and the other for animal health, which would oversee the establishment of a unified colonial agricultural service and the network of central research stations. The total cost of the scheme, including the service, the central council, and one central research station was estimated at £127,000 per year, of which the Empire Marketing Board agreed to contribute roughly one fifth. The remaining expenses were to be paid out of a central fund financed by an assessment of one quarter of a percent on the annual revenues of each colony.[57]

In submitting the proposal for approval to the various colonies and protectorates, Amery asked for their cooperation, stressing that the success of the scheme rested on their readiness to accept the financial liabilities set out in the report.[58] The response was mixed, revealing the real differences and cleavages within the "dependent" empire despite Amery's attempt to forge a single, distinct identity. The scheme was accepted wholeheartedly by the African dependencies, but ran into trouble in the West Indies.[59] Officials in Barbados, the Windwards, and Trinidad were apprehensive about having to contribute both to the new plan and the Imperial College of Tropical Agriculture, which they felt was unfair given that the college benefited the whole empire and not just the region. Both no specific terms of reference, but was given the power to co-opt additional members, to be appointed initially for three years.[70]

In his opening statement as chairman, Ormsby-Gore outlined the many aims of the new council. Its most immediate task would be to review the general proposals for the unification of the colonial agricultural services in light of the criticisms submitted by the various colonial governments. In addition, it was to work closely with the Imperial Bureaus of Entomology and Mycology and the new Imperial Agricultural Bureaux in order to ensure a greater dissemination and interchange of scientific information. It would also need to work closely with the EMB and its Research Grants Committee in screening funding applications for colonial agricultural research projects. There would have to be periodic reviews of agricultural research work and development in the colonies, and a further general review of the proposed chain of central research stations. Ormsby-Gore finished by reminding members of the "science for development" mission, concluding "that the future prosperity not only of the Colonies themselves but also of Great Britain depends very largely on the increasing application of the results of scientific research to the problems of agricultural development. Though there are still large gaps in our knowledge which we must endeavor to find the best means of filling, there is a vast amount of experience not merely in the British Empire but in other Colonial Empires, and in the Universities and research centers of the whole world, which can be rendered more available to producers and workers if only that experience was more generally appreciated and widely known."[71]

Despite such optimism, Ormsby-Gore's expectations for the CAC were only partially fulfilled in the years following his and Amery's departure from the CO in June 1929. The council's most immediate task was to review the recent proposal for the unification of the colonial agricultural service in light of the criticisms submitted by the various colonial governments. In the face of continuing opposition from Malaya and apprehension from Ceylon and the West Indies, it was clear that the larger proposal would have to be abandoned. A subcommittee of the CAC suggested a partial unification in which only certain colonies, perhaps confined to the African dependencies, would initially take part.[72] The venture would be modest, providing funds for a small central pool of specialists and perhaps also for officers stationed in smaller colonies. The subcommittee's report was considered together with the findings of the interdepartmental Colonial Services Committee, chaired by Sir Warren Fisher, at the second Colonial Office Conference, held in June 1930. The conference concurred with the Fisher Committee, concluding that a single colonial service and the unification of each of its special branches were imperative in order to improve recruitment and efficiency. Over the next decade, the gradual "unification" of the different branches of the colonial civil services began, starting with the administrative service in 1932.[73] However, in light of the Depression and the associated retrenchments of staff and salaries in many colonies, full implementation of the Lovat scheme for the colonial agricultural service was ruled out. The idea of a common funding pool to finance the service was abandoned, and there was no attempt to divide the service into administrative and specialist wings or to introduce a standardized system of classification and grading. The unification of the veterinary and agricultural services in 1935 was, in the end, more of a symbolic gesture than an actuality.[74]

Unification was not the only casualty of the Depression. The broader design for greater coordination of "imperial science" through the formation of new mechanisms and central organization in London began to dissolve in the early 1930s as well. Not only were colonial governments and legislatures weary of greater central control over local officers and revenues, but they also found themselves without the financial resources to contribute in any substantial way to a common pool. In Britain, government retrenchment in response to the Depression drastically reduced support for organizations funding scientific research in the name of empire development. Most notably, the Empire Marketing Board was disbanded in March 1933, and, as a result, expansionary plans for an empire-wide chain of central research stations came to a grinding halt.[75] Even the future of the EAARS at Amani was in doubt in the 1930s, as staff layoffs and budget cuts threatened its survival. The turn of events had a significant impact on the actions of the CAC. In the sharply altered conditions of the early 1930s, the council's role was reduced to one of giving advice on agricultural problems referred to it by the colonial governments—a far cry from the earlier blueprints of the "science for development" zealots.

Despite the setbacks, the council managed to carve out a niche as a colonial research advisory body, responding on an ad hoc basis to proposals and problems as they arose rather than in any planned and programmatic manner. For the first few years, the CAC worked closely with its patron, the EMB, whose interest in funding the body was premised on the need for expert advice in vetting grant applications from the colonies.

in small experimental plots.[86] A broad foundation in plant genetics and physiology, they believed, would provide the basis for logically planned and executed hybridization programs that would produce new, higher yield, disease resistant crop varieties.[87]

The strong orientation toward investigation of fundamental principles, rigorous field experiment methods, and statistical analysis at Cambridge was particular significant, given, as noted above, the special role this institution played in the development of research and training at the ICTA. It is also worth noting that the CAC's early membership included some of the most prominent members of the British scientific establishment, including Sir E. John Russell, the director of Rothamsted, and T. B. Wood and Frank Engledow from the Cambridge School of Agriculture. Not surprisingly, the logic and organization of imperial science was also patterned on the principle of pure science and fundamental research. The primary focus of tropical agricultural science was on extensive investigations and surveys into primary export crops, to be conducted largely by the proposed network of central research stations. The stations were to concentrate on problems requiring more prolonged research than could be carried out by local agricultural department staff, or which affected more than one colony within the empire. They were also to be organized on a regional basis, with each center focusing on the primary export crops of that area; sugar, cocoa, citrus fruits, and Sea Island cotton in the West Indies; coffee, sisal, and cloves in East Africa; palm oil, cocoa, and rice in West Africa; and so on.[88]

The ICTA's program, for instance, was modeled in close consultation with Engledow, who was commissioned in 1929 by the EMB to write a special report on the future needs of the college as a teaching and research institution.[89] Engledow advocated a similar approach to that of Cambridge, where the School of Agriculture was virtually synonymous with the Animal Nutrition and Plant Breeding Institutes, and where staff of the institutes also served as teaching faculty at the school. Following Engledow's visit, greater emphasis was placed on the college's contributions as a center for long-range research into problems connected with the region's main cash crops. The Botanical Department, for example, concentrated on the genetics and cytology of sugarcane, cocoa, and bananas with the aim of discovering the methods of selection, propagation, and crossing that produced the best yields, quality, and uniformity; the Department of Chemistry and Soil Science under Professor Frederick Hardy developed an important program in crop ecology, soil profiles, and pedology.[90] Postgraduate training was also systematized, with students specializing in three fields in addition to presenting a piece of original investigative work.

Next to the ICTA, the East Africa Agricultural Research Station at Amani was the most significant regional agricultural research center developed between the wars. The EAARS received particularly close attention from CAC members, many of whom considered Amani to be a model for future research stations. Following recommendations put forward by the Lovat Committee and the 1927 Imperial Agricultural Research Conference, the EAARS was reestablished after years of neglect as a center for basic, long-range agricultural research for East Africa in 1927. The station's new director, William Nowell, and his staff were to operate independently of local departments of agriculture, supplementing their work by conducting investigations of the broadest and most fundamental type, such as the surveying and classification of soils, climate, fauna, and vegetation; the introduction, breeding, and distribution of plant varieties; and the incidence and distribution of crop pests and plant diseases.[91] The proposed emphasis on basic research did not, however, impress local authorities and legislative councils. After its first annual report for 1928–29, for example, the station was heavily criticized by local colonial governments for putting too much stress on the study of underlying principles, and not undertaking work on problems of more immediate practical importance, such as soil erosion. Colonial authorities in East Africa felt the station needed to demonstrate its usefulness to contributing governments. Investigators at Amani, on the other hand, argued that it was necessary to build up a "groundwork of knowledge" based on prolonged studies before they could tackle the more direct questions to do with land use or soil and plant improvement.

The wrangling over what sort of problems researchers should concentrate their efforts on was referred back to the CAC, and Nowell was invited to London to discuss the matter in February 1930. The debate prompted one CAC member, William Furse, the director of the Imperial Institute, to write a strongly worded letter defending the principle of fundamental research at Amani. Despite serious reservations expressed by Furse and others, the council concluded that in order to secure the full support of the East African dependencies, a certain amount of work on practical problems had to be undertaken. To promote greater cooperation between scientists and policymakers, the CAC recommended that a Conference of East African Directors of Agriculture be held at Amani in exchange of information throughout the colonies. The CAC cooperated, for example, with the Imperial Agricultural Bureaux (IAB), which were formed in 1928 as clearinghouses for the collection, collation, and exchange of information among scientific researchers working throughout the wider empire. Each bureau was attached to one of the major existing research institutes in Britain. The Imperial Bureau of Soil Science (IBSS), for example, was housed at Harpenden under the supervision of Rothamsted's director, Sir John Russell. As head of the IBSS, Russell stressed the value of soil surveying and mapping as a preliminary to all resource development planning, and resisted the pressure on soil scientists to produce practical results while they had insufficient knowledge of the fundamental nature of the soil. When in 1931 the council was asked to consider how investigational work on the problems of lateritic soils in Sierra Leone might be continued and extended, it turned to Russell and the IBSS to prepare the scheme. The bureau proposed that a team of soil chemists be sent out to conduct a West African soil survey. The survey, it was estimated, would take three years to complete at a cost of £14,100. The EMB agreed to cover the costs of the coordinator, but the salaries and expenses of the four chemists would have to be paid for by the West African governments. In the face of financial constraints caused by the Depression, however, all four colonial governments rejected the plan. The scheme was shelved—but, arising out of Russell's concern for soil classification, a much more ambitious proposal would emerge. In 1935 the British Association for the Advancement of Science appointed a committee to work in conjunction with the IBSS and the Royal Geographical Society to assemble a soil map of the entire British Empire.[104]

Probably the most crucial way the CAC helped facilitate greater cooperation and exchange of scientific knowledge was through the periodic visits made by Stockdale to the different regions of the colonial empire in order to meet with local agricultural officers. Between 1929 and 1937 Stockdale made eight tours, visiting West Africa in 1929, South and East Africa in 1930–31, the West Indies in 1932–33, Jamaica in 1933, Malta and Cyprus in 1934, Palestine and Trans-Jordan in 1935, West Africa in 1935–36, and, finally, East Africa in 1937. Often, Stockdale's visits coincided with regional agricultural conferences, which were instrumental in hammering out common research priorities and disseminating new ideas and information. Stockdale played a pivotal role at these conferences, often acting as chair. After each tour he prepared a full report outlining his recommendations, which was then circulated and debated at council meetings.

The purpose of Stockdale's first tour, for example, was primarily to attend the Second West African Agricultural Conference, which met in Accra in 1929 to discuss issues directly related to the cocoa and palm oil industries. During the conference, Stockdale drew attention to the dearth of research being conducted on declining yields in the Gold Coast cocoa belts.[105] There was forceful criticism of the Agricultural Department for not being in touch with the agricultural practices and problems of the colony, and for having no data on which to base its extension and propaganda work. In particular, Stockdale and the council decried the fact that a colony which had millions of acres of land under cocoa cultivation, and which had enjoyed such phenomenal prosperity in the earlier part of the century, had taken no steps to safeguard the industry's future.[106] Stockdale felt collaboration was needed with scientific workers in other cocoa producing regions, especially Nigeria and Trinidad, where thought had already been given to the future.

Investigations on cocoa began at the ICTA in 1930 with the setting up of a five-year Cocoa Research Scheme, funded by the cocoa producing colonies, including the Gold Coast, and chocolate manufacturers in Great Britain. The main focus of the research was on increasing yields and on rehabilitating the cocoa industry in Trinidad, which was facing a serious threat from the spread of "witch broom" disease.[107] One of the findings that appeared to be of primary importance for the younger, West African industry was that cocoa plantations seemed to reach their peak in production at about twenty-five to thirty years, after which they steadily declined unless rejuvenation schemes were implemented. Rejuvenation involved replacing old, disused, and nonbearing pickets with special high-yielding types, or simply cutting down mature trees and replacing them with new ones. This research seemed to indicate that the time was fast approaching when rejuvenation of cocoa fields in the Gold Coast would be necessary.[108] By the time Stockdale returned to West Africa in 1935, he was convinced the cocoa industry was under threat. His report commended the recent reorganization of the Agricultural Department, but stressed that experimental investigation would have to form the basis of the new system.[109] Stockdale felt immediate steps were needed to set up a number of well-equipped and -staffed experiment stations and at least one special cocoa research station to carry out investigations on new and high-yielding varieties and on methods of disease treatment. The new varieties and improved methods could then be made available to farmers through concerted extension and propaganda work before the industry started to decline and the welfare of the colony was put in serious danger. The government finally responded to the agricultural adviser's pressure by founding the Cocoa Research Station at Tafo in 1937, which would later form the nucleus for the West African Cocoa Research Institute (WACRI) after the war.[110]

Stockdale's regional tours also generated support for pan-colonial scientific collaboration and coordination. Following suggestions made at the 1929 West African Agricultural Conference, it was decided to hold periodic conferences of all colonial directors of agriculture, the first of which was organized by the CAC at the Colonial Office in 1931.[111] The purpose of the conference was to compile a review of colonial agricultural development and research in preparation for the second Imperial Agricultural Research Conference, to be held in Australia in 1932. Among the resolutions passed were one favoring the continuation of the Colonial Agricultural Scholarship Scheme; a plan for Kew Gardens to compile information on the introduction of economic crops and varieties; an endorsement of the principle of agricultural "cooperation" as a means of improving the inspection, grading, financing, and marketing of smallholders' produce; and an expression of support for the introduction of animal husbandry into agriculture through mixed farming systems. In addition to the conferences of colonial directors, the CAC also sent delegates to represent the colonies at such imperial and international scientific fora as the British Commonwealth Scientific Conference in 1936 and the United Nations Conference on Food and Agriculture in Hot Springs, Virginia, in 1943.[112] In a very real sense, then, Stockdale and the CAC played an important part in the growing globalization and authority of colonial scientific knowledge in the 1930s. Their support for regional and colonial scientific networks helped to facilitate the exchange and synthesizing of ideas, which, as will be seen in chapter five, became linked with the new development and environmental agendas of the late 1930s and 1940s.

In his memoirs, Leopold Amery proudly detailed his contribution as colonial secretary to the field of research, declaring, "there is no part of my work during those years to which I look back with greater satisfaction."[113] That the promotion of scientific research was the one bright star in the sky for the constructive imperialists of the 1920s was, in many ways, a reflection of how little they managed to achieve in the face of continuing obstinacy from the Treasury and resistance from critics who opposed a full-blown program of imperial development at the expense of local rights and interests. Colonial governments remained reluctant to adopt a more interventionist approach based on direct state enterprise, while private capital lacked the financial resources and access to labor to transform the tropical "estates" along capitalist lines. Against this background, to embrace a policy of scientific and technical assistance was a compelling alternative, one that gave the appearance that something was being done.[114] Emphasis was placed on the ability of science to supply the requisite knowledge of new economic resources and colonial conditions, and to eradicate tropical diseases and control agricultural pests. With the rejection of tariff reform for a second time in 1923, moreover, the debate over imperial policy shifted to different fields of battle, such as empire settlement and "non-tariff preference."[115] The promotion of imperial science and research was seen as an attractive and relatively cost-effective means of forging closer intra-imperial economic and political cooperation.[116]

Yet, as the evidence presented here highlights, the extension and coordination of colonial agricultural research and technical services was an uphill climb, punctuated with recurrent setbacks and fiery debate. Indeed, one may legitimately wonder why it took so long for scientific colonialism, with all its perceived promise, to be put to full effect. Beneath the facade of steady organizational progress, colonial science policy, like the wider debate over development itself, was riddled with rivalries, opposing agendas, and structural constraints that limited and deflected the imperial government's modernizing project. The political rationale for using colonial development projects, including research, to deal with Britain's declining economic position and chronic unemployment woes was substantially undercut by the acceptance of imperial preferences at the 1931 Ottawa Conference and the resulting Import Duties Act of 1932. The most notable causality, as mentioned earlier, was the disbanding of the Empire Marketing Board in 1933. In addition, as the case of agricultural research makes clear, the push for greater imperial coordination and central direction, through the formation of new mechanisms and organization in London, was challenged and constrained by the growing autonomy of "colonial science" on the peripheries. In the 1930s and 1940s, talk of imperial coordination gradually gave way to the notion of imperial, and even international, scientific cooperation and exchange of information.[117]

This, however, does not fully explain the enigmatic character of the alliance between science and colonialism in the early twentieth century. Although colonial agricultural and medical research and practice were touted as important "tools of empire" and domination, their application often posed difficult questions and intractable problems regarding the nature and legitimacy of colonial power. This contradiction needs to be stressed, not only for what it tells us about the tensions of empire between the wars but also because it foreshadows many of the dilemmas that would later plague and ultimately cripple the more extensive postwar colonial initiatives.

Amery's vision of imperial science was effectively challenged in two important respects. First, as the next chapter examines, his more expansive ambitions were held in check by counter-development agendas, which gathered strength after the First World War. Colonial reformers and philanthropic groups, bolstered by mounting international criticism and concern for the principle of imperial trusteeship, began to shift and redirect the terms of debate away from its Chamberlainite roots toward an emphasis on what was termed the "human side" of development. Less attention was given to the material exploitation of the "tropical estate," and more to the problems of health, education, and social welfare.

This overlapped with and fed into a second challenge to the Chamberlainite doctrine, one that came from the frontline trenches of colonial rule. As we have seen, local administrators were beginning to fathom that the introduction of new cultivation methods and practices into local agricultural systems inevitably entailed upsetting the prevailing economic and social conditions of the colonies, thus incurring the risk of unleashing unforeseen and unwanted consequences. Similarly, metropolitan visions of a new medical and sanitary order, whether it was "health segregation" or the extension of rural health services, were also received with skepticism, not only because of the costs involved but also because such schemes demanded a level of state power and ability to coerce that simply did not exist. All of this made clear to officials both in the colonies and at the CO that developing the economic and human potential of their tropical empire was not going to be a quick and easy mission; it would have to be carried out methodically over the long term. The need for greater knowledge and exchange of information about actual conditions was recognized. But as the fruits of this new research began to amass, critical questions were raised about many of the preconceptions inherent in earlier imperial constructions, contributing, in time, to serious challenges to and an ultimate reworking of the colonial development mission. By the late 1930s, as chapter five explains, the meaning of development was being redefined, brought on not so much by "high policy" decisions at the center as by the momentum of events in the colonies themselves.

CHAPTER 4

The "Human Side" of Development

Trusteeship and the Turn to "Native" Health and Education, 1918–35

ALFRED MILNER AND LEOPOLD AMERY may have renewed hopes within government and scientific circles for a more rational approach to colonial development in the years following the First World War, but as with Chamberlain before them, their vision of colonial progress ignited heated controversy among liberal reformers and humanitarian critics who saw instead a recipe for continued exploitation of tropical lands and peoples. In the interwar milieu such criticism could not be ignored. The setting up of a Permanent Mandates Commission under the authority of the newly formed League of Nations in 1919 marked an important milestone in international relations.[1] Under the mandate system, Britain and France were awarded administrative responsibility for the lion's share of the former German colonies and territorial remains of the Ottoman Empire, but they were also held accountable to the commission for their administrative practices in these territories. This meant, for the first time, that the actions of the European colonial powers became subject to international scrutiny and would be measured against the principle of trusteeship now the movement of Europeans into non-European areas threatened to displace or eliminate the indigenous peoples. Such rhetoric enjoyed great popularity in the 1920s among those who, like the Round Table Group, favored the further expansion of white settlement in the highlands of East and Central Africa.[27] As earlier advocates had done before them, they were defending the rights of white settlers to occupy and make use of the land in the name of development. But this was more than simply a rhetorical device to legitimize white colonization. In many ways, the discourse of emptiness and underpopulation spoke to the fundamental problem faced by almost every regime in the early decades of colonial rule in Africa and elsewhere, of gaining control over indigenous labor in a context in which the possibility of resistance or migration by potential workers was greatly enhanced by the resilience of precolonial social organization and continuing access to land as an alternative source of livelihood.

What made these concerns even more pressing, however, was a perceived reproductive crisis among the indigenous populations of many colonial territories. Severe population declines had, for example, followed Western imperial expansion in the Pacific and other territories in the late nineteenth century, while in the interwar years colonial reformers, missionaries, and health experts became increasingly alarmed over the high mortality rates and signs of depopulation in many parts of sub-Saharan Africa.[28] Concern about high infant mortality rates and stagnating population levels in the wake of the influenza epidemic of 1918 prompted governments in West Africa to open the first maternity homes and child welfare clinics. The Gold Coast under Guggisberg led the way with the creation of dispensaries and government centers for child welfare staffed by women medical officers, while in Sierra Leone the government launched a campaign to promote child welfare through the registration and supervision of midwives.[29] The official and semiofficial views tended either to blame the decline on the apparent lassitude, ignorance, and backwardness of indigenous peoples and "tribal" customs, or else to see it as a consequence of cultural degeneration or a "*malaise* of the stock" under the onslaught of European occupation.[30] This concern regarding degeneration and sterility was particularly strong in colonial medical and public health circles in the 1920s and 1930s, which lobbied for the extension of mother and child welfare services and for the introduction of women's training programs in mothercraft, dietetics, and domestic science in an effort to check high maternal and infant mortality rates.[31] Indeed, in the 1920s and 1930s, disease was defined as one of the greatest obstacles to the development of the productive forces of the tropics, and its cure or prevention through increased medical and public health intervention was seen as potentially the strongest demonstration of the utility of science to economic imperialism, and, at the same time, of the benefits of Western civilization.

As the attention of medical and sanitary experts in London was increasingly drawn to the problems of health and demographic decline, policy priorities emanating from the CO's medical committees were substantially revised. Interest in "sanitary segregation" steadily faded in the 1920s, while the need for maternity homes, child welfare clinics, midwifery supervision, sanitary education in the schools and through public lectures, health treatment programs, and rural dispensary systems was pushed to the fore. This did not, however, result in any coherent or concerted effort by the imperial government or colonial states to promote health among indigenous peoples, any more than the earlier reforms succeeded in making sanitary segregation anything more than an official dream. Throughout the interwar period, as metropolitan advisers complained time and again, colonial medical policy and practice remained sporadic and piecemeal, lacking any long-term systematic planning or direction.[32] Colonial health resources remained concentrated in the towns and hubs of colonial capitalism, spreading only fitfully to the rural areas and to those groups marginal to colonial enterprises. Provision of rural health care services remained largely dependent on the voluntary efforts of missionary societies, while local governments continued to subordinate public health and sanitation expenditures to other, more pressing needs. Colonial governors consistently argued that they lacked the necessary funds and qualified staff to do more, but the unwillingness of authorities to extend health care systems more broadly was also a result of their reluctance to interfere too heavy-handedly in indigenous social customs, as well as the refusal of colonial medical professionals to admit indigenous medical practitioners into the service on a more equal footing. More rigorous plans to promote rural African health and welfare would have to wait until the late 1930s and 1940s, when mounting material crises and social upheaval on the colonial periphery would prompt officials and experts in Whitehall to reconsider the gulf between the rhetoric and reality of colonial medical and public health efforts. It is important to note, however, that the turn in the interwar years toward the problem of native health was, as the next section explains, part of a wider shift in imperial rhetoric calling attention to the "moral and material advancement" of colonial peoples.

reserves. In an effort to turn criticism to more "constructive" ends, Oldham prepared a confidential "Memorandum on Native Affairs in East Africa" for the parliamentary undersecretary of state, Edward Wood, in which he maintained that "A policy which leaves the native population no future except as workers on European estates cannot be reconciled with Trusteeship. Nor can it, in the long run, conduce to the economic prosperity of the East African Protectorates. The chief wealth of these territories is the people, and, on a long view, the cardinal aim of policy must be to maintain tribal life, to encourage the growth of population by combating disease and promoting sanitation and hygiene, and to develop by education the industry and intelligence of the population."[43] Oldham insisted that only if sufficient land was guaranteed for Africans in the reserves could it be assured that those who left to work for Europeans did so by their own will as free labor.

In response to its critics, the colonial government issued a statement on future development policy in Kenya, announcing in May of 1922 its commitment to what came to be known as the "dual" policy of parallel development of both the European settlement areas and the African reserves. This meant, on paper anyway, that the colonial government was obligated to increase expenditures on social services and on programs aimed at stimulating African agricultural production in the reserves. This was followed by a 1923 white paper which declared that Kenya was primarily an African territory in which "the interests of the African natives must be paramount and that if, and when, those interests and the interests of the immigrant races should conflict, the former should prevail."[44] Far from being an altruistic statement, however, the Devonshire Declaration, as it was known, was intended mainly as a way of circumventing the Indian Crisis, which had erupted in 1922 over demands for the equal treatment of South Asians in the colony.[45] The declaration nevertheless highlighted the rhetoric of trusteeship as never before, placing additional pressure on the imperial government to show it was making efforts to promote the rights and welfare of indigenous peoples. In the midst of the crisis, Parliamentary Undersecretary of State William Ormsby-Gore affirmed in the Commons that the first duty of the imperial government was to the "moral and material advancement of the native population."[46] Just what he meant by this statement in practice was not entirely clear, but what was certain was that the issue could no longer be ignored.

Over the course of the next few years Oldham would seize the opportunity, thinking the problem through more thoroughly while at the same time pushing forward his plans for a more "constructive" native policy. Oldham drew on the ideas of Dr. Thomas Jesse Jones, the American expert on "negro" education in the southern states and well-known propagandist of the Hampton-Tuskegee doctrine of educational adaptation.[47] As educational director of the American philanthropic foundation, the Phelps-Stokes Fund, Jones headed the highly influential African Educational Commission to West, South, and Equatorial Africa in 1920–21, and, together with Oldham, was instrumental in arranging a second Phelps-Stokes commission to visit East Africa in January 1924 in the wake of the Devonshire Declaration.[48] Jones's educational theory was rooted in a sociology of race differences, both physical and mental, that implied a special type of education was necessary for non-Europeans. As such, Jones scorned literary education as inappropriate. Instead, he pressed for the adaptation of education to the needs of the rural black community, which he defined as sanitation, health training, improved housing, and increased industrial and agricultural skills. In pressing for a program of rural betterment, Jones consciously avoided controversial views that might be construed as challenging the supremacy of southern whites, such as the demand for black voting rights. Jones saw the same "Simples of Education" as even more appropriate for Africa. As he told guests at an honorary dinner given by the British government in 1925, "The simple structure of the African village offers an easier problem for educational research. Here needs are almost acute and insistent in the demand for correction and guidance. Even casual observation reveals four conditions that are so obviously essential to community welfare as to merit the rank of 'fundamentals in education.' To avoid the appearance either of pedantry or scientific abstraction, these necessities of sound community life are called the 'simples' of education. They are, first, sanitation and health; second, agriculture and simple industry; third, the decencies and safeties of the home; and fourth, healthful recreations."[49] Oldham seized upon Jones's depoliticized, reconciliatory approach as the way to make trusteeship a practical reality, especially in Kenya, which for Oldham was the "test case" of empire.[50] He felt that the only way to make trusteeship work was with the cooperation of the local settler community, and the best chance to secure their goodwill was through the Phelps-Stokes approach.[51] But Oldham was not content to simply court settler favor. He also hoped to use the commissions to his advantage in London, seizing the opportunity to readdress the principle of trusteeship and colonial relations in Africa in general.[52] Bolstered by the first commission's findings, Oldham approached the CO early in 1923 to discuss the subject of African education and was invited by Ormsby-Gore to prepare a memorandum for the next African Governors' Conference.[53]

Oldham's memorandum, "Educational Policy in Africa," challenged the imperial government to honor its recent claim in Parliament that its first duty was to the moral and material advancement of the African population.[54] This should be done, Oldham claimed, by allocating greater resources to education. He suggested that policy in India as laid down in the *Educational Dispatch* of 1854 provided the best framework for constructive policy by allowing the expansion of a national system of education based primarily on the voluntary efforts of private agencies, generously supported rather than directly managed and controlled by the state. What had to be avoided, however, was the type of education previously encouraged in India, which, according Oldham, had been responsible for producing an indigenous class lacking in reverence and respect for either traditional or colonial authority. To correct this, he strongly endorsed the recent Phelps-Stokes report, which recommended that education in Africa be practical and oriented to the needs of the community. The memo also emphasized the importance of Christian religious instruction as a way to remedy defective tribal beliefs and instill a moral sense of citizenship. Moreover, it called for the creation of advisory boards, not only in the various colonies but also in the form of a central body at the Colonial Office to oversee policy guidelines, research, and the collection and dissemination of information. Such a body, according to Oldham, would provide an effective vehicle for consultation and cooperation between government and private agencies. It would also, Oldham's memo stressed, be an important step toward widening the scope of colonial development itself: "In the administration of our large African Empire considerable provision has been made for the scientific study of the best means of developing its material resources . . . Very little has yet been done to devote serious thought to the best means of developing its human resources and advancement of the peoples for whom we claim to act as trustee. The mind of the nineteenth century was predominantly occupied with the development of material resources. The present century is happily witnessing an increasing recognition of the importance of the human factor in the life of mankind."[55]

The colonial secretary, the Duke of Devonshire, gave his assent to a permanent standing committee, and six months later the newly formed Advisory Committee on Native Education in Tropical Africa (ACNETA) held its first meeting on 9 January 1924 in London. As the composition of experts who initially made up the committee suggests, the body was designed to act as the corporate voice of the establishment on colonial education issues.[56] Notably absent was any form of African representation or consultation.[57] Initially, the ACNETA operated in an ad hoc manner, responding to specific problems in various colonies as they arose, while the committee's secretary, Hanns Vischer, embarked on an information-gathering tour of East Africa as part of the second Phelps-Stokes commission. With the return of Vischer in July, however, attention turned to articulating a more definitive statement on colonial education in tropical Africa.[58] Although it was Lugard who originally offered to prepare the policy statement, by the time the final draft was endorsed by the committee in March 1925 it had gone though a series of revisions, with Oldham active throughout, offering numerous suggestions and recommending changes.[59] Indeed, the 1925 white paper, *Education Policy in British Tropical Africa,* bore the recognizable mark of Oldham's pen, restating the case that he and the Phelps-Stokes commissions had been urging on individual African governments for the past five years.[60] The memorandum was careful to place the issue of greater state involvement in education within the context of the contemporary debate. It noted that while economic development had given local administrations access to larger revenues, there had emerged at the same time the recognition of the government's responsibility as trustee for the moral advancement of the indigenous population. The paper was clear on this point: "The rapid development of our African Dependencies on the material and economic side demands and warrants a corresponding advance in the expenditure on education. Material prosperity without a corresponding growth in the moral capacity to turn it to good use constitutes a danger. The well-being of a country must depend in the last resort on the character of its people, on their increasing intellectual and technical ability, and on their progress."[61]

In light of these concerns, the paper recommended several broad principles to serve as the basis for a sound educational policy. The central tenet of the new policy was the "adaptation" and development of education "along native lines." This meant that

> Education should be adapted to the mentality, aptitudes, occupations and traditions of the various people, conserving as far as possible all sound and healthy elements in the fabric of their social life; adapting them where necessary to changed circumstances and progressive ideas,

> as an agent of natural growth and evolution. Its aim should be to render the individual more efficient in his or her condition of life, whatever it may be, and to promote the advancement of the community as a whole through the improvement of agriculture, the development of native industries, the improvement of health, the training of the people in the management of their own affairs, and the inculcation of true ideals of citizenship and service. It must include the raising up of capable, trustworthy, public-spirited leaders of the people, belonging to their own race.[62]

The memorandum left the meaning of adaptation vague and loosely defined. Under this general catchphrase, the paper stressed the importance of training indigenous teachers and establishing a system of visiting instructors as a way of bringing new ideas and improvements to rural village schools. It advocated the use of vernacular languages and the preparation of vernacular textbooks. The education of women and girls was seen as integral to the whole question since, as the memo suggested, "[t]he high rate of infant mortality in Africa, and the unhygienic conditions which are widely prevalent make instruction in hygiene and public health, in the care of the sick and the treatment of simple diseases, in child welfare and in domestic economy, and the care of the home, among the first essentials."[63] Yet, the policy also laid stress on the technical training of skilled artisans and mechanics with a good knowledge of English to work in government workshops, and on vocational apprentices for the various technical departments. It stressed the importance of the "discipline of work" and the formation of habits of industry as part of the "foundation of character." Finally, the memo also noted the need for training for those required to fill administrative posts. Thus, "As resources permit, the door of advancement, through higher education, in Africa must be increasingly opened for those who by character, ability and temperament show themselves fitted to profit by such education."[64]

The 1925 memorandum was the single most important document on education in Africa produced by the advisory committee. As a set of general principles based on the adaptational model of education it continued to inform government thinking up to the Second World War and well beyond.[65] Over the course of the next decade much of the committee's time was spent sketching out the details of the 1925 white paper through a series of further memoranda. This stream of committee paperwork included reports on the improvement of the education staff, the place of the vernacular in African education, the role of biology in tropical Africa, grants in aid of educational institutions, the education of African women, and the teaching of domestic science and its application in the colonies.[66] Despite the flurry of official pronouncements in London, however, the colonial record in Africa and elsewhere in the 1920s and 1930s reveals a system plagued by widespread confusion of purpose and lack of coherency. The actual practice of education was torn in many directions, with the result being an educational program characterized by improvisation, vacillation, and sheer expediency.[67] On a certain level, of course, inertia followed from the fact that policy was not matched by a guarantee of financial resources from the imperial government. As with colonial medical and sanitary services, education continued to be treated as a recurrent social cost that was the responsibility of local colonial governments to provide for out of their own revenues. Advisory committee pronouncements remained just that: advisory objectives and guidelines, not mandatory directives. This reluctance to endorse the state's role more fully meant education in the colonies remained largely dependent on the efforts and commitment of local authorities and social agents: enthusiastic governors and directors of education, missionary societies, and, more than anything else, Africans themselves, whose demand for educational facilities led to the rapid growth of so-called bush or hedge schools and increasing pressure for higher education.

Yet the shortcomings of colonial education cannot be attributed entirely to a lack of resources or commitment. Even colonies such as the Gold Coast, which was prepared to allocate significant sums in laying the foundations of an education system, have been criticized for their essentially reactive and incoherent approach to educational development.[68] Such criticism suggests that in the interwar years the problems in enacting policies regarding colonial education, as with the other specialist services, went much deeper than the shortage of funds and educational staff. The confusion and paralysis reflected the enigmatic aims of the policy itself, and, beyond this, the makeshift nature of the colonial project as a whole. The 1925 policy statement papered over what in reality were highly contested aims and designs. Government departments, missionaries, European settlers, and, most crucial of all, the demands of Africans themselves pulled the practice of colonial education in divergent directions. Far from the modernizer's tale of gradual and steady expansion of educational facilities, colonial authorities found themselves compelled to contain the process—a process set in motion by colonial intervention, but which officials were increasingly unable to manage and control. The result was often one of expediency and complex improvisation.

This did not mean, however, that colonial education was rudderless or without intention. It is clear that Oldham and his band of old colonial hands and liberal reformers were consciously steering African education in a particular direction. With the lessons of past colonial blunders in India and elsewhere firmly in view, these so-called friends of the natives were fashioning an alternative education that would "fit" Africans for an agrarian vision of development, one which would foster happy and prosperous rural communities firmly tied to the land and would avoid producing a surplus population of rootless and discontented youths. And in this sense, the significance of the 1925 white paper went well beyond its immediate purpose in laying the basis for British colonial education policy in the interwar period. In many ways, it formed the nucleus for a whole new way of thinking about the nature of colonial development and the type of scientific investigation necessary to address the emerging problems of welfare of indigenous peoples. Once again, Oldham's handiwork was central in broadening the mandate of education, and once again it was the intractable question of white settlement and capitalist enterprise in Central and East Africa, where the contradictions of colonial rule were perhaps most apparent, that served as his foil. Here, the demands of mining capital and settler farming for cheap migratory labor threatened to destabilize the reformers' visions of Africa as a land of self-sufficient village communities. Here, African migratory labor threatened to spill over the boundaries set for what colonial officials and experts considered to be the "real" Africans: those living in containerized "tribes" under the authority of "native" chiefs and councils. Here, in Oldham's mind, the "detribalized native" constituted a troubling and deepening problem that needed to be managed and ordered before it was too late.

Even as he was busy crafting the 1925 white paper, Oldham was already beginning to mull over the larger question of colonial development and its relation to African welfare. As he explained to the archbishop of Canterbury in October 1924,

> It is becoming increasingly clear to me that, on a long view, the interests of the native population, of European settlers and of the Empire, are fundamentally the same. The ultimate limiting condition of the economic development of East Africa is the question of population. The interests both of this country and of the local settlers are entirely de-

> pendent in the end on an increasing, healthy, industrious and contented population. If any of these conditions are ignored economic development is inevitably arrested. I believe we shall make more progress by demonstrating the truth of this view than by appearing to advocate native rights in antagonism to policies of economic advance. *The opposition is transcended if one takes a sufficiently large view of the problem.*[69]

Oldham laid out his new forward-looking policy in a memorandum written to Ormsby-Gore in response to the report of the 1925 East African Commission.[70] He praised the report for drawing attention to the potential wealth of East Africa, but he also reminded Ormsby-Gore of the parallel and equal obligation to promote the "physical, mental, moral and social advancement of the native inhabitants." He suggested the want of African population was the most crucial and dominating fact confronting colonial governments in carrying out this twofold responsibility. From an economic standpoint it was clear that continued exploitation of the territories' resources depended on an abundant supply of African labor. On this issue, the settler community could wholeheartedly agree, yet it appeared the African population in many parts of East Africa had stagnated or even declined since the onset of European occupation. The lack of a natural increase of population, according to Oldham, could be traced to the spread of disease following conquest and to the disturbance of traditional life as a result of contact with a more advanced civilization. He pointed in particular to the adverse effects of male labor migration in undermining the social stability of the family and community and imposing an excessive burden on women to provide the subsistence needs for both adults and children.

Oldham therefore supported the implementation of a constructive policy aiming at a healthy increase of the indigenous population. To this end, he called for the scientific treatment not just of the problems of economic development, but of the advancement of colonial peoples as well. Attention and resources had to be devoted to the "human side" of development in order to promote the health and industry of the people and protect their way of life from disintegration. He suggested that a rational, scientific treatment of the problems of East Africa would mitigate the conflict between the interests of settlers and others involved in exploiting the region's natural resources, on the one hand, and the welfare of the African peoples on the other. And he warned:

> Unless Government is prepared to accept responsibility of regulating the rate of development in accordance with the supplies of labour that are available, or that can be made available . . . There is bound to come . . . a rupture of the equilibrium between the demands of development schemes and of European enterprises and the existing supply of labour. When that point is reached either development schemes must be held up and European enterprises allowed to go bankrupt, or recourse must be made to methods of obtaining labour which are injurious to native welfare and which can therefore only have the effect of reducing the population and making the situation worse. The means of averting such a disaster is to be found in a far-sighted policy based on an accurate knowledge of all essential facts.[71]

To this end, Oldham proposed that a portion of the East African Transport Loan be spent on research and education, arguing that the resolution of East Africa's political and economic problems would, in large measure, depend on the training and development of its indigenous population. He called on the imperial government to take greater financial responsibility for education in the widest sense of the term, including not only formal schooling but also the introduction of improved agricultural methods, encouragement of labor-saving techniques and use of better tools, and improvement of sanitary and public health services in the reserves. Such a progressive approach to development, Oldham argued, would produce a healthy and abundant population, which, in turn, would mean a greater supply of voluntary African labor. In essence, Oldham argued that a policy of rural betterment in the reserves would not only lead to the advancement of "native welfare" but also induce population growth, the lack of which had been the primary obstacle to the further development of the region's natural resources and the main source of conflict between settlers and indigenous peoples. He argued, moreover, that in order to circumvent African demands for more land and to stabilize the policy of land segregation, the African reserves had to be made more habitable and productive through a program of rural improvement and education that was adapted to local conditions and practical in nature.[72]

The allocation of imperial funds for education and other social services in the colonies was an unprecedented move which, barring some profound crisis or transformation in government thinking, senior staff at the Treasury would have found impossible to swallow. As it was, Amery was forced to accept major concessions before the Treasury would allow the passage of the East African Loans Act of 1926—concessions that so diluted the recommendations of the East Africa Commission that Ormsby-Gore doubted whether it was even worth proceeding.[73] Without imperial funds there was little that metropolitan specialists like Oldham and the members of the ACNETA could do beyond offering ad hoc advice on problems that arose on an administration-by-administration basis. As Oldham noted in reference to Northern Rhodesia in 1930, the committee could only "state the conditions"; it could not impose the rules.[74]

But the spotlight that Oldham and others succeeded in placing on the development of the human resources of the tropics had, nevertheless, revealed a fundamental flaw in colonial administration prior to the 1930s: colonial governments knew very little about the local inhabitants whose welfare they claimed as trustees to protect and advance. Oldham recognized the problem very clearly, advocating, as previously noted, greater investment of resources not just in education widely defined, but also in research. In his memorandum on the East Africa Commission he outlined a scheme for a general survey of conditions in East Africa, the most immediate task of which would be to obtain accurate estimates of population and of the proportion of males that could be safely withdrawn from the reserves for employment outside without destroying the life of the "tribe." He drew attention to the great need for scientific and technical experts, especially anthropologists, to make inquiries regarding the conditions of life, customs, and beliefs of various social, religious, and ethnic groups and to gather information about the health and physical efficiency of indigenous peoples and their capacity for greater industrial organization and production. The survey would also consider a whole range of other questions, from the best economical use of existing labor to questions of health and education, to investigations of soil fertility, crop production, and autochthonous methods of agriculture. Obtaining such information through local scientific and technical departments was the crucial preliminary step to framing an imperial policy that would deal with the problems of East Africa in a "scientific spirit."

Oldham's call for a more scientific approach to colonial problems struck a chord. Whatever differences there were between he and Amery over the priorities of imperial policy, both fervently believed in the need for further research and expert knowledge to help manage the problems of colonial development. For Oldham, the common quest by both constructive imperialists and liberal reformers for a scientific solution to imperial problems seemed to offer a way forward.[75] In the late 1920s, for problem of "adapting" traditional social institutions and "mental cultures" to modern requirements. The aim, in other words, was to help Africans adjust to the new, modern order of things, while at the same time cultivating and preserving their own unique "mental heritage" and cultural distinctiveness as much as possible.[91]

With the aid of generous funding from the Rockefeller Foundation, the institute embarked in 1931 on an ambitious five-year plan of research on Africa, directing its program to the fundamental problem of effects on social cohesion "arising from the interpenetration of African life by ideas and economic forces of European civilization."[92] The goal of colonial administration, according to the IIALC, was to prevent the breaking apart of African societies into "a mass of unrelated individuals deprived of the restraining influences of tribal law and customary sanctions" by fostering the development of "healthy, progressive organic" communities.[93] The institute could assist colonial governments by scientifically and objectively studying the processes of social and economic change affecting the indigenous way of life. Such processes, it stressed, could not be viewed in isolation, but would have to be studied in relation to the whole sociological background of the people, taking into account customary uses of land and resources and systems of kinship and ownership. Most crucial of all, the effects of male migrant labor and the resultant growth of individualism on clan and tribal organization as well as on family, gender, and generational relations needed careful examination. To this end, the institute planned to use most of its resources for fellowships to support specially trained investigators who would work closely with governments, missionaries, and African representatives to carry out inquiries on selected African communities.

The aims of the institute followed closely the ideas popularized by Bronislaw Malinowski, the pioneer of the functionalist school of anthropology in Britain.[94] Indeed, an impressive number of participants in Malinowski's seminar at the London School of Economics in the 1930s would go on to do important fieldwork in Africa, often funded by the IIALC.[95] Malinowski urged his new breed of functional anthropologists to go beyond the study of remote or lost "tribes" and focus instead on the "anthropology of the changing native." Investigating the problem of "culture contact," as it was termed, involved exploring the central question of how European influence and Westernization affected indigenous communities. This new approach, Malinowski argued, could fulfill important practical functions. The task of the anthropological expert was to study in an analytical spirit such questions as the changing demographic patterns of "primitive tribes" and the importance of fundamental "tribal" institutions such as the family, marriage, and educational agencies in molding the character and social nature of the individual. Research of this sort could help prevent or ease the suffering and social crises affecting areas where Western capitalism was pressing.[96]

The new functionalist research model, with its stress on the "adaptation" of culture and the imperatives of social cohesion, promised to make its most valuable contribution to colonial education. Education, it was argued, was "bigger than schooling"; it involved the upbringing of the individual by the whole community. If the process of "detribalization" and the disintegration of African communities were to be arrested or slowed, then African education would have to broaden its scope and look beyond the provision of formal schooling.[97] Education would have to be thought of more as "community education." This practical anthropology was just the sort of applied research model Oldham had in mind. As administrative director of the IIALC from 1931 to 1938, Oldham used his influence to ensure that Malinowski's students received the lion's share of the institute's fellowships.[98] Colonial Office staff, however, soon grew and remained weary of Malinowski's band of research workers: they were highly conscious of the fact that many of them were not British nationals, and fearful that they would be "let loose at random on Africa" and might engage in subversive activity. Once again Oldham's networking with the CO and with senior colonial administrators was crucial to allaying fears and securing their cooperation. In the case of Godrey Wilson and Monica Hunter, for example, Oldham met with the governor and the commissioner of native affairs for Tanganyika while they were on leave in England, ensuring their acceptance of the scheme even before a formal request for permission was made to the CO.[99]

Anthropologists were not the only group of experts probing for new ways of thinking about the social ills thrown up by colonial capitalism. Demands for a more positive form of trusteeship and for greater imperial responsibility for colonial peoples, especially in the area of social welfare, were heard from an increasingly broad range of actors as the interwar years wore on, many of them academic researchers or prominent critics connected in some way to what may be called the London lobby. As Deborah Lavin and others have observed, London was buzzing in the 1930s with groups studying and thinking about Africa, and in the wake of the Depression the Colonial Office invited many of the key figures

CHAPTER 5

View from the Field

Rethinking Colonial Agricultural and Medical Knowledge between the Wars, 1920–40

EVER SINCE THE TIME of Joseph Chamberlain, the structural problems of chronic high unemployment and a depressed and declining core of staple industries *in metropolitan Britain* had driven the colonial development agenda. Development was primarily thought of as a question of "opening up" the presumed natural riches of the tropics to serve the commercial interests of British trade and industry. The problem, it was believed, was largely one of distribution. Early colonial state efforts focused overwhelmingly on building transportation and communication infrastructures and capturing and redeploying indigenous labor power toward the development of primary resource and agricultural industries for the export market. There was a persistent fundamental assumption that what was good for British trade and industry would also have a beneficial effect on the moral and material advancement of colonial peoples. Such unbridled optimism for British colonial rule as an agent of social and economic progress was sorely tried in the years between the wars. Questions of trusteeship and responsibility for the welfare of indigenous populations, as we have seen, gained saliency in the 1920s through the agitation of colonial reformers and humanitarian lobbyists like Joseph H. Oldham. The onset of Depression in 1929 reinforced this trend, shattering the illusion of a harmony of interests. Earlier justifications for colonial development lost their appeal in the wake of the apparent failure of policies premised on an export-led economic recovery.[1] In Britain, government economic policy moved toward abandoning the once sacred principle of free trade, while in the colonies it became increasingly clear that social and economic conditions were getting worse, not better, as the decade wore on. Deteriorating conditions in the colonies, critics charged, were evidence of years of complacency, neglect, and exploitation. New attitudes were taking shape which by the end of the decade would usher in a far-reaching process of colonial reform, symbolized by the passing of the Colonial Development and Welfare Act of 1940.[2]

The 1940 act championed a strong welfarist agenda.[3] It was imagined as an attempt not to replicate the material conditions present in the imperial heartland of industrial Britain, but rather to mitigate the contradictions thrown up by private capitalist enterprise and colonial rule in the empire. At the center of the revolution in metropolitan thought were the perceived loss of productive force and the rise of a relative surplus population throughout much of the colonial empire. In stark contrast to earlier discussions of the problem of development, in which, as we have seen, an abundance of untapped resources and a want of population were held to be the greatest obstacles to the rapid social and economic transformation of the tropics, by the late 1930s the pendulum had dramatically swung toward addressing the socioeconomic and environmental dilemmas gripping many colonial regions. There was a growing realization among officials and experts that they were facing a new set of problems: the limited and fragile nature of tropical environments and the appearance of surplus population in many areas. It was only after visions of impending ecological devastation and overpopulation instilled a new sense of urgency that imperial state interventions in the name of development were seriously contemplated and initiated with the aid of Treasury funds.

The new coordinated policy aimed to create stable and sustainable agricultural communities that were protected from the vagaries of the world market. Development had come to mean planned, state-directed projects for improving agriculture, particularly local food production, and providing community welfare and social services in an attempt to allay the misery of growing unemployment and poverty and the flow of rural emigration, not in Britain, but *in the colonies themselves.* The specialist advisers and metropolitan experts enlisted as part of the CO's advisory network in London occupied a pivotal place in this redefinition of development principles. But the new development priorities and policy paradigm of the late 1930s and early 1940s did not spring fully formed from their own imaginations. It is important to note that much of the "new thinking" and impetus for change originated from below: from the myriad of local technical officers and research practitioners operating on the ground, who were beginning to survey and gather data on local conditions and carry out field trials at various research and experiment stations peppered throughout the colonial territories. As the results of their investigations permeated colonial policy debates—through regional and pan-colonial conferences, government reports and memoranda, and publications in scientific and imperial journals—many earlier assumptions about and images of tropical development were brought into question and dramatically altered. The various local reassessments making their way up to London took on added significance as economic hardship and social turmoil swept across the empire in the 1930s and 1940s, generating a new sense of urgency and of need for government planning and action. This chapter looks at the crucial years between 1920 and 1940, when many of the fundamental assumptions of earlier imperial doctrines were overturned and the discourse surrounding tropical development began to be radically redefined.

Rethinking Tropical Environments: From Natural Abundance to the "Menace" of Soil Erosion

Perhaps the most far-reaching revision of colonial knowledge that occurred between the wars involved the long-standing belief in the apparent natural exuberance of the tropics. Before the 1920s, distress over the tropical environment had most often been voiced by colonial foresters, who linked climatic changes and land desiccation to shifting cultivation methods practiced by indigenous agriculturalists, which they believed were leading to the rapid destruction of primary forest cover.[4] In India, as Richard Grove has shown, concern regarding the deterioration of natural resources and the environment was strongest among the surgeons and medical botanists of the East India Company Medical Service, and later the Indian Medical Service, who tied extensive deforestation to the increased likelihood of drought, famine, and social disorder.[5] Largely through their lobbying, the Raj created in 1865 a state forestry service that would go on to combine commercial exploitation with extensive programs of reforestation. Indian-trained foresters were often recruited to produce reports, conduct surveys, and help establish new forestry departments in colonies throughout South Asia, the Caribbean, and sub-Saharan Africa in the late nineteenth and early twentieth century.[6] But despite the trepidations of colonial foresters, most observers in the early part of the twentieth century continued to exult over the potential economic bounty of tropical environments. The CO's Committee on the Staffing of Agricultural Departments in the Colonies, for example, reported in 1920 that "[i]t is not generally realized how rich a harvest may be reaped if the agricultural resources of the Colonies and Protectorates are properly developed, or how small a part of the Empire has been so far subjected to such development. These colonies, situated as they are in the tropical and semi-tropical zones, are full of all manner of valuable economic products, and though the development of agriculture in recent years in many places has been rapid, a vast amount of country which has so far only been lightly 'scratched over' still remains to be developed."[7]

Images of rich tropical harvests waiting to be reaped were kept alive throughout the 1920s in the reports and public addresses delivered by William Ormsby-Gore, who as late as 1927 was telling members of the Royal Colonial Institute that "Hitherto in West Africa we have been enabled to base our development upon the existing bounties of nature . . . Except in the palm belts and the cocoa growing areas, and in a few areas of Northern Nigeria, what strikes the visitor first . . . must be not how much use is made of land, but how little."[8] In a similar vein, the 1925 East Africa Commission spoke of the "dormant possibilities" of that region as the future granary of the empire, hoping to breathe new life into the dream of transforming the great alluvial plains and plateaus of High Africa into islands of white settlement. "East Africa is a wonderful country," Archibald G. Church informed members of the Royal Geographical Society. "The greater part of it is amazingly fertile, and only needs the development of its transportation services to make it one of the most productive areas in the world."[9] In the early decades of the twentieth century the presumption of tropical abundance also permeated the agendas of most colonial governments, which were, as we have seen, preoccupied with turning these territories into vibrant producers of agricultural commodities and natural resources for the global export market.

This, it was increasingly argued, demanded more research and more scientifically trained officers in areas such as plant genetics and breeding, plant physiology, and plant pathology. "The whole future of successful agricultural development," Ormsby-Gore suggested, "is bound up with the work of the soil chemist, the plant breeder, the entomologist, and the mycologist. We have to remember that in the tropics the successful introduction of new crops and the improvement in the quality and yield per acre of existing crops often depends on years of spade work at the laboratory and the experimental farm, the practical application of which cannot be hurried."[10] And indeed, as explained in chapter three, colonial technical and research services and staff, especially in agriculture and agriculture-related fields such as forestry, surveying, and veterinary medicine, expanded considerably in the interwar period despite the retrenchments and cutbacks of the Depression years.[11] Their efforts were supplemented by the more long-range investigative work of the East Africa Agricultural Research Station (EAARS) at Amani, the research departments and commodity research schemes of the Imperial College of Tropical Agriculture (ICTA) in Trinidad, and countless other commodity research stations and institutes located throughout the colonies. In addition, many colonies established their own special research departments and centers such as the Tsetse Survey and Research Department in Tanganyika, the Sleeping Sickness Service in Nigeria, and the Henry Scott Agricultural Laboratories in Kenya.[12]

The model for much of this organization, as noted earlier, was the nationwide system of agricultural research established in Britain under the auspices of the 1909 Development Commission. The metropolitan orientation toward investigations of fundamental principles, rigorous field experimental methods, and statistical analysis had a strong bearing on the development of British tropical agricultural science, especially, as we have seen, through the Cambridge School of Agriculture's close connection with and influence upon the teaching and research program at the ICTA in Trinidad. Not surprisingly, colonial agricultural research priorities tended to replicate the technical influences and biases of the ICTA in Trinidad and, in turn, the leading agricultural schools and research institutes in Britain, especially Cambridge. Colonial agriculture departments and research centers before the mid-1930s were driven by a commodity-based research agenda preoccupied with the introduction and improvement of cash crops such as sugarcane, cotton, rubber, coffee, palm oil, and tea; the breeding and introduction of higher-yielding and more drought-resistant crop varieties; and the study and eradication of insect pests and plant diseases.[13] Building on the experiences of the nineteenth century, British tropical agriculturists initially remained heavily biased in favor of plantation enterprises and settler estate farming; in areas of peasant cultivation, preference was given to the introduction of "sound" European agricultural practices such as planting in pure stands, deep plowing, crop rotation, and clean weeding. Agricultural officers and researchers were preoccupied with improving yields, although much attention was also directed to the problems of quality control and genetic uniformity of crops grown, especially by indigenous farmers whose crop husbandry practices, methods of land management, and attitudes to government advice were often singled out for blame.[14] Preference also tended to be given to the introduction of exotic crops and plant breeding materials, over research into local varieties.

But colonial agricultural officers and other natural resource personnel were also increasingly aware of the need for collecting information on local environmental conditions and for experimental trials on the most ideal varieties and best methods to apply to those conditions before the practical work of agricultural extension and improvement could take place.[15] The Nigerian Agricultural Department under O. T. Faulkner and later James Mackie led the way in investigating indigenous farming systems and methods of cultivation. The greater attention Faulkner and Mackie gave to local agricultural practices reflected, in part, the phenomenal growth of peasant commodity production of both food and cash crops, not just in tropical Africa but throughout much of the colonial empire in the early decades of the twentieth century.[16] Such unprecedented expansion persuaded both the CO and many local colonial authorities of the potential value of peasant agriculture and of the need to pay greater attention to indigenous systems of husbandry. Faulkner and Mackie, however, were also highly critical of earlier colonial efforts to promote cotton cultivation. The disappointing results of introducing new, exotic crop varieties and European practices such as deep plowing, clean weeding, and continuous systems of rotation without careful regard for the variability of local climatic and soil conditions and the dynamics of local farming systems were not soon forgotten. Faulkner and Mackie's approach instead stressed continuous, systematic research and field experimentation leading to sound extension work and the introduction of improvements that built upon existing practices.[17] Under Faulkner and Mackie a vigorous program of research work began in the 1920s, including a prolonged program of experimentation with mixed farming in Northern Nigeria, which led by the end of the decade to the initiation of a pioneering extension campaign to integrate a system of animal husbandry, using cattle as a source of manure and traction, into local crop-farming systems to improve efficiency and maintain soil fertility.[18]

Faulkner and Mackie may have been pioneers in examining local farming systems and climatic and soil conditions, but it is important to realize that their approach both mirrored and influenced wider trends in British and indeed international research currents, which increasingly moved toward more quantitatively rigorous methods of field experimentation and statistical analysis in the interwar period. Engledow's investigations on cereal yields at Cambridge in the 1920s, for example, pointed to the need for understanding the spatial relations of plants in the field, examining the effect of spacing on plant growth and performance, and documenting what happened in the field throughout the growing season in relation to plant population densities and interplant competition. Such investigations pointed away from the purely technical side of plant breeding toward studying the wider relationship between environmental conditions and agricultural practices and their long-term effects not only on plant nutrition but on the soil as well—which, as we will see, became the central preoccupation of much of Engledow's subsequent career.[19] In other words, the influence of the new botany, with its emphasis on studying the physiology of plants as living organisms affected by their total environment, and the growing importance of the field as much as the lab as a site of investigation, not only led to fundamental research on heredity and genetic uniformity but also gave rise to new interest in crop ecology, plant geography, and the specificity of local environmental factors like climate and soil fertility.[20]

Similar trends shaped tropical agriculture, where the overriding focus on commodity research, while perhaps culpable for its neglect of food crops, had the effect not only of prompting studies on wider questions of agroclimatology but of drawing researchers into the problems and difficulties faced by primary producers, especially the need for more effective extension methods and services for smallholders.[21] In his influential report on agricultural extension methods, published in 1949, Charles Lynn discerned a notable shift in attitudes about indigenous methods of agriculture and an increasing use of the term "extension" in the reports and statements of agricultural departments in the late 1920s.[22] It was increasingly recognized that the application of knowledge based on research done in Britain, Trinidad, and elsewhere had to be worked out locally on central and district experiment stations and substations, and that too hasty an extension of new methods ran the risk of undermining the confidence of local communities and leading to resentment and an erosion of trust. Nor, it seems, was this greater attention to local variability peculiar to the British. Suzanne Moon has recently argued, in the case of the Dutch East Indies, that the department of agriculture from its inception in 1905 adopted a localized approach, conducting, for example, demonstration and field trials across the island of Java in order to gain a greater understanding of the diversity of social and ecological conditions that affected rice yields and to develop crop varieties and techniques that best suited local conditions. Like Faulkner and Mackie, the department seems to have favored a more evolutionary model, running long-term empirical investigations on the social and environmental impact of introducing new crops or cultivation practices before recommending any changes to indigenous farming systems.[23] In French West Africa, the failure of grandiose projects for *colonisation indigène* in the 1920s and 1930s led colonial agronomists after 1945 to develop a more subtle approach to agricultural development, one premised on extensive field trials and careful studies of local soil types based on indigenous knowledge and classification systems. They also turned to the local practice of allowing nomadic pastoralists to graze their cattle on farmers' fields in the dry seasons as a way of maintaining soil fertility.[24]

Indeed, in her study of the African Research Survey, Helen Tilley identifies three fundamental and overlapping trends in the activities of most British colonial agricultural departments in the 1920s and 1930s: an increasing emphasis on research into indigenous methods of cultivation and local crop varieties; a new awareness of the dangers of soil erosion and the importance of soil investigations; and the increasing incorporation and application of concepts from the emerging science of ecology.[25] As more research on tropical environments and indigenous farming systems was carried out by technical personnel and research officers employed in the colonial professional departments or in the network of pan-colonial commodity research stations and institutes, and as the results of their investigations began to be exchanged and compiled through periodic conferences or written up in departmental bulletins and annual reports, CO reports and memoranda, and imperial scientific periodicals like *Tropical Agriculture* and the *Empire Journal of Experimental Agriculture*, many of the earlier images of and assumptions about the "tropics" which had informed previous colonial development doctrines would be challenged and brought into question. By the end of the 1930s this new awareness of the complexity and diversity of tropical environments would make itself felt in "high policy" debates and would lead to a considerable reworking of the colonial development agenda at the center of empire in Whitehall, largely through the efforts and activities of the CO's specialist advisers and advisory bodies.

A common thread running through much of the new research was an awareness of the limits of tropical fertility. Frequently, colonial foresters, who as noted earlier had long been concerned about the biological consequences of deforestation on climate change and rainfall reduction, sounded the initial alarm, warning that further losses of forest reserve could lead to droughts and soil erosion, and, in turn, to declining yields, food shortages, and ultimately social unrest.[26] But colonial forestry was in many ways the Cassandra of the colonial sciences, its dire predictions ignored—or subordinated to the opinions of nonforestry government advisers and officers.[27] In the 1920s, however, desiccationist discourse began to have a wider impact on imperial debates as attention was drawn to the closely intertwined problems of vegetation disturbance, deteriorating soil conditions, and declining crop yields of important cash and food crops such as bananas in the Caribbean and coffee in East Africa.[28] Desiccationist warnings also gained a larger audience through the emerging and broadening field of tropical ecology. A new "systems approach" to ecological research took shape in the 1920s and 1930s through the influence of Arthur Tansley, Thomas Ford Chipp, Charles Elton, and other prominent members of what Peder Anker has dubbed the "Oxford School of Imperial Ecology," which saw the new science as a powerful administrative tool not only for the study of the environment but for the improved management of the material and human resources of the empire as well.[29] The British Empire Vegetation Committee's 1926 handbook, *Aims and Methods in the Study of Vegetation,* edited by Tansley and Chipp, was particularly influential not only in laying out a general outline of ecological methods and their application but in drawing attention to the need for a more complete botanical survey and classification of the empire's "vegetational assets." The first step was for ecologists and other resource experts to gather this information through proper land reconnaissance methods and the preparation of ecological surveys for each territory, which would aid in the rational management and development of the vast natural stores of Britain's imperial estate. The new ecological framework would form the basis for much of the research carried out by colonial scientific practitioners in the 1930s and 1940s.[30] The aim of much of this research was to ground the imperial mission on a technical basis through the use of systematic planning, coordination, and resource management, with development conceived as a scientific problem and ecology held out as the model, thus helping to elide controversial political, racial, and class issues surrounding the practice of imperial trusteeship.[31] But in the hands of local colonial scientists, the new ecological approach could also lead to innovative and critical analyses of the nature of the problems facing colonial regimes that often challenged widely held earlier beliefs about tropical environments and indigenous land use practices.[32]

One of the most pressing imperatives, as both the British Empire Vegetation Committee and many local colonial investigators recognized, was the need for proper geological, topographical, botanical, and soil surveys to ascertain the empire's real productive potential in terms of agriculture and other natural resources. Duncan Stevenson, the deputy conservator of forests in British Honduras, for example, noted in 1925 that no serious attempt had up to that point been made to determine the actual extent of the colony's agricultural and forestry resources, primarily because it had always been assumed that the tropics were vast storehouses of untold wealth and that the main problems of development were access and distribution, which the expansion of transportation and communication networks would resolve.[33] The need for greater knowledge and regional surveys was also one of the main themes of the East Africa Commission, which drew particular attention to Tanganyika's southern plateau as an area of potential European settlement and cereal farming. All that was needed was the construction of a railroad connecting Tukuyu in the southern highlands to the central line at Dodoma.

Ironically, the interwar drive to survey the tropics would reveal an unexpectedly somber picture of the empire's economic potential. In response to the East Africa Commission, for example, the senior district engineer for Tanganyikan Railways, Clement Gilman, was sent to conduct a topographical survey of the southern highlands in order to determine the best possible route for the proposed railway line. But Gilman, who had a background in geology and geomorphology as well as displaying great sensitivity to local climatic variations and soil conditions, questioned the possibility of successfully developing the region at all, laying stress in his report on the difficulties presented by the topography, climate, and limited population.[34] He observed that, although the southern plateau was well watered and contained many perennial streams, it was of a lower latitude than the Kenyan highlands and experienced a more inclement climate, with frequent, forceful thunderstorms. There were no forests on the rolling uplands, and mosquitoes and Spirillum ticks were prevalent. Gilman also noted the great climatic variation across the region, with the lands beyond the Mporoto chain, for example, experiencing dramatically lower rainfall. In fact, Gilman felt it had not been sufficiently realized to what extent the semiarid and indeed uninhabitable condition of much of the land, given the absence of a permanent and readily accessible water supply for human domestic use, acted as a geographically determined barrier to development in the region. He suggested that far more accurate and reliable studies of the region's vegetation and soils were needed before its potential as a producer of food and cash crops could be properly assessed.

Gilman's findings are significant not only for their influence in tempering earlier generalizations about the agricultural potential of East Africa but also because of their impact on the work of Geoffrey Milne, a young soil chemist at the EAARS with whom Gilman corresponded regularly about soil conditions in the region. Milne was one of the most innovative colonial soil scientists of the 1930s. His studies of basic types of East African soils and development of new units for mapping soils, like the "soil catena," were regarded as landmarks in the development of soil classification, gaining international recognition after the war.[35] Milne was deeply impressed by Gilman's emphasis on the interrelationship of topography, climate, and soil and their influence on vegetation types and plant growth. He was also inspired by the director of Rothamsted, Sir John Russell's insistence on the necessity of creating detailed soil survey maps before any agricultural development schemes were undertaken, in order to avoid disappointments and problems due to insufficient knowledge.[36] In 1935, Milne undertook a soil reconnaissance tour of the Central, Western, and Lakes Provinces of Tanganyika which became the basis for much of his Provisional Soil Map of East Africa, which Gillman described as "a grand synthesis of rock and soil, vegetation and man."[37] What distinguishes Milne's approach is his argument that simple vegetation or soil surveys did not go deep enough, and that what was needed was a comprehensive "soil-vegetation complex" or "land type" classification survey which integrated plant ecology with pedology and addressed wider questions of proper land use.[38] His work displays a deep sense of the importance and diversity of locality, and of the complex succession of soil types that could be found in any district, based on its climate, geomorphology, and topography. Local soil differences like level of soil acidity were closely related to the distribution and variation of vegetation types. Land types, in turn, were viewed ecologically as distinctive "habitats," which could be extended to include humans as an environmental factor impacting the land. The soil surveyor, therefore, had to be acquainted with physical geology, climatic effects, vegetation and plant selection, knowledge of crops and cultural practices, and man-made patterns of social land units based on historical anthropological research. The goal, for Milne, was a field classification map that would serve as the foundation for a more rational utilization of the land in which particular problems, such as introducing new crops or the logistics of population resettlement, could be resolved. And, though he was very much oriented to the practical applications of his work, Milne also stressed the importance of pursuing fundamental laboratory research on soil characteristics in tandem with fieldwork, and the necessity of undertaking extensive field experiments before any measures, such as local soil conservation programs, were actually carried out.[39]

Milne's research at Amani also contributed in important ways to a reassessment of both European and non-European agricultural practices in the tropics. One of the most serious mistakes of previous surveys that studied the natural vegetation as an indicator of the agricultural potential of an undeveloped country, he argued, was that they did not take into account the complex relationship between plant cover and soil conditions, and therefore failed to carefully examine the soil's true character before development began. The discovery of rich, luxuriant tropical forests or vast expanses of natural grasslands was taken as evidence that the soils beneath were highly fertile and capable of supporting permanent agricultural use with intensive cropping without much regard for conservation. This assessment did not take into account that a soil and its plant cover have evolved together over a long period of time, and that the sudden removal of the natural vegetation for the purposes of human land use could radically, and sometimes fundamentally, alter soil conditions.[40] Although humid tropical forest soils were initially highly fertile, after only a few years of cropping their productivity rapidly declined until they had to be abandoned. One of the most salutary lessons of tropical forest exploitation occurred not too far from Amani in the East Usambara highlands, where German companies began planting coffee in the late nineteenth century. In terms of rainfall and elevation, conditions there seemed ideal for growing Arabian coffee, which was indigenous to African rainforest, and yet yields rapidly declined and by the First World War most of the estates had been abandoned or planted with alternative crops.[41] Debate ensued about the cause of the diminishing yields, but without resolution. In the late 1920s, Milne began studying the laterized red earth soils of the region, noting that the topsoils were surprisingly rich in nitrogen content and organic matter, but that the base-absorptive power of the soil was largely dependent on that organic matter. If it was removed or lost, the remaining mineral material was incapable of supplying plant nutrients or retaining those supplied by artificial manures. The forests of the East Usambara Mountains, Milne argued, did not simply grow "in or on the soil"; they were participants in a working system, having grown up slowly along with the soil that nourished them, and were maintained largely through a self-contained circulation of plant nutrients derived from fallen leaves, branches, and trunks. Milne's account stressed not only the complex balance between vegetation and soils but also the dangers of large-scale exploitation using single-crop plantings based on outdated assumptions about the natural exuberance of tropical forests.

Milne was also critical of the "plant more crops" campaigns that were initiated in the Shinyanga District of Sukumaland in Central Tanganyika in the 1930s. This area had been cleared of all woody vegetation so that it could be used as common grazing lands, and had at one time or another also been used for cultivation, yet Milne found that the land had not been damaged by the inhabitants and their cattle to nearly the extent suggested by most accounts. Its hard, level surfaces had resisted trampling and wind and water erosion, with the result that the soil over most of the cleared areas was still intact. But he warned that there was a real risk of serious land degradation if intensified agriculture stimulated by the planting campaigns proceeded in advance of working out the necessary improvements in agricultural and stock-owning practices. In the meantime, he felt they should be left in strict abeyance.[42] As the above examples suggest, much of Milne's work involved critically rethinking earlier opinions about not just European farming but indigenous land practices as well. For example, to the common denunciation of Shamba burning as superstitious and destructive of humus, Milne offered an alternative account, noting that low-intensity running fires cleared the land of trash that harbored pests and fungi, partially sterilizing the soil and producing mineral residues that were retained on the soil. "If the native is going to be convinced by our 'modern enlightenment,'" Milne observed, "then we must be very sure that the objections we advance do actually hold good."[43]

Milne was not alone in drawing upon ecological concepts to broaden his understanding of the relationship between vegetation and soils and its bearing on land utilization. A similar approach was pioneered by Colin Graham Trapnell, J. Neil Clothier, and William Allan in Northern Rhodesia in the 1930s and 1940s.[44] Trapnell, Clothier, and Allan were critical of the inadequate knowledge of local climatic and soil conditions and traditional farming systems that characterized previous government approaches. Earlier officials tended to regard all African farming systems as wasteful and destructive, to be replaced by often unsuitable European methods of "improved agriculture." As a result, the native reserves that were demarcated in the 1920s failed to take into account the amount of land needed to practice shifting cultivation on relatively poor soils. Within a few short years the reserves were suffering from overcrowding which was leading, in some cases, to severe deterioration of natural resources and food shortages. This led Trapnell and his associates to develop specific methods for estimating population carrying capacity relative to specific areas and systems of land usage. These were used in assessing the urgency of land problems in the reserves and planning resettlement schemes that transferred "surplus populations" to less congested Crown lands and former estate lands. In estimating the carrying capacity of a given area the first step was to carry out an ecological survey based on the vegetation-soil classification methods developed by Trapnell in the 1930s (which were similar to the classifications used by Milne in his East African soil map), clearly defining and mapping different land types based not only on vegetation and soil characteristics but contour and topography as well.[45] Trapnell's surveys also contained detailed accounts of the two main African cultivation systems of Northern Rhodesia—chitemene systems and soil selection systems—which he found to be both remarkably diverse and complex, and in many cases impressively well-suited to their challenging environments. The ecological surveys of the soils, vegetation, and agricultural systems of northeastern and northwestern Rhodesia carried out by Trapnell, Clothier, and Allan in the 1930s and early 1940s became the basis for the territory's subsequent land use planning and policies after the Second World War.[46] Similar ecological survey and land use planning methods, as we will see in chapter 7, were taken up by agricultural staff in other colonies in the 1940s, forming the basis for numerous land resettlement and "progressive" or master farmer schemes following the war.

What fueled the extension of ecology as a colonial science more than anything were the problem of soil erosion and the need for broader investigations of soil conditions in the face of signs of escalating land degradation brought on, in part, by the commercial expansion and intensification of colonial agriculture. The marked expansion of peasant smallholder cultivation both of cash and food crops in the 1920s was followed in the early 1930s by the collapse of prices for raw materials and primary products during the Depression, which had a devastating impact on much of sub-Saharan Africa, the West Indies, and elsewhere.[47] Colonial governments, dependent as they were for the bulk of their revenues on export sector earnings, responded by reducing staff, cutting investment on public works projects and social services, and looking for new ways to mobilize previously untapped or inefficiently utilized resources. Authorities looked to peasant development with new eyes, estimating the number of additional acres that might be brought under cultivation and the amount of much-needed revenues it would generate.[48] In Ceylon, for example, increasing attention was directed to the potential of peasant colonization in the thinly populated dry zone as a way of relieving land hunger in more congested areas and expanding cultivation to achieve self-sufficiency in food supply.[49] In the West Indies, C. Y. Shephard, an agricultural economist at the ICTA, began a series of extensive surveys on peasant economics, while the West Indian Sugar Commission of 1929–30, chaired by Sydney Olivier, raised once more the problem of overspecialization of the plantation sector and the need to diversify agriculture through the creation of agricultural extension services in order to improve the productivity of peasant cultivation.[50] In Tanganyika, as noted above, the government initiated a "plant more crops" campaign aimed especially at expanding cotton cultivation, while in neighboring Kenya efforts were made to stimulate African cash crop production for the export market, on the assumption that African farmers could operate more easily under the lower profit margins than the faltering European estate sector.[51]

But even as they did, administrative and technical officers were troubled by signs of land degradation and fears that such rapid expansion would lead over time to serious social and ecological consequences. The frantic push for increased commercialization and expansion of agricultural production was blamed for encouraging "selfish individualism" and an over-taxing of the soil. Objections were perhaps most vocal and politically charged in settler-dominated colonies like Kenya, but throughout much of British colonial Africa and elsewhere the sharp rise of land under cultivation in the interwar years roused unsettling concerns about soil depletion, land utilization, and indigenous crop and animal husbandry practices, and contributed to the emergence of a new conservationist ethos among colonial agricultural officers.[52] In South Africa, a major step toward raising a more widespread concern about erosion was taken by the South African Drought Investigation Commission of 1922–23.[53] The commission's report drew attention to the settler pastoral farming areas of the mid-Cape, arguing that the veld's destruction was due to exhaustion of the vegetation and erosion of the soil brought on by burning, overstocking, the kraaling of sheep and cattle, and intensively farming for short-term economic gain. In British colonial Africa similar fears were aired, primarily in reference to African agricultural methods. In Lesotho, growing concern over signs of gully erosion along roads, paths, and cattle tracks prompted the 1934 Commission of Inquiry into the territory's financial and economic position to recommend a coordinated national anti-erosion campaign to reclaim the areas most severely affected. In Nyasaland, European officials and planters were by the early 1930s targeting the shifting system of millet cultivation in combination with the growing problem of congestion in the highland areas of the Southern Province as the causes of rapid soil erosion, deforestation, and poor husbandry. In Kenya, droughts in the late 1920s caused worry over possible land degradation, while at the same time officials became concerned about the problem of overstocking. By 1930 reconditioning schemes were initiated in Baringo District by the provincial administration. In Uganda, falling cotton crop yields in certain districts, particularly Teso, were noted by the Empire Cotton Growing Corporation by the late 1920s, prompting fears of declining soil fertility and leading the director of agriculture, John D. Tothill, to compile a series of soil surveys. In Tanganyika, meanwhile, the densely occupied central and northern areas of Sukumaland were also showing signs of declining crop yields and losses in soil fertility. Local administrative and agricultural officers such as Brian Hartley and Norman Rounce began carrying out field studies on traditional forms of soil conservation as well as introducing anti-erosion measures and such changes in husbandry practices as contour ridging to reduce runoff, early planting, the use of manure and "famine" crops, and the distribution of new seed varieties.[54] By 1929 the director of agriculture in Tanganyika, E. H. Harrison, was describing soil conservation as a top policy priority, and in 1931 a Standing Committee on Soil Erosion was set up to monitor the situation.[55] The newly formed committee decided to approach Amani, encouraging the station to conduct studies into the causes and processes of land degradation. In 1932 attention was also drawn to the problem of shifting cultivation at a conference of Central and East African soil chemists held at Amani, where it was agreed that soil surveys using standardized methods of analysis, similar to Milne's soil classifications, should be made in preparation for soil maps of the whole region.[56]

A different but no less urgent set of climatic and agroecological questions caught the attention of local technical officers working in West Africa, where alarms over drought, desiccation, and "savannization" were first raised in the 1920s. British and French foresters and botanists in West Africa had long pointed to the need for forest reserves along the northern margins of the forest belt to act as curtains against the southward spread of the so-called derived savanna zone. In the 1920s, however, there was greater apprehension concerning this transitional zone, as officials feared it was becoming progressively drier through the farming and bush-burning practices of local inhabitants.[57] Foresters and agricultural scientists, also disturbed by evidence of drought and declining yields of important tree crops such as cocoa, coffee, and kola, expressed fears that through excessive deforestation the region might be losing its "forest climate."[58] In the Gold Coast, for example, A. S. Thomas argued that the crucial factor determining the growth and cultivation of cocoa in the region was the amount and intensity of rainfall, or "degree of wetness" received in a given area during the dry season. Cocoa was indigenous to the "wet tropics," but in many parts of the Gold Coast it had to survive periods of intense dryness, which he felt was causing the plant to acquire a "deciduous habit" of losing its leaves and might be an important factor in reducing its resistance to insect pests and disease.[59] He also thought the relatively dry conditions of the country might explain why cocoa yields were initially very high, but the life of the tree often very short. Thomas's views were supported by Herbert Moor, a forestry officer in Accra, who argued that the dry northeast harmattan wind acted as a strong desiccation agent in the cocoa-growing zones. The adoption of deciduous habits, susceptibility to "die-back," falling yields, and premature die-off were all warning signs that Gold Coast cocoa trees were under great stress due to insufficient rainfall. What is more, Moor suggested that conditions were getting worse due to the effects of bush-burning by shifting cultivators, along with haphazard and excessive deforestation of the transition forest belt and its replacement by savannah. Moor felt that the erection of forest shelter belts and shade cover would help, but that such remedial measures were already too late to save "another prosperous tract of country [which] is in the process of degeneration to comparatively infertile savannah."[60]

Moor originally presented his findings at the Second West African Agricultural Conference held in Accra in 1929, calling for expanded forest reserves and shelterbelts as a means of retaining moisture and increasing rainfall. The CO's agricultural adviser, Frank Stockdale, attended the conference to discuss, among other issues, the concerns raised about cocoa cultivation.[61] In the event, Stockdale rejected Moor's desiccation analysis, drawing attention instead, as we have seen in chapter three, to the need for increased research through the setting up of a central cocoa research station and collaboration with investigators at the ICTA in Trinidad, who were grappling with the problem of rehabilitating that colony's cocoa industry as a result of declining yields brought on by the spread of "witch broom" disease.[62] The similarity of developments in the regions, especially the common pattern of premature die-off, suggested to Stockdale that the problem was systemic, and that further investigation of new, higher-yield varieties and methods of disease treatment offered greater hope for a solution. However, the appearance in 1937 of a new disease of cocoa, "swollen shoot and die-back," prompted the CO to send H. A. Dade to investigate. He drew the conclusion that it was a soil deficiency disease, due to unfavorable environmental conditions caused by a lack of shade and forest shelterbelts. For Dade as well as local foresters like Moor, the "swollen shoot phenomenon" appeared to be the result of a secondary infection in trees planted in unsuitable conditions for cocoa. The real problem, they felt, was drought die-back due in large measure to the increase in desiccation brought on by the clearing of protective bush and original shade trees.[63]

Despite the concerns of local colonial research officers in the early 1930s, however, the problem of soil fertility and deterioration was not extensively reviewed in London by the CAC until the end of 1937. In fact, before the mid-1930s, soil conservation hardly registered in metropolitan and imperial research currents. At the ICTA in Trinidad, for example, the postgraduate training program for colonial agricultural candidates gave only cursory attention to the subject.[64] It was only briefly noticed, if at all, at various imperial and colonial agricultural conferences in the early 1930s.[65] Yet, from the middle of the decade on, the problem of soil erosion increasingly caught the attention of officials and experts in London as the concerns of local authorities were lifted onto the imperial stage. A kind of erosion scare began to grip officials, academics, and specialists, facilitated by the exchange of ideas and transmission of knowledge through such forums as the Royal African Society, which held a special dinner to discuss the problem in 1937.[66] A resolution was passed insisting "that immediate steps should be taken for the adoption of a common policy and of energetic measures throughout British Africa in order to put an effective check upon this growing menace to the fertility of the land and to the health of its inhabitants."[67] The sense that humans were reaching the limits of unfettered exploitation of the land was compounded in the 1930s by a series of droughts over large areas of the United States, Australia, and sub-Saharan Africa. The dust storms and drought that swept across the Great Plains in the 1930s are but the most devastating and well-known of these climatic events, which inspired a global debate about the limits of material progress and the capacity of the earth to sustain human existence. Many observers suggested that these areas, under both climatic and human pressures, were drying up and becoming uninhabitable.[68]

The concerns of local researchers in West Africa like Thomas and Moor took on even greater significance after outside experts drew attention in the mid-1930s to the fact that the region of the Sahara and the western Sudan was becoming increasingly arid, and that the populous countries between the Sahara and the humid belt bordering the Gulf of Guinea appeared to be in the process of degeneration into an inhospitable wilderness. One of the foremost British forestry experts who popularized this view was E. P. Stebbing, a professor of forestry at the University of Edinburgh and former officer in the Indian Forest Service.[69] According to Stebbing, the Sahara was advancing at an average rate of one kilometer a year. The most seriously affected area was Northern Nigeria, where the view of the future could not have been more wretched: "The people are living on the edge, not of a volcano, but of a desert whose power is incalculable and whose silent and almost invisible approach must be difficult to estimate. But the end is obvious: total annihilation of vegetation and the disappearance of man and beast from the overwhelmed locality."[70] In response to Stebbing's dire warnings, a joint Anglo-French Boundary Forest Commission was appointed in 1936, which found little evidence that the desert was actually encroaching.[71] The problem, as Dudley Stamp later explained, was that Stebbing read the West African landscape as a forester, not as a plant ecologist or climatologist, and therefore presupposed a state of climax forest vegetation that was being progressively degraded, whereas most of Nigeria was in fact made up of new vegetative covers formed by shifting cultivation and bush fires.[72] The commission nevertheless concluded that West Africa was threatened, not by an invasion from the desert without, but by a man-made desert generated from within, largely due to the uncovering of light soils for cultivation. In response, Stebbing shifted his emphasis to human-induced change, arguing that greater population pressure was leading to soil erosion and land spoliation.[73] The culprit, according to Stebbing, was the "native," whose "improvident habits" of shifting cultivation and great herds or flocks of cattle, goats, and sheep were hacking, burning, and grazing the forests into extinction. With excessive forest destruction came "sand invasion," the dying-out of natural vegetation as existing water supplies dried up, and, eventually, the transformation of the savannah forests into complete desert. Such gloomy predictions of desiccation and loss of productive resources led G. V. Jacks and R. O. Whyte of the Imperial Bureau of Soil Science to write their influential world survey of soil erosion, entitled, significantly, *The Rape of the Earth.*[74] Stebbing's findings also influenced the French forester André Aubréville, who coined the term "desertification" to describe the man-made process of land degradation in the West African Sahel.[75] They were not alone. Increasingly, British colonial periodicals and agricultural journals in the mid-to-late 1930s and 1940s were filled with articles and discussion on the dangers of soil erosion and ecological degradation. Many warned of a pending colonial environmental crisis.[76]

The first concerted move by the Colonial Advisory Council on Agriculture and Animal Health in London came in October 1935, when the council called for the appointment of full-time soil erosion officers in each of the Central and East African territories. Then, in early 1937, Stockdale made a tour of East Africa with the object of examining current research work at Amani, and, more generally, studying the problem of soil erosion.[77] Upon Stockdale's return and for the remainder of the decade, the CAC was increasingly taken up with the problem of soil erosion and conservation. The council felt soil erosion had become a serious problem and should be brought to the notice of all local colonial governments, impressing upon them the need for anti-erosion measures. By the end of 1937 the number of official reports and memoranda prepared or received by the CO on the subject of erosion had grown substantially. The most serious losses, it was agreed, were occurring in East Africa, but concern was also expressed regarding other territories, including Basutoland, Ceylon, Cyprus, Jamaica, Northern Nigeria, Palestine, and St. Vincent, prompting the CAC's Standing Committee on Agriculture to suggest a general review of the problem.[78] The council met in December 1937 and recommended that soil erosion be treated as a "major question of policy." Following the council's advice, the colonial secretary, Ormsby-Gore, sent out a circular to all colonies requesting that separate annual reports on soil erosion be prepared and submitted jointly by the administrative and technical departments, beginning with the year 1937. In these reports, he expected a review of measures and progress already made, as well as a summary of research work and legislation. The replies would then be examined by the CAC and compiled into a general review for the year. Future reports would supply updated accounts, supplementing what had already been provided.[79]

By 1938, then, the problem of soil erosion had become a matter of central concern for the specialist advisers in London. Every colonial administration was expected to be alive to the potential threat and to report its progress in "combating" soil loss to the CO in London. By the late 1930s the CAC had emerged as the leading imperial think tank in the campaign to preserve the empire's dissipating "natural capital." Ostensibly, the goal was still to make the dependencies more productive and to ensure the basis of future capitalist accumulation, but gone was the buoyant talk of vast "undeveloped estates" and the need to stimulate the stagnant demographic growth of the inhabitants. Science and technical expertise were as important as ever, but now the mission was to "save the soil" from burgeoning human and stock populations and the "improvident habits" of colonial peoples. Science, as Libby Robbin suggests, "became the voice of reason and restraint, of management for a long-term yield."[80]

The shift in the council's terms of reference reflected the cumulative concerns of local colonial agriculturists working in the field, as well as changing trends in imperial and, indeed, international scientific research currents, but the main arbiter of the new policy was the CO's agricultural adviser, Frank Stockdale, who emerged as the leading expert on soil erosion in the British empire. Unlike most of his fellow CAC members, whose agricultural and scientific background was based in Britain and who dealt with a range of crops and livestock primarily found in northern temperate countries, Stockdale's career had been almost exclusively formed in the tropics as part of the colonial agricultural services. It was in Ceylon, in particular, that Stockdale began to take an active interest in soil conservation work. Ceylon was part of what was known as the "moist" or "wet" tropics, where the felling of trees on hillsides and near watersheds had long been recognized to contribute to erosion.[81] The clearing of forests and vegetation exposed surface soils, which were carried off by rivers or washed down slopes by intense, heavy downpours. Under Stockdale's direction between 1916 and 1929, the agricultural department devoted considerable time and energy to the prevention of soil loss and the maintenance of soil fertility in the island's hill regions, where the clean-weeding practices favored by the tea and rubber estates were found to be a main cause of depletion. Drawing on the experiences of the Dutch in Java, the department experimented with the use of contour drains, silt pitting, low contoured stone walls, and, especially, ground covers, which latter were found to be the most efficient way of minimizing erosion by preventing run-off while at the same time serving as green manures to increase fertility.[82] Stockdale continued to be a strong advocate of soil conservation even after his appointment to the CO. In 1929, for example, he spoke at the annual meeting of the Empire Cotton Growing Corporation, where he called for the appointment of a general officer to deal with the whole question of the prevention of erosion in the colonies. Such an officer, he felt, should spend considerable time studying the different measures of control and ways of improving soil structure and fertility that had been effectively adopted, not just within various British colonies but also in countries outside the empire as well.[83]

In essence, this is exactly what Stockdale did in the years that followed, using his many tours of the empire to gain a wider knowledge of and perspective on the problem. His tour of East Africa in 1931, for example, included an excursion to South Africa to learn more about the pioneering work being done there. Stockdale also kept informed of international efforts, attending the 1939 Conference on Tropical Agriculture held in the Netherlands and, most importantly, touring North America in 1937 to see at first hand the work of the U.S. Department of Agriculture's Soil Conservation Service (SCS), which was set up in 1935 under the direction of Hugh Hammond Bennett to combat the devastation wrought by the infamous North American Dust Bowl.[84] The American example heightened colonial, and indeed global, awareness of the dangers of soil erosion, especially among soil and forestry experts in Southern and East Africa.[85] It also made a deep impression on Stockdale and other CO advisers, who seemed awed by the SCS's emphasis on scientific planning and coordination, extensive use of technical experts, large-scale interventions, and unilateral state action.

Stockdale's extensive experience and touring in the colonies and elsewhere undoubtedly made him more aware of and responsive to the problems being raised by local agricultural officers. Although he certainly shared the views of other council members regarding the importance of fundamental and commodity-based research, he had a much greater sense

many native communities, a very different and more complex picture of the relationship between local population and resources was frequently uncovered. Concern about population growth and the consequent pressures on land, food resources, and social facilities, as well as about increasing urban migration and social turmoil, was perhaps most pronounced in India, where after a half-century of devastating droughts, famines, epidemics, and declining life expectancy population figures began to steadily rebound, climbing from just over 300 million in 1921 to nearly 400 million by the census of 1941.[94] Sir John Megaw, the director general of the Indian Medical Service, and later president of the India Office Medical Board and medical adviser to the secretary of state for India, believed that India's population had doubled in a generation, rising dramatically since 1930. The country, he believed, was on the brink of a Malthusian crisis which required only a succession of bad seasons to ignite.[95] Outside India, trepidation about surplus population was initially expressed in overcrowded areas of the empire such as the West Indies, where restrictions on immigration to the United States added to chronic fluctuations in unemployment in the region's sugar industry, or in settler colonies, where the demarcation of reserves and alienation of land for whites put an effective brake on the territorial expansion of indigenous communities.[96] An increase in male labor migration in the wake of the Depression, which was seen as the cause of mounting insecurity and social breakdown in rural areas as well as vagrancy and crime in the newly formed urban centers, also generated fresh anxieties within missionary and anthropological circles.

In the early 1930s, for example, Joseph H. Oldham's boundless energy and attention were drawn to the movement and concentration of population brought on by the phenomenal growth of the mining industry in the Northern Rhodesian Copperbelt. With the discovery of valuable copper ores, a once sparsely populated region had been transformed almost overnight into a great industrial mining complex with a large urban African population. As Oldham noted:

> One prodigious consequence of the introduction of these new forces into Africa is the migration of population in search of employment, which is the outstanding fact in the life of the continent to-day . . . The problem is unduly simplified if we shut our eyes to the fact that the native peoples are growing. They are receiving education. They are absorbing new ideas. They are becoming articulate. They are developing their

> own aspirations and ambitions . . . Of hardly less significance are the first beginnings of trades-union organization, and of recourse to modern methods of industrial conflict such as the strike. The economic issues arising out of the presence of cheap labour have cut clean across the racial division . . . European civilization in Africa has to contend with two forces which it may find itself unable in the end to control. The one is the ravages of insect life and bacteria . . . The other is the untameable impulses of man himself.[97]

Indeed, the unprecedented social and economic changes taking place in the Copperbelt were potentially explosive: a large migratory male workforce; poor and inadequate housing and sanitation in the newly built company towns; next to nothing in the way of schools, hospitals, and other social services; a de facto color bar between white and black labor which kept wages for African workers abysmally low.

As secretary of the International Missionary Council (IMC), Oldham proposed that the council's Department of Social and Industrial Research undertake a thorough study of the conditions of African industrial life in the Copperbelt. The department's director, John Merle Davis, was put in charge of a commission of inquiry whose report was published as *Modern Industry and the African.*[98] One of the crucial sociological problems identified by the commission was rapid population growth: "It is significant that the total number of Natives for the Protectorate has increased 32.9 per cent during the decade ending 1931. A number of factors are responsible for the phenomenal growth, all of which are connected with the superimposition of European civilisation . . . Should the conditions favouring this effective Native fecundity continue at the same rate we might reasonably expect the population to double in from thirty-five to forty years."[99] Moreover, the opening of the copper mines and the demand for labor brought with it a great movement of population, especially male migrant workers, which it was feared was leading to a breakdown in customs, the spread of nonindigenous infectious diseases, and a dependence on wage employment which, in times of economic downturn, would lead to destitution, crime, and violence. The link between "native fecundity," unemployment, and subversion could not have been made more clear. In many ways, the commission was ahead of its time. Its call for greater labor stabilization through government-missionary cooperation in the provision of social welfare and community planning services at the mines prefigured postwar initiatives.

As the IMC's investigation of African mining communities indicates, the sudden appearance of "surplus population" in the 1930s had as much to do with growing concerns over colonial poverty and social upheaval as it did with natural increases. Between 1929 and 1933, the problem of labor shortage, which had consumed nearly every African colony in the first three decades of the century, was transformed into the new problem of *excess* labor.[100] The collapse of export prices during the Depression triggered a sharp fall in overall wage employment and wage rates not just in tropical Africa but throughout much of the colonial empire. Despite this, rural workers continued to enter the labor market in increasing numbers, prodded by the depressed state of small commodity production and the need to generate income to pay colonial taxes. In the aftermath of Depression, as world prices began to climb once more, the demand for labor increased, but employers persisted in keeping wages low. The result was a drop in real incomes and erosion in living standards of wageworkers as food, imported goods, and urban rents became more expensive. The most visible signs were the swelling ranks and escalating unrest of the unemployed and underemployed in many of the empire's important centers of production and communication, especially in the "tropical slums" of the West Indies. But population pressures were also surfacing in rural areas, where alarms were being raised over the fragmentation of holdings, rural indebtedness, and land shortages. Official concerns over rising population pressures, land shortages, overcropping, overgrazing, soil erosion, and the fragmentation of holdings in the African-occupied areas of Central and East Africa testify to the fears of rural landlessness and unemployment and to the fact that increasing social divisions were emerging in the African reserves—divisions in which wage labor played an increasingly vital but complex role as a source of income for rural households.[101]

It is with this backdrop of growing landlessness, the threat of rural emigration, and rural and urban unemployment in mind that missionary reformers, anthropologists, nutritionists, public health officers, agricultural advisers, and others grew increasingly alarmed over the signs of surplus population among urban and rural communities. The "insistent pressure of the march of events" from below, as one CO adviser described it, gave a new sense of immediacy to the findings of researchers and technical staff on the peripheries.[102] Experts and authorities were particularly disturbed by the degree of "detribalization" among "native" workers, many of whom, as the economic slowdown of the early 1930s revealed, were either unable or unwilling to return to their former villages in times of dearth.

It was this "unwanted drift" of "derelict labor" that officials feared would provide fertile ground for subversive doctrines and anticolonial agitators. In anticipation of the more comprehensive intellectual framework of the late 1930s and 1940s, as we shall see, they urged the introduction of rural stabilization programs to strengthen indigenous ways of life and arrest the flow of migrants leaving the countryside to seek work in the towns.[103]

To this consternation over surplus population, as Michael Worboys has shown, was added the related "discovery" of colonial malnutrition and undernourishment.[104] As chapter four outlined, colonial public health officials, alarmed by the spread of communicable, infectious diseases, the prevalence of unsanitary living conditions, and mounting evidence of high maternal and infant mortality rates in many colonial areas, began in the 1920s to direct more attention to the problems of health and welfare. Under the rationale primarily of boosting indigenous population rates and increasing the productivity and efficiency of labor, they lobbied for the creation of rural dispensary and auxiliary clinical services, better housing, water supplies, and waste disposal in villages, the extension of mother and child welfare services, and the introduction of women's educational training programs.[105] It was out of this general concern for ill-health and demographic decline that research on diets and physique, food security, and malnutrition initially began. The first important study to draw attention to the subject of colonial nutrition was made by Dr. John (later Lord) Boyd Orr, the director of the Rowett Research Institute in Aberdeen, and John Gilks, the director of medical and sanitary services in Kenya, who in 1927 carried out a comparative investigation into the health and diet of two East African ethnic communities, the Maasai and the Kikuyu.[106] According to Boyd Orr and Gilks, East African "natives," far from being "noble savages" unaffected by malnutrition and the various prevalent diseases that afflicted the peoples of "civilized" countries, were in fact plagued with the same bad dietary habits and poor physique—and were arguably even worse off, having an infant mortality rate much higher than that of Britain and suffering from other tropical diseases such as malaria, yaws, and illnesses due to intestinal parasites. Their work caught the attention of Lord Balfour and Leopold Amery, and as a result a new subcommittee of the Committee of Civil Research was set up to investigate dietetics in East Africa, including a full survey of the health and diet of the Maasai and the Kikuyu under the direction of Gilks.[107] The final report, which was published by the Medical Research Council in 1931, found that malnutrition among "native tribes" was largely due to inadequately balanced diets, especially the lack of animal proteins, fats, and calcium among agricultural peoples like the Kikuyu, and it suggested that the problem was undoubtedly common to many different areas of the tropical dependencies.[108]

Boyd Orr and Gilks's study served as an important stimulus to further nutritional research. It was recognized that more intensive studies of "native dietary" undertaken in different regions were urgently needed before further generalizations could be drawn. One group of specialists eager to heed the call was the rising generation of social scientific investigators funded by the International Institute of African Languages and Cultures (IIALC), who seized upon the field of nutrition as an ideal practical application for the new functionalist anthropology being pioneered by Malinowski at the London School of Economics. The IIALC, it will be recalled, was instrumental in providing fellowships for a number of qualified researchers, many of them students from Malinowski's LSE seminar, to carry out fieldwork in Africa with the aim of studying the impact of Western cultural and economic forces, and especially the effects of "detribalization," on the cohesiveness and stability of traditional African societies. The sociological and psychological aspects of health, diet, and nutrition were among the subjects to be investigated.

Undoubtedly the most influential analysis of African diets and food consumption before the war was carried out by Audrey Richards, Elsie Widdowson, and Lorna Gore Brown on the Bemba people of Northern Rhodesia.[109] The Richards team found that with one or perhaps two good meals a day, the Bemba millet-based diet, including little or no animal protein or raw green vegetables, amounted to no more than half the calorie intake of the average European and was deficient in fat and vitamin C. There were also marked seasonal variations, with critical food shortages occurring two or three months of the year in many districts, during which people were forced to rely on wild fruits, grubs, insects, and honey to stave off hunger.[110] Even more striking were the pronounced variations in food consumption patterns found both across villages and at the individual and household level within villages. It appeared that food shortages were more serious, communal meals less common, and kinship ties weaker among households in villages closer to towns, where the high percentage of men away at work led to shortages of labor for making gardens. Marginal groups such as old widows, deserted wives, and youths looking for work, who were less able to draw upon kinship networks for access to resources and support, were more likely to suffer from inadequate diets and were far more vulnerable to shortages in food supply. From her assessment of the data, Richards concluded that the integration of the region into a market economy and the resulting high rates of male absenteeism due to labor migration were leading to increasing social differentiation of individuals and households, and that this was the source not only of greater social insecurity but also of the breakdown of conjugal ties and kinship relations that held traditional African village communities together. The pattern of economic and social change in Bemba communities, as Henrietta Moore and Megan Vaughan have recently argued, was far more complicated than the picture that emerges from Richards's account, but what is important for us to note is that her study was both informed by, and, in turn, helped to reinforce the emerging colonial development and environmental narratives of the 1930s and 1940s.[111]

The Richards team had provided, as was noted by Raymond Firth, one of Malinowski's staunchest supporters at the London School of Economics, a potential model for future field studies, in which the value of the biochemical analyses of diets offered by nutritional science could be greatly enhanced by taking into account the way in which the production, preparation, and consumption of food was shaped by cultural systems and preferences, social and sexual relations, kinship obligations, and ritual barriers.[112] Such studies could be expected to pay real practical dividends by helping to determine the nutritional standards and material well-being of specific communities and how they might be improved. In 1934, at the suggestion of Firth, the IIALC appointed a small committee to consider the possibility of cooperation between anthropologists and other scientists in the study of "native dietectic" problems, under the premise that improvements in health should begin with the provision of adequate and well-balanced diets.[113] The committee pressed for more long-range research on specific villages as well as general dietary surveys carried out on a regional basis, since the few studies that had already been completed demonstrated the importance of local variations in soil, climate, and native food crops. Such interdisciplinary studies could best be carried out by special teams of metropolitan experts: biochemists, medical officers, agriculturalists, veterinary officers, and anthropologists, who would be sent out to work closely with local research officers and laboratory facilities in the field. But above all, shortages of food supply and dietary deficiencies demanded a combined and coordinated approach involving many different branches of government, since the crux of the problem, as Boyd Orr situation of global food supply, in which departments of agriculture in the advanced countries were restricting food production or curtailing imports in an effort to deal with the apparent problem of overproduction, while departments of health and medical advisors in these same countries were raising alarms over the serious underconsumption of the same foods. Such incongruity pointed to the urgent need for a greater coordination of government agricultural and health policies, and, further, the need for a large-scale inquiry to determine the world food position. Bruce's proposal for "a marriage of agriculture and health" was accepted by the assembly, which decided to set up a Mixed Committee on Nutrition under the chairmanship of Lord Astor.[122] The committee's report represented an important turning point as it endorsed the principle of optimum (rather than minimum) dietary standards to promote "positive health," and urged member states to set up national nutrition committees to study its recommendations.

The work of nutritional, medical, and social researchers like Boyd Orr, Gilks, and Richards, together with pressure from League of Nations initiatives, finally motivated the CO to take the problem of malnutrition in the tropical colonies more seriously. In November 1935, shortly before the League's Technical Commission on Nutrition held its first meeting in London, Bruce's economic adviser, Frank MacDougall, sent a paper to Gerard Clauson, head of the CO's Economic Division, entitled "Can Improved Nutrition Solve the Agricultural Problem?"[123] In it MacDougall drew upon the findings of Boyd Orr, emphasizing the importance of "protective foods" and pointing to the strong correlation between low income and inadequate diet, poor physique, poor mental alertness, and lower resistance to infection. Malnutrition appeared to be widespread among the working-class population of more advanced countries like the United Kingdom and the United States, and was the "common fate" of the bulk of humanity living in India, China, and Africa. Drawing upon recent discussions at the League's Assembly, MacDougall highlighted the current paradox of overproduction but underconsumption, and the need for an international plan for "a marriage of agriculture and health." It was a not-so-gentle reminder that the current widespread discussion of nutrition and public health and its close connection with agricultural problems would soon force governments around the world to take some form of action to improve nutritional standards and the general welfare of their people.

MacDougall's memo came as no surprise to CO officials, who were well aware of the political repercussions of the heightened awareness over nutrition and were in fact already planning their own "monumental circular despatch" on the subject.[124] In 1936, a circular was sent out by the colonial secretary, J. H. Thomas, requesting all colonial governments to provide information on the nutritional status of their indigenous populations, and a Standing Committee on Nutrition in the Colonial Empire was established under the auspices of the Economic Advisory Council (EAC) to review the replies and compile an empire-wide survey of colonial malnutrition.[125] Many of the members of the Committee on Nutrition were either recruited from or would later serve on the CO's standing advisory committees on agriculture, education, and medicine.[126] The replies to the secretary of state's circular made it abundantly clear that the problem went beyond simple "dietetics." Undernourishment was described as the most pressing issue, while poverty, ignorance, and, in certain territories, increasing population were identified as the chief culprits. Nutrition was perceived as less a public health or medical concern than, as Boyd Orr had observed, an agricultural and educational problem linked to a myriad of other problems—including the expansion of cash crop production for export at the expense of subsistence food crops, declining yields due to soil erosion, wasteful husbandry practices, high infant mortality rates, backward customs, and the lack of education and ignorance of women. On the recommendation of the committee, Dr. B. S. Platt was chosen to undertake nutritional surveys in the colonial empire, the first of which was carried out in Nyasaland in 1938. Platt's suggestion that specially trained officers be provided for all the dependencies to coordinate the activities of the many departments involved was also acted on, and by 1946 the first nutrition officers were being stationed in various colonies.[127] From the late 1930s, then, colonial malnutrition and undernutrition became a widely debated issue within colonial and scientific circles—and nutrition, along with soil erosion, became the key variables around which a number of interconnected crises were framed within the shifting colonial discourse of development.[128]

By 1940 the central assumptions that grounded colonial development doctrine had been turned on their head. No longer did development signify the "constructive" exploitation of the "imperial estates" through the expansion of transport and communication facilities and tropical export products. Tropical agriculture had become less geared toward fundamental and commodity-based research tied primarily to the plantation industries and much more involved in the development of agriculture, local

Plate 4. F. L. Engledow, Drapers' Professor of Agriculture, Cambridge, 1930–56. Copyright Godfrey Argent Studio. Royal Society, IM/GA/WS/1464

Plate 5. Sir Frank Stockdale, agricultural adviser to the secretary of state for the colonies, 1929–40. F. A. Stockdale, T. Perch, and H. F. Macmillan, *The Royal Botanic Gardens, Peradeniya, Ceylon, 1822–1922* (Colombo: H. W. Cave, 1922)

Plate 6. James Mackie (*center*), director of agriculture, Nigeria, 1936–45. From the collections of the Bodleian Library, University of Oxford, Mss. Afr. s. 1996 (2), p. 2

Plate 7. Experimental field trials, Kericho District, Kenya. Copyright Kenya National Archives. From the collections of the Bodleian Library, University of Oxford, Mss. Afr. s. 1717 (81), fol. 16

Plate 8. Citemene cultivation, Northern Rhodesia. From the collections of the Bodleian Library, University of Oxford, Mss. Afr. t. 41, p. 8

Plate 9. Communal bush-burning, Gold Coast. From the collections of the Bodleian Library, University of Oxford, Mss. Afr. s. 1812, p. KM 3/3 21

Plate 10. Ground cover and stone terracing on rubber estate, Ceylon. From the collections of the Bodleian Library, University of Oxford, Mss. Brit. Emp. s. 311 (11), item no. 3

CHAPTER 6

View from Above

The Consolidation of Knowledge and the Reorganization of the Colonial Office, 1935–45

> Every year it becomes more and more manifest that health, agriculture and education in the Colonial Empire are very closely bound together; that they are not so much three separate subjects as three aspects of the same subject, social and economic welfare, and that each aspect must be considered in relation to the others.
>
> —*Malcolm Macdonald, Secretary of State for the Colonies, 1938*[1]

BY MOST ACCOUNTS, the years just before and during the Second World War constitute a turning point in the history of metropolitan policy toward the colonial empire.[2] The problem of development was revised substantially from the earlier visions of Amery and his fellow "Constructive Imperialists." Development thinking in the wake of the Depression swung sharply away from the Chamberlainite "Imperial Estates" doctrine, with its emphasis on augmenting British power and wealth and reducing British unemployment, toward a policy which can best be described as an attempt to ameliorate colonial conditions. Officials and experts in London were confronted in the late 1930s by an empire that appeared to be in the grips of a series of dramatic social, economic, and ecological crises; the administration was shaken out of its usual slumber to find the very legitimacy of the colonial project challenged from both above and below. The most notable outcome of this reckoning was the much vaunted Colonial Development and Welfare (CD&W) Act of 1940, which along with subsequent acts committed the British government to providing direct subventions of imperial aid to help improve colonial living standards. The acts were a landmark not so much for the actual sums disbursed as for the acceptance of the principle first articulated by Chamberlain nearly half a century earlier, that development was something that could be *made to happen* by means of state agency, planning, and intention.[3] The motivations which lay behind this novel stance had less to do with philanthropy than the conscious need to intervene to forestall and manage what one senior CO official later candidly described as a "social revolution."[4]

But perhaps more than anything, the years between 1935 and 1945 opened up a whole new space for the expert—a space in which the members of the CO's advisory bodies figured prominently. The CO's advisory network provided an important forum not only for those with close official connections but also for academics and prominent critics interested in colonial affairs. As outside experts separate from the traditional geographic departments, they were to a certain extent free from the bureaucratic routine and procedure of the old CO structure. Out of the advisory discussions and committee work of the late 1930s and early 1940s emerged a wide consensus that colonial agricultural, educational, and health knowledge and expertise needed to be mobilized and the corresponding technical services better coordinated in order to bring about a comprehensive scheme of rural community development. The reports and memoranda produced by the CO's advisory experts would provide the blueprints for the new integrated approach that, as we will see in the final two chapters, underpinned the conceptual framework not only of the early postwar colonial development drive but of much international development and environmental policy and state practice ever since.

CO Advisers, Neo-Malthusian Crisis Narratives, and the Concept of Rural Community Development

The concerns about land degradation, soil erosion, overcrowding, and malnourishment analyzed in the previous chapter were compounded in the 1930s by the fallout of the Depression and the anxieties it generated both in Britain and the colonies. By 1932 world industrial production had been reduced by two-thirds what it was in 1929. In Britain, production of iron and steel dropped by half, shipbuilding came to a grinding halt, and by the summer of 1932 some three million men and women were unemployed and almost seven million people were on the dole.[5] The British national government responded by going off the gold standard, which sent the pound tumbling and threw European currency markets into chaos. Governments tried to deal with the crisis by erecting tariff barriers and import quotas, causing world trade to fall by 60 percent between 1929 and 1932. The total value of British exports fell by half. The collapse of international markets hit colonial economies even harder, leading, as noted earlier, to a dramatic drop in earnings from exports and sharp declines in wage employment and wages rates.[6] African peasant producers were also affected, especially in West Africa where higher prices for import goods motivated farmers to expand the volume of exports as a way of maintaining living standards, but to no avail. Even when prices began to slowly recover after 1934, the market for primary goods remained sluggish and was hampered by another major cyclical downturn in 1937–38, which yielded a serious deterioration in livelihood and real income both for commodity producers and wage earners. Many plantation and mining companies adjusted to the prolonged crisis by scaling down operations or even liquidating their holdings, forcing thousands of migrant laborers out of work and rendering them unable to pay their taxes. The loss of tax and customs revenues placed tremendous strain on colonial governments, which responded by cutting staff, vital services, and public works and searching for new sources of income through "plant more crop" campaigns and the intensification of peasant agriculture.

Local indigenous reaction to the erosion of living standards took many forms: urban protests, spontaneous demonstrations, strikes, riots, and commodity "hold-ups." The first major disturbance was the massive strike in the Northern Rhodesian Copperbelt in 1935. Organized without the aid of trade unions, it spread by way of religious organizations like the Watch Tower Movement, dance societies, and mass gatherings of both miners and others. In West Africa, meanwhile, miners, railwaymen, and public sector workers led the way in forming unions and organizing strikes and other forms of militant action in Sierra Leone, Gold Coast, and Nigeria.[7] In 1939 a wave of strikes and work stoppages spread across British colonial Africa, including the railwaymen's strike in the Gold Coast, the Dar es Salaam dock strike, and the Mombasa general strike. The increasing labor organization and action in Africa took on added significance in the context of the West Indian disturbances of 1935–38. There, reaction to growing unemployment and underemployment, low wages, and widespread urban and rural deprivation made itself felt in a series of strikes, riots, and social protests that broke out in island after island, beginning in 1935 with riots on St. Kitts and culminating with an island-wide rebellion in Jamaica in May and June 1938.[8] A similar chain of events unfolded in other sugar-producing islands such as Mauritius, where in 1937 the cut in cane prices set off the worst outbreak of violence and rioting in the island's history as laborers walked off the job on many estates in support of small planters.[9] Colonial unrest continued and in fact intensified during the 1940s under the hardships and severities of the war. A further wave of strikes and protests spread across the vital ports and communication centers of empire, culminating in a colony-wide movement in Nigeria in 1945.

For Britain, the Depression brought an historic end to free trade doctrine with the passing of the Import Duties Act and the embracing of the principle of "imperial preference" for colonial products at the Imperial Economic Conference in Ottawa in 1932. These measures allowed the empire, especially the dominions, to play an enhanced role in Britain's trade in the second half of the 1930s, and arguably brought about a speedier and less turbulent recovery compared to some of the other Western industrialized countries.[10] But despite the relatively smooth recovery "at home," the Depression still left an indelible mark on the British psyche and imagination, perhaps best captured by George Orwell's gripping portrayal of working-class life in the industrial north in *The Road to Wigan Pier.*[11] One trend, already under way but unequivocally confirmed during the Depression, was a striking decline in the country's birth rate and a consequent reduction in the average size of the British family. The proportion of families with five or more children fell from 27 percent before World War One to fewer than 10 percent in the 1930s, provoking a depopulation panic among pro-natalist groups. Indeed, in the writings of prominent metropolitan intellectuals, authors, and scientists like Enid Charles, A. M. Carr-Saunders, Julian Huxley, and H. G. Wells, the "population problem" was seized upon as the most serious and fundamental question facing Britain and the empire. Magnifying this fear of depopulation was, as noted in the previous chapter, the sudden discovery of "surplus" population in many of Britain's colonial possessions, the significance of which was made graphically clear by the wave of strike action and social unrest that swept across the empire in the late 1930s and 1940s. To the general consternation over increasing numbers, Huxley and Carr-Saunders added the disturbing specter of racial conflict and "swarming," noting that birth rates of British and other "white nations" were declining at the same time that population was beginning to shoot up among nonwhite peoples of the empire, auguring a fundamental shift in the demographic balance of world power.[12] A kind of neo-Malthusian crisis narrative began to creep into public discourse as well as official colonial reports and correspondence, which would lead many observers and scientific advisers by the end of the decade to urgently press for aggressive state intervention into colonial societies.

Already by 1935, Sir Daniel Hall, the former chief scientific adviser at the Ministry of Agriculture and director of the prestigious John Innes Horticultural Institute, was drawing attention to the marked increase of population in many parts of the empire, especially the eastern regions of tropical Africa.[13] According to Hall, many African "tribes" had arrived at a dangerous impasse into which they were sure to drift if something was not done. The price of noninterference would be high: land shortages, famine, desertification, breakdown of order, widespread social unrest, and eventually war and the disintegration of indigenous cultural organization. The problem was that traditional shifting cultivation systems were unable to cope with changing demographic conditions. Adding to the crisis was the tragedy of overstocking. Possession of large numbers of cattle and other livestock was the basis of the bride-price and therefore of marriage and kin relations, but it was also, according to Hall, leading to widespread overstocking, a reckless indifference to the quality of the stock or its commercial value, and the overgrazing and destruction of the soil itself. Hall also linked such practices to the unbalanced and inadequate diet, which was, as Boyd Orr and Gilks's investigation had indicated, a major contributor to poor physique, prevalence of disease, and malnutrition among many African communities.

"It is difficult to exaggerate the gravity of the situation," Hall asserted, "but what is to be done?" He suggested that "the European, having arrived in Africa and having exercised that major interference . . . must go on with the work and become responsible for the organization of a new society." Custom and a traditional, conservative mental outlook, mired as they were in ignorance, were seen as the immediate obstacles to be overcome. The alternative was not to be a policy of exploitation of resources for the benefit of Britain, but, rather, a wholesale plan for the amelioration of the conditions of the indigenous peoples themselves. The aim was to raise living standards not through a continuing concentration on export crops, which, Hall felt, was partly to blame for the mounting crisis, but through boosting the production and supply of food crops. Moreover, it was to be premised on the application of modern science and rational planning. The first step was the improvement of cultivation through the introduction of a fixed agriculture, which would allow a more intensive and continuous method of production. The European system of crop rotation was preferred, although the merits of mixed cropping were also acknowledged, but in either case Hall suggested the introduction of composting methods such as those pioneered by Albert Howard at Indore to improve soil fertility. The inclusion of leguminous crops such as peas and beans as part of the rotation and the introduction of meat, milk, and other dairy products through mixed farming practices would also lead to a more balanced diet and do much to ameliorate the dismal living conditions of rural Africans. While Hall was open to the use and improvement of indigenous food crops as well as introduced varieties as part of an adequate dietary, it had to be based on research and experimentation through the establishment of plant-breeding stations.

More than anything, the problem was one of education. Government action must be preceded by a massive "campaign of enlightenment" to be initiated with the cooperation of the chiefs. The whole "mental outlook" of the African had to be transformed. The shift to fixed agriculture would involve extensive changes in land tenure and custom. It required the formation of individual property or tenant rights, which would inevitably disrupt the whole social framework of the village community. What was needed most was a great expansion of rural extension services and the mobilization of the missions to supply African instructors and demonstrators. Colonial agricultural staff and scientific officers, in turn, could provide important investigations of indigenous food crops, research on the adaptation of composting to local practices, and a sustained plant-breeding program. The bulk of the extension work, however, would continue to rest with the district officers, whose influence over local chiefs and Native Councils was key to the whole "betterment" campaign. Such a far-reaching program called for a great expansion of both European and local staff, additional research, and substantial capital expenditures—expenditures which went beyond the immediate resources of the colonies themselves and could be met only by liberal financial subventions from the British government.

Although agricultural advisers at the CO tended to be more cautious and less draconian in tone, they echoed many of Hall's arguments.[14] An indication of how the changing intellectual terrain of the 1930s was affecting discussions on colonial agricultural strategy can be seen from the proceedings of the Conference of Colonial Directors of Agriculture held at the Colonial Office in July 1938. The conference was attended by a wide cross-section of experts in tropical agriculture and other related fields, as well as the members of the CAC and its standing committees on agriculture and animal health. Four main subjects dominated the agenda: soil erosion, nutrition, mixed farming, and land settlement. From the *Report and Proceedings of the Conference* it is clear that an integrated and multifaceted approach was being framed in which the preservation and improved management of the colonies' natural resources was given top priority. Given the deteriorating resource base and burgeoning indigenous population, it was argued, further improvements in the social and material welfare of the colonial peoples could be achieved only through the "proper" development of land industries and the creation of self-sufficient agricultural communities. Development, according to the conference report, meant the preservation and improvement of the fertility of agricultural lands and other natural resources through the adoption of such practices as green manuring, strip cropping, mixed rotational farming, and reforestation; greater concentration on food production and on improving the nutritional standards of colonial people through the consumption of meat, fruits, and green vegetables; the introduction of land settlement and cooperative credit and marketing schemes; and a system of agricultural education and training designed to fit the majority of people to an agricultural way of life.[15]

The new integrated strategy was primarily fashioned by the CO's agricultural advisers, Stockdale, Dr. Harold A. Tempany, and Professor Frank L. Engledow, who were instrumental in drafting the key policy statements on colonial agriculture of the late 1930s and early 1940s.[16] Two issues of central importance emerged out of these policy discussions. In the first place, it was clear that the introduction of anti-erosion measures and other agricultural improvements was intimately tied up with questions of land utilization and land tenure. While dissenting from earlier views of indigenous peoples as lazy, unskillful, and negligent, and acknowledging that their agricultural practices were often rooted in sound knowledge based on experience, Stockdale and Tempany nevertheless argued that the expansion of population under British rule had brought with it unprecedented problems which could best be solved by the modification of existing practices to suit changing conditions.[17] The practices most favored were those being pioneered by James Mackie in Northern Nigeria as part of a system of mixed rotational farming in which both animals and crops would play an enhanced role. Only in this way, it was claimed, could a more "intelligent" use of livestock be introduced that would reduce overstocking, maintain soil fertility through organic manuring, and meet nutritional requirements through the introduction of meat and dairy products into local diets. All this, however, implied the replacement of shifting cultivation and communal land tenure by a more settled and individualist agricultural system. Customary patterns of land use involving a multiplicity of widely separated small plots, with grazing pastures often miles away from arable ground, possessed the advantage of allowing everyone a fair share of the best lands, but they were not conducive to "sound" farming practices, which required compact farms. Some degree of redistribution and consolidation of fragmented holdings would be necessary. The predicament, as one agricultural officer stationed in Tanganyika pointed out, was "how to deal" with those unwilling to allow the reallocation of land already in their possession. Short of coercion, all he could suggest was to encourage the more "progressive" farmers to give up good soil in exchange for large consolidated farms of moderate fertility and the promise of government assistance, or to resettle families onto new lands with the proviso that they adopt the new farming methods.[18]

The other point quickly seized upon was that such changes depended on far-reaching government involvement in indigenous crop and animal husbandry practices, the success of which would depend upon close collaboration of administrative officers with the staff and specialists of the various technical departments. The need for a planned approach to agricultural development in this respect was first addressed by Professor Engledow in 1937, in a memo in which he argued that the time had come for colonial agricultural policy to be carefully thought out along broadly conceived lines. Only in this way would it be possible for agricultural research programs to be framed in accordance with local area needs and coordinated with government policy as a whole. Engledow's proposal provided the first rough outlines of a coordinated program of work for the colonial agricultural and veterinary departments.[19] It called for an overall review of the social, political, and economic conditions in each colony, as well as separate programs of executive and investigational work to be drawn up by the director of each department. The executive side would include schemes of agricultural development and education, while the investigational work would focus on inquiries into plant diseases and soil problems and the study of new crops, livestock improvement schemes, and new methods of cultivation. The proposal also encouraged greater coordination and cooperation between the administrative and technical departments, and between specialists and other agricultural officers, in order for technical work to be brought into step with the general policy of development. Along the same lines, it was suggested that colonial agricultural policy should be regionally integrated through such interterritorial organizations as the East Africa Agricultural Research Station at Amani. Engledow's report was endorsed by the CAC in June 1938 and issued as a circular despatch by the colonial secretary, Malcolm MacDonald, in August 1939.[20] With this despatch, according to Engledow, the work of the Colonial Advisory Council on Agriculture and Animal Health in London was reformulated. Instead of being an advisory body that dealt with colonial problems as they arose, it was to act as a central coordinating agency in matters of agricultural development policy common to all colonies.[21]

The CO's agricultural advisers were not alone in calling for a more coordinated and planned approach to rural development and welfare in the colonies. In the mid-1930s members of the CO's Advisory Committee on Education in the Colonies (ACEC) also began to take stock of the many lessons learned and ideas arising from various local, village-based educational experiments initiated since the publication of the 1925 white paper on education, and to extend those principles as the basis for a new "community development" strategy. The result was Joseph H. Oldham's 1935 *Memorandum on the Education of African Communities,* which argued that the success of any rural reconstruction scheme would ultimately rest on the education and participation of the whole community, with the village elementary school serving as the center of inspiration and "community advance."[22] It was assumed that the basis of African life was and would remain agricultural, and that, therefore, the aim of education should be "the growth of rural communities securely established on the land," producing crops primarily for their own subsistence. What was needed was a comprehensive program of rural reconstruction and community betterment—one that would demand the close integration of education policy with economic development planning, on a colony-wide basis, and the coordination and cooperation of all the different agencies responsible for social welfare work, especially the Church and missions. Knowledge imparted in the school, for example, would need to be correlated with efforts by the agricultural department to improve the economic life of the community, and with the medical department's organization of community health campaigns.

Behind the eloquent pining in the 1930s for "community development" lay the desire to counteract the effects of "detribalization" and rural migration. Far from being a romantic delusion, it was a rather desperate attempt to contain the contradictions and ameliorate the worst effects of capitalist transformation. In the face of social ills thrown up by rapid economic change, missionaries, district commissioners, and anthropologists and other colonial experts all sought to recreate the supposed organic order of "traditional" village life by invoking the ideals of "community" welfare, as against "excessive individualism" and the vagaries of the world market. "Investment in rural education is a wise investment," one colonial educationalist argued, "since it creates a reserve of contented and progressive peasantry which nothing less than a catastrophe can diminish in value . . . any programme for social amelioration that is linked onto world economy or which depends for its sinews on the Territory's export trade, is bound to suffer from the baffling rise and fall of world prices. But a programme that has in view the steady improvement of the peasant's lot by way of increased food production, better health and greater understanding is largely independent of world conditions and, since its vitality lies within itself, it can survive whatever economic blows the world may unwittingly bestow upon it."[23] Thus, the problem was the disintegrating effect of Western colonialism on indigenous people's cultures. The solution was centralized planning, conjoined with interdepartmental coordination.[24]

Colonial Office educational advisers recognized that the new emphasis on a coordinated, community-based approach would involve far-reaching changes in education. Dr. William H. McLean, who along with Oldham sat on the ACEC in the 1930s, was particularly impressed by the latter's emphasis on relating colonial education to economic development and to the work of other government departments such as public works and the agricultural and medical services.[25] What was needed was a long-term policy; this, in turn, would demand greater coordination of the CO's machinery for dealing with the various lines of long-range planning of social and economic development.[26] In 1934 McLean put together a memorandum on the need for colonial development planning and coordination, in which he suggested that some form of effective contact or liaison between the various CO advisory committees was needed. In the wake of Oldham's memo on rural communities, McLean prepared a second memo on the question of advisory committee coordination at the CO.[27] He noted that recent work by the ACEC demonstrated the need for greater cooperation by and among all agencies aiming at economic and social improvement in the colonies, and as a way of facilitating this he proposed the formation of a central coordinating committee—to consist of the heads of the General Department and the new Economic Department, the chief specialist advisers, and representatives of the three main CO advisory committees on agriculture, education, and health—which would deal with matters of general policy. Most senior staff in the general and geographic departments were opposed to the scheme. One of them went so far as to suggest that it be "humanely put to death."[28] The plan was supported, however, by the educational secretaries, Hanns Vischer and Arthur Mayhew, and Gerard Clauson, the head of the Economic Department, who felt that the recently raised issue of colonial nutrition was a case in point. Stockdale was more cautious, suggesting that a start be made by setting up an ad hoc committee of agricultural, medical, veterinary, and educational experts to examine the problem of malnutrition and its relation to health and agricultural development.[29]

In the face of CO inaction, members of the ACEC decided to form their own standing subcommittee on social and economic development, with McLean as chairman.[30] McLean's subcommittee was one of the first official forums connecting the burgeoning network of CO specialists and advisory staff.[31] Strong links were forged, for instance, between the McLean subcommittee and another important imperial coordinating effort, the Economic Advisory Council's Standing Committee on Nutrition in the Colonial Empire. As noted earlier, many colonial advisers began to see nutrition as an issue of urgent importance demanding, by its interdisciplinary nature, greater collaboration among agriculture, education, and health experts.

One product of this new collaboration was the report on the welfare of women and children in the colonies prepared by Dr. Mary Blacklock. The "discovery" of colonial malnutrition and its consequences for the welfare of indigenous peoples had led to renewed demands for additional expenditures on midwifery and childcare services and the education and training of African women. Dr. Blacklock, who served as the first woman medical officer in Sierra Leone before being appointed as the only female medical specialist on the Colonial Advisory Medical Committee, was one of the leading figures in this campaign. With the aid of a Levelhulme Fellowship she toured China, Hong Kong, Malaya, Burma, India, Ceylon, and Palestine in 1935 to survey medical efforts being carried out for the welfare of women and children. Her report reflected much of the current thinking within colonial and scientific circles on problems of welfare.[32]

Blacklock was highly critical of the one-sidedness of most colonial medical work, which she saw as not only male-dominated but specifically geared to serve the adult male members of local populations. She expressed grave dissatisfaction with the lack of attention given to and progress regarding the welfare of women and children, particularly as concerning the high rates of infant and maternal mortality, the lack of education for girls and of training of local nurses, and the dearth of knowledge about social conditions affecting women. She attributed the poor health of women partly to their own ignorance and the backwardness of their customs, and partly to the effects of poverty and deleterious living conditions, chiefly poor housing, inadequate sanitation and water supplies, and malnutrition. However, the major causes of these backward conditions, according to Blacklock, were the ignorance and prejudice of many medical administrators, and the imbalance of expenditure within the colonial medical service itself. Many male medical officers did not appreciate the valuable and even crucial work done by women doctors and medical officers. Moreover, the separation of the curative from the preventive or health branches, and the strong bias in favor of curative work and institutions, especially large hospitals, had led to a neglect of rural welfare work.

What was needed was a more unified and integrative approach aimed at improving the health and welfare of women. This would involve better-planned systems for the development of local services of well-trained nurses, midwives, and health visitors; an increase in the number of medical women; and expanded opportunities for the training of girls to become medical assistants and doctors. Welfare work needed to be incorporated as closely as possible with other medical practices by having welfare centers either situated near to main hospitals and dispensaries or combined with medical dispensaries to form composite health centers in rural areas. Blacklock praised the work of the missions, whose voluntary efforts in some places accounted for the only welfare work being done, and she cited the Jeanes system of village "uplift" as the kind of model to emulate. It was essential to have close cooperation between the various government departments, especially health and education authorities. Blacklock argued that the education of women was key to the social welfare of the whole community, since it was they who mainly looked after the home and raised the children. Greater expenditure needed to be devoted to training women teachers, while domestic science and mothercraft were advocated as the backbone of a special type of education designed to prepare girls for the "noble occupations" of wifehood and motherhood. Adult women's education was essential as well, and for this she recommended women health visitors, women's societies modeled on the cooperative principle, health exhibitions, and various other types of propaganda.[33] In a follow-up memo prepared by Dr. Philippa Esdaile, who was a reader in biology from the University of London, it was proposed to link dietetics to the teaching of "domestic science" as a way of bringing the new science of nutrition and its practical application to bear on every rural household.[34] Overall, Blacklock's report was at times highly critical of the previous colonial record and its structural and gender biases, and it marked an important shift in colonial medical discourse that helped lay the groundwork for a stronger population-based and preventive approach to colonial health.[35] But the most salient features of her analysis were her portrayal of women's health problems more as a question of official neglect than of material poverty or poor living conditions, and her consequent vision of the solution primarily in terms of enhanced expert knowledge and better-designed and better-coordinated state programs.

This was the gist of the general findings of the Sub-Committee on Social and Economic Development as well. The degree of coordination between departments and other agencies advocated by the ACEC and the Committee on Nutrition, according to McLean, went far beyond the traditional sphere of any one subject, necessitating a general review of government social services and the preparation of development programs.[36] An integrative and coordinated program was needed for each colony, covering all projected work for the next five years. This would ensure greater continuity and balance in the expansion of social services and avoid disruptions in times of economic crisis. The key was to utilize colonial resources, especially land, more efficiently and productively in order to raise revenues for social expenditure.[37] The execution of rural programs was to be aided by what the subcommittee described as "community education." What is striking for a document prepared under the auspices of the ACEC is how little it had to say about formal colonial schooling.[38] Following the principles laid out earlier by Oldham, and supported by anthropological opinion, "community education" was defined as a sweeping program of village welfare and "uplift" in the rural areas: better housing and town planning, better water supplies, sanitation and drainage; greater focus on domestic food crops to increase food supply, including the introduction of new varieties necessary for a more balanced diet and improved health; a land utilization plan taking into account its population "carrying capacity," maintenance of soil fertility, and the prevention of overstocking through improved agricultural practices and anti-erosion measures; the organization of markets through the introduction of systems of cooperation; and, finally, a better-adapted curriculum which sought to instill an agricultural and "rural bias" in the teaching of all village school subjects.[39]

The work of the CO's specialist advisers and bodies was reinforced by the proposals of the Departmental Committee on Colonial Development and by the publication of the West India Royal Commission's recommendations in 1939. These reports, together with the added urgency and weight of the war, precipitated the issuing of the 1940 white paper and the subsequent adoption of the Colonial Development and Welfare Act.[40] The 1940 act promised up to £5 million a year for ten years for development and welfare projects. This sum was substantially enhanced by the second CD&W Act of 1945, which pledged £120 million (amended to £140 million in 1950) to be made available over the full period of the next ten years. Too much can be made of the financial contributions provided by the imperial government under the CD&W acts, since, at the end of the day, far more of the revenues for development were raised by the colonies themselves.[41] Nevertheless, they signaled the CO's growing commitment to a new state-sponsored colonial mission, which encouraged metropolitan advisers and experts at the Colonial Office to become even more imaginative and inventive in their thinking.

In light of the 1940 act, the CO's education advisers, at the urging of Sir Fred Clarke and Dr. Margaret Read of the University of London's Institute of Education, decided to revisit the whole question of "community education" and mass literacy.[42] The result was the 1944 report *Mass Education in African Society*, which emphasized the need for a campaign approach whereby teams of experts would target small localities and draw on the "real cooperation" of the people. It also recommended the appointment of "Mass Education Officers" to spearhead the new initiative.[43] A further Sub-Committee on Education for Citizenship in Africa was established in 1947, followed by a CO conference held in 1948 to discuss mass education and ways of encouraging initiative in African society.[44] Finally, in 1949, a new advisory body, the Mass Education (Community Development) Committee, was formed, which defined community development as "a movement designed to promote better living for the whole community, with the active participation and, if possible, on the initiative of the community but if this initiative is not forthcoming, by the use of techniques of arousing and stimulating it in order to secure its active and enthusiastic response to the movement."[45] Included among the range of development activities at the district level was anything that would make local villages better places in which to live: better crops and livestock, better water, health, housing, and roads, infant and maternity welfare, and education of both adults and the young. Community development officers were appointed to work together with representatives from the technical departments and the provincial commissioner to stimulate and develop ideas that could be implemented at the district level in consultation with local councils. The community development model would resonate beyond the CO's advisory network, becoming enshrined in the framework of such international organizations as the newly formed United Nations Educational, Scientific and Cultural Organization (UNESCO), whose first director-general was one of the architects of the CO's scheme, Sir Julian Huxley.[46]

Meanwhile, the CO's agricultural advisers were also busy drafting the blueprints of the new postwar order. In 1940, following the recommendations of the West India Royal Commission, Frank Stockdale was appointed as the first comptroller for development and welfare in the West Indies. The proposals for agricultural reform in the West Indies developed by Stockdale and his team of experts followed closely the ideas laid out by Professor Engledow, who had served as the agricultural adviser to the commission. These proposals provided the basis for an overall reorientation of colonial agricultural policy.[47] Among the many problems confronting the islands, Stockdale considered the following to be most crucial: the increase of population throughout the whole area; shortage of employment for this increasing population; the previous wastage of natural resources; and the dearth of suitable land for agriculture. In his report as comptroller for development, Stockdale put the "long-term problem" squarely, arguing that "[i]n schemes of development, the highest priority should be given to those [measures] which will help to improve conditions in the rural areas . . . The greatest asset of the West Indies is its land, and it is the land which must continue to be the major and primary basis of employment."[48] This was the underlying message of the West India Royal Commission as well, which after many months of hearing evidence arrived at the conclusion that "the problem of the West Indies is essentially agrarian."[49] The Depression had exposed the vulnerability of primary-producing economies and the instability of international markets. These factors, in conjunction with rapid population growth and the demand for better living conditions, dictated a change in economic policy away from the traditional bias toward the sugar industry and the heavy reliance on export-oriented production, toward an agricultural policy that focused on diversification, increased productivity and efficiency, and self-sufficiency in local food supplies. Engledow was even more emphatic, arguing that a readjustment of the whole agricultural system was needed, amounting to what he described as an agricultural revolution.[50] While not ruling out export crops, Engledow saw the first object of agricultural development as one of reducing the risk of dependence on imported foodstuffs, by making the colonies as self-sufficient as possible through homegrown essential foods—for the reality was, as he told members of the CAC in 1940, "90 per cent of male inhabitants of the West Indies must look forward to an agricultural life."[51]

What was true of the West Indies, it seems, was even more the case in the rest of the colonies. At least this was the view of those who were drafting colonial agricultural policy in the early 1940s. The move toward a policy aimed at increasing the production of foodstuffs for local use was reinforced by the imperatives of the war, which caused drastic shortages of imports, foreign exchange, and shipping, forcing the colonies to rely on their own resources as much as possible. For the CAC, the war presented a unique opportunity to restructure colonial agricultural policy so that local food sufficiency would become the ultimate long-term goal of each territory.[52] "Agriculture," as the 1943 memo on the Principles of Agricultural Policy began, "is by far the most important industry in the Colonial Empire, and upon it rests the material well-being of Colonial peoples."[53] Agricultural and veterinary policy was to be coordinated with other areas of government initiative, including irrigation, forestry, nutrition, public health, and education. The object of such coordination was "to establish agriculture as a way of life . . . to make the individual realize that he is a member of community, to encourage him to take part in community activities, and to discourage aimless drift from the land into urban areas. This involves the encouragement not only of co-operatives in the narrow and technical sense of the term, but also of a co-operative spirit in the daily life of the farming community." The provision of welfare services for rural communities and the instilling of a "rural bias" in the schools would do much "to make rural life more attractive." In other words, the intent was to create stable and self-sufficient agricultural communities. The first objective, therefore, was to preserve and improve the fertility of agricultural lands and other natural resources, taking into account the future rate of population growth and its social and economic consequences. The paramount purpose of the soil, it was agreed, was the production of food, especially "protective foods" of high nutritional value, in order to shelter local communities from the vagaries of the international market and to reduce the prevalence of disease due to malnutrition. To this end, a system of mixed, rotational farming appropriate to local soil and climate conditions, and a continuous program of agricultural research in each colony, were recommended.

The CO's memo on agriculture, further revised in light of resolutions made at the UN's Conference on Food and Agriculture, served as the blueprint for W. A. Robertson, the CO's newly appointed forestry adviser, who with Engledow and Tempany's guidance prepared a statement on the "Principles of Forest Policy in the Colonial Empire" in September 1943.[54] According to the statement, forestry and agriculture were bound in alliance to ensure, in the face of shifting systems of cultivation and increasing populations, not only the continued production of food crops and grazing lands but also preventive measures against ecological crisis or disaster. Forest resource management, therefore, had to be correlated with agricultural and pastoral policy as well as with health and education. Forests were to be held in trust by the government as "forest estates," ensuring that sufficient area was retained to provide protection from erosion and flooding and to conserve water supplies, while at the same time supplying a "sustained yield" of timber, fuel, and other forest products to meet the needs of the local population. Only after these objectives had been achieved were forest exports to be permitted.

The interrelated problems of forestry and agriculture were thus to be treated as two sides of the same coin: the proper utilization of the land. Indeed, as all of the CO's natural resource advisers recognized, the future prosperity of the colonies would depend ultimately on the proper use of the soil and the effective coordination of all government agencies involved. There was a sense that development in the past had proceeded along ad hoc lines, based on preconceived and illusory ideas and inappropriate practices that had been applied without modification. Capitalist enterprises had been undertaken without proper checks and safeguards and with the view of obtaining maximum profits in the minimum time. There had been excessive concentration on exports and a lack of attention given to the production of foodstuffs for local consumption. Within administration, there had been a subordination of the technical services and an unwillingness to see the interconnectedness of different departments' work, with each expanding independently and haphazardly. Sudden changes and a lack of continuity in policy, combined with outbreaks of disease, malnutrition, soil erosion, and deforestation, had been the unfortunate and often unforeseen result. All this pointed to the fact that research and an adequate background of knowledge were the essential prelude to development. Preliminary investigations of locality and the physical characteristics of each territory were crucial and had to precede the land-planning stage in order to avoid development failures. The work of each department had to be planned long-term and coordinated with other departments, as well as linked to long-range research being carried out on a regional basis. What was needed most of all was a "sound" policy of land utilization and rural development, one that was suited to local conditions and which took into account the full effect of any changes on a colony's welfare and environment.[55] Both the CO and colonial governments overseas were thus to be instilled with a "new spirit" of comprehensive planning, and ready to take action on an unprecedented scale when the war was over. In practical terms, as the next section relates, the new integrated approach required a vast expansion and reorganization of imperial machinery and the preparation over the next several years of strategic, long-range plans for the execution of the postwar development mission.

Revolution from Above: The Colonial Office Gears Up for the New Metropolitan Initiative

The growing enthusiasm for central planning and coordination displayed by specialist advisers and experts at the CO in the late 1930s and early 1940s was indicative of broader trends in British public and official attitudes about state involvement in society. As we saw in the debate over malnutrition in the 1930s, social activists such as John Boyd Orr pushed for state-sponsored collective feeding programs, food subsidies, and state-supported nutritional research to help alleviate hunger, poverty, and disease among the poorer classes. Indeed, in the wake of the Depression many on the Left and in the Labour Party championed "constructive" state power and action as the answer to the country's mounting social bill. But the turn toward state welfare provision, as Jose Harris has argued, went far beyond party politics and tactics.[56] British idealism, with its roots in the late-Victorian political thought of T. H. Green as well as Edwardian philosophers such as Bernard Bosanquet, Edward Urwick, and J. H. Muirhead, came to permeate not only public administration and academic circles but also a large segment of the socially active middle classes in the interwar years, providing a shared vision of state and society. Its emphasis on inculcating active citizenship and enhancing the organic sense of community and corporate consciousness as the goals of social reform would help lead to the reconceptualization of the state as a positive and benign force for the public good.

The boundaries of state action were further transformed and revolutionized by the Second World War, which demanded the total mobilization of the nation for the war effort. The entire economy was put under government direction, from the allocation of material to rationing to price controls. Social services such as school meals, public nurseries, and emergency hospitals were extended as never before. The war also marked an important, if contradictory, watershed for the colonial empire: on the one hand it inaugurated a more ameliorative development and welfare strategy, but it also delayed the full implementation of that policy and generated new pressures for the exploitation of colonial resources for British imperial and national interest. In order to save foreign exchange and deal with the shortage of shipping, colonial governments were to reduce consumption to a minimum by raising taxes and cutting imports.[57] Initially, this reinforced the prewar aim of encouraging production of local foodstuffs to bolster self-sufficiency and restrict the need to buy imported essential goods.[58] But after the loss of the Far Eastern dependencies in 1942, the demand for raw materials from the remaining empire increased, prompting a massive expansion in the exploitation of tropical resources, especially in the African colonies. They were instructed to intensify all forms of local production, both of foodstuffs for local consumption and of raw materials for the war effort.[59] These demands led to the imposition of various control measures: exchange controls, restrictions on trade with non-sterling countries, import and export quotas, price controls, and rationing.[60] State marketing control boards were set up to organize bulk purchasing of surplus colonial commodities, and forced labor practices were imposed to conscript workers for the wartime drive to maximize local mining and agricultural production.

As the war continued, thoughts turned to planning for the postwar social reconstruction of Britain. The appointment of an interdepartmental committee under William Beveridge was the key catalyst. Beveridge's *Report on the Social Insurance and Allied Services,* published at the height of the war in 1942, called for the extension and consolidation of Britain's prewar social services into a comprehensive scheme of public insurance and state benefits. Sweeping state intervention was needed to create a universal system of security for all Britons, from cradle to grave, against social hardships beyond their control.[61] All citizens would be entitled to a minimum subsistence based on full employment, social security, family

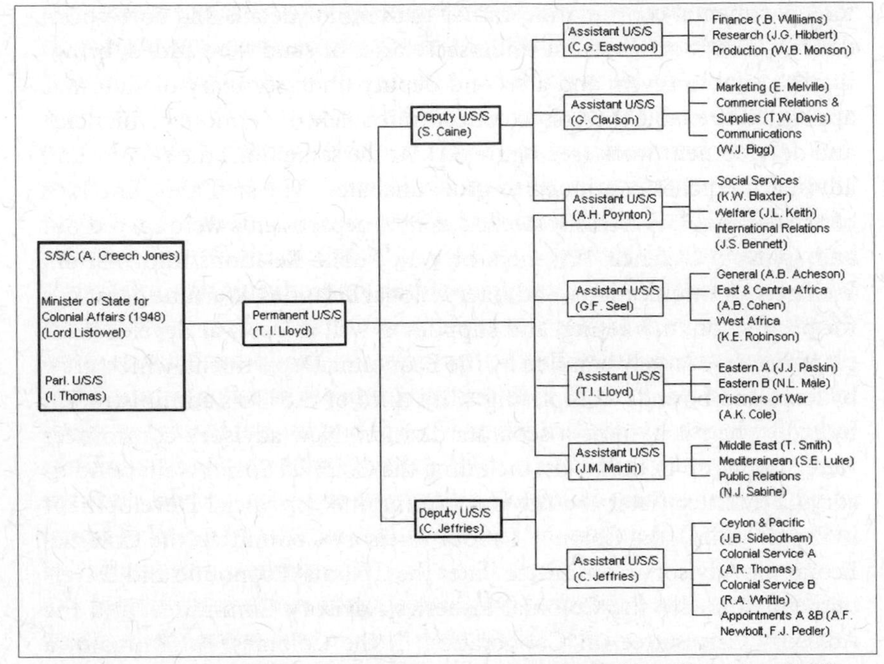

Figure 6.1. Colonial Office organization, c. 1947. Charles Jeffries, *The Colonial Office* (London: Allen and Unwin, 1956), 208–9; J. M. Lee, *Colonial Development and Good Government* (Oxford: Clarendon Press, 1967), 104; Anne Thurston, *Records of the Colonial Office, Dominions Office, Commonwealth Relations Office and Commonwealth Office* (London: HMSO, 1995), chap. 7.

Development and Welfare Organization (WIDWO), set up in 1940 as a regional body of experts and administrative officers under the supervision of Sir Frank Stockdale, who as comptroller was initially charged with administering the proposed West Indies Welfare Fund and with working out long-term development and welfare programs in consultation with, but independent of, local colonial administrations.[66] Stockdale and his team of experts were responsible only to the secretary of state for the colonies, not to the West Indian governments or to the Treasury, an arrangement which it was hoped would avoid the mistakes of the past by allowing them to circumvent local vested interests and by giving them wide discretionary powers over the approval and supervision of grants. But before the organization could even get off the ground, the fund was merged into the general CD&W scheme, leaving it with no resources to administer and no executive authority. It was decided to retain the WIDWO nevertheless, as an expert body which would advise the secretary of state on any schemes submitted by the West Indian governments for financing under the CD&W Act, while the actual administration of grants would remain the responsibility of local colonial governments. During the war, the organization's main task was to carry out surveys and preliminary studies, to determine costs, and to assist local administrations in working up five-year plans in preparation for grant applications under the CD&W Act. Over time, Stockdale's team of technical experts also took on an increasingly active and important role as traveling consultants, especially for the smaller colonies that did not have an adequate technical staff of their own.[67]

Although Caine commended the Stockdale team for what they had managed to achieve in the West Indies, he felt it wasn't enough: it was still largely a projection of the existing CO advisory machinery for the screening of locally initiated schemes, much like the CAC or the CDAC, since it had neither the staff nor the authority to formulate and submit proposals of its own. What was really needed, he believed, was a new "habit of initiative" at the center, through the creation of some kind of colonial development board or government-controlled company, along the lines of the Tennessee Valley Authority, which would have substantially greater discretionary power than any previous body to freely and directly work out new schemes and send out technical investigators to report on development possibilities, without kowtowing to local colonial interests or the parsimony of Treasury officials. Other senior CO administrative staff remained more cautious, however, supporting a more active role for Whitehall but also sensing the importance of carrying local opinion and taking into account knowledge of local circumstances.[68] In the interest of bridging the gap between center and periphery, a number of innovations were agreed upon: the establishment of a separate planning section within the CO; the building up of the CO's technical staff to enable regular visits by experts to smaller colonies in need; the creation, wherever practicable, of regional organizations along the lines of the Stockdale model, but with greater executive power; and, finally, the setting up of special territorial planning machinery or development councils with the aim of preparing long-range plans for each colony. The findings on social and economic planning were summarized in a paper produced by the newly formed Colonial Economic Advisory Committee (CEAC), which was circulated to all colonial governments in April 1944. Then, in 1945, the secretary of state sent out another important circular requesting, in light of the passage of the Second CD&W Act, that all overseas territories prepare comprehensive ten-year development plans. To help with the compilation, negotiation, and settlement of these plans, Sir Frank Stockdale was appointed the CO's adviser on development planning and a new class of development officers was established. By 1952 over 370 had been appointed to assist individual governments with the implementation of those plans approved and financed under the CD&W Acts.[69] One of the first to respond was the Ugandan administration, which had set up a Development and Welfare Committee during the war and in 1946 hired E. B. Worthington as a special adviser to prepare the territory's first ten-year development plan.[70] Worthington's plan for Uganda served as an important model for other colonial governments.

The reorganization of the CO and the introduction of new state planning practices were not the only innovations of the 1940s. The new colonial development initiative also stimulated an unprecedented demand for new knowledge and new forms of scientific organization. As Worthington had pointed out in the case of Uganda, one of the main factors limiting development efforts was a lack of fundamental information about the country.[71] The need for some kind of central direction of colonial scientific research had long been recognized by authorities in London, who, as we have seen throughout, periodically revisited the idea of greater imperial coordination, only to come up against the recalcitrance of local officials and the growing autonomy of local scientific communities. The case for extending the facilities and scale of research work and for taking a more comprehensive view of all scientific efforts was put forward once again by the 1938 *African Survey* under Lord Hailey, who recommended the provision of substantial funds from the British Treasury to finance African research and the establishment of an African Bureau in London to act as a clearinghouse of knowledge and assistance to those pursuing research on problems arising from the material and social development of the continent.[72] Hailey's idea of a special fund was taken up and extended by the CO, which under the CD&W Act of 1940 made £500,000 a year available for research for ten years; this was doubled to £1 million a year under the 1945 Act. A special Colonial Research Committee was formed during the war, with Hailey as its first chairman, to advise on research grant expenditures under the CD&W Act as well as to examine research needs across a wide range of fields. This led to the establishment of several new advisory bodies at the end of the war to assess research priorities and advise the secretary of state on the provision of funding within their respective fields: the Colonial Social Science Research Council;

the Colonial Agriculture, Animal Health and Forestry Research Committee; the Colonial Medical Research Committee; and the Colonial Economic Research Committee.[73] A separate Research Department was also established within the CO in 1945, taking over responsibility for all areas of research both in the colonies and in fields relevant to colonial questions. By the end of the decade some 329 research schemes totaling £6 million had been financed under the CD&W Acts.[74]

Similar to trends in development planning, the CO also sought to assert greater control over the use of resources and trained scientific personnel by organizing research on a regional basis, as a way of supplementing the already substantial network of agricultural research and experimental stations located throughout the overseas colonial territories (see table 6.1). The most notable development in this direction was the setting up of the East African Agricultural and Forestry Research Organization and the East African Veterinary Research Organization at Muguga, near Nairobi, after the war. The plan for the new East African research organizations came at the urging of the CO's key agricultural advisers, Engledow, Tempany, and Dr. H. H. Storey; much of the initial capital came from grants provided under the CD&W Acts. Similar to their predecessor at Amani, the new organizations were intended to concentrate on long-range basic or fundamental research in agriculture, forestry, and veterinary science, while the territorial departments were responsible for the technological review of basic research under a range of soil, vegetation, and climatic conditions and for the introduction of practical measures based on the conclusions reached.[75] Similar schemes patterned on the East African model were planned for West Africa, while in the West Indies it was proposed that all long-range agricultural research be centralized at the ICTA, where a separate West Indian Agricultural Research and Forestry Organization might possibly be established.[76] Along the same lines, regional institutes for sociological research were also supported or established through the Colonial Social Science Research Council: the West Indies Institute of Social and Economic Research in Jamaica, the Institute of Social and Economic Research for West Africa at Ibadan, and the Rhodes-Livingstone Institute in Northern Rhodesia.[77]

Finally, the new colonial mission would not have been possible without the remarkable expansion, beginning in the late 1940s, of colonial technical departments and personnel in the field. During the war, recruitment to the Colonial Service dropped dramatically, while at the same time many existing officers were called up or volunteered for duty, or else

Table 6.1. List of colonial agricultural research stations, 1951

Barbados	Groves Sugarcane Station; West Indies Sugar Cane Breeding Station (BS)
Borneo	Rice Padi Experimental Station (ES)
British Guiana	Rice ES, Livestock ES, and Sugar ES, Department of Agriculture (DOA); St. Ignatius BS, Lethem; Ebini Station; Forest ES, Mayasuri
Cyprus	Central Experimental Farm, Morphou; Government Stock Farm, Athalassa; Deciduous Fruit Station (DFS), Galata; DFS, Trikoukkia; Viticulture and DFS, Saittas; Tobacco Station (TS), Akrades
Fiji	Principal Agriculture Station (AS), DOA; AS, Sigatoka
Gambia	Cape St. Mary ES; Masembe ES; Yoroberi Kunda ES
Gold Coast	Kpeve Research Station (RS), Togoland; Asuansi RS, Cape Coast; Pokoase RS; Tamale RS; Zuarungu RS; Babile RS; Kumasi Forestry RS; Nungwa RS; Pong-Tampale RS
Jamaica	Forestry Dept, Kingston; Bodles AS, Old Harbour; Hope Agricultural Sub-Station (ASS); Grove Place ASS; Irwin ASS; Orange River ASS, Highgate; Caenwood ASS, Hope Bay
Kenya	Kitale ES; Njoro Plant BS; Ol Joro Orok ES; Molo ES; Coffee RS, Ruiru; Matuga Coast Investigation Station; Thika High Level Sisal RS; Kabete Veterinary Lab; Kabete Pasture RS; Maivasha ES
Leewards	Friar's Hill AS
Federation of Malaya	Federal ES, Port Swettenham; Federal ES, Serdang; Federal ES, Cameron Highland; Telok Chengai Padi ES; Titi Serong Padi ES; Kota Bharu Padi ES; Pulau Gadong Rice ES; Forest Research Institute, Kepong; Rubber Research Institute, Kuala Limpur; Veterinary Dept
Mauritius	Central ES, Reduit; Pamplemousses ES; Hermitage ES; Barkley ES; Government Dairy Station, Curepipe; Palmer BS; Richelieu TS
Nigeria	Oil Palm RS, Benin; Moor Plantation, Ibadan; Samaru ES, Zaria; West African Institute for Trypanosomiasis, Kaduna
Northern Rhodesia	Lunzuwa AS; Kambowa AS; Lusaka Government Farm; TS, Choma; Msekera TS, Fort Jameson; Forest Dept, Ndola; Veterinary Dept, Mazabuka
Nyasaland	Agricultural RS, Lilongwe; Bvumbwe ES
Sarawak	Tarat AS; Kanowit AS; Rantau Panjang Rice Station; Bangkita AS
Seychelles	Union Vale; Grande Anse Estate; Val Rice RS
Sierra Leone	Rice RS, Rokupr; Njala AS; Musaia Animal Husbandry Station
Singapore	Botanical Gardens
Tanganyika	Ngomeni Sisal ES; Lyamungu Coffee RS; Ukiriguru ES; Ilonga ES; Veterinary RS
Trinidad	ICTA, St. Augustine
Uganda	Kawanda ES, Kampala; Serere ES, Soroti; Bukalasa Farm, Bombo; Bugusege; Ngetta; Kyembogo
Western Pacific	British Solomon Islands AS
Windwards	Mount Hope Experimental Cocoa Station; Ashenden Experimental Cocoa Station
Zanzibar	Kizimbani ES; Hanyegwa Mchana Sub-ES
E.A. High Commission	EAAFRO, Nairobi; EAVRO, Kabete; East African Tsetse and Trypanosomiasis Research and Reclamation Organization, Tanganyika
West African Council	West African Cocoa Research Institute (WACRI), Tabo

Source: CO 908/9 Colonial Agricultural Research Committee, Papers and Minutes, 1951, CAR (51) 48, List of Agricultural Research Stations, October 1951.

were interned or invalidated, with the result that by the end of the war there was a substantial shortage of staff. Renewed recruitment began in 1944, aimed initially at demobilizing armed forces personnel, followed by a full-scale recruitment campaign launched in June 1945. More than eleven thousand new appointments were made between 1944 and 1953, pushing the total size of the Colonial Service to historic levels. By 1954 it reached its peak of eighteen thousand, of which, significantly, over 85 percent held posts in the professional departments and services such as education, health, and natural resources (see table 6.2).[78]

In the aftermath of the Depression and the widespread social protests and labor unrest that ensued throughout the colonial empire, a fundamental shift or "conversion" in official thinking took place at the CO in London. Building on the cumulative efforts of colonial agricultural, educational, and health researchers in the field, especially revelations of new and pressing problems such as soil erosion, malnutrition, and surplus population, metropolitan advisers and experts successfully persuaded key senior authorities of the need for a more systematic and "constructive" policy of colonial reform. Much of the cumulative knowledge of the previous ten or fifteen years was mobilized and synthesized into the new doctrine of rural community development enshrined in the CD&W Acts and the various policy statements emanating from the CO's network of advisory committees. The Second World War acted as a further catalyst for new metropolitan visions of state planning and engineering both in Britain and in the empire overseas. Imperial financial assistance was substantially

Table 6.2 Colonial Service recruitment, 1944–53

	1944	*1945*	*1946*	*1947*	*1948*	*1949*	*1950*	*1951*	*1952*	*1953*	*Total*
Administrative	8	103	475	344	211	158	267	210	166	108	2,050
Scientific[a]	66	175	477	368	446	585	593	538	542	309	4,099
Other Expert[b]	80	205	420	310	321	381	361	322	369	474	3,243
Other[c]	38	121	301	121	165	243	325	270	255	97	1,936
Total	**192**	**604**	**1,673**	**1,143**	**1,143**	**1,367**	**1,546**	**1,340**	**1,332**	**988**	**11,328**

a. Includes appointments in Agriculture, Chemistry, Engineering, Fisheries, Forestry, Geology, Marine, Medicine, Meteorology, Surveying, Veterinary, and other/miscellaneous scientific posts.
b. Includes appointments predominantly in Education but also Architecture and Town Planning, Civil Aviation, Commerce and Industry, Cooperation, Dentistry, Development Officers, Economics, Labor, Mining, and Social Welfare.
c. Includes Auditing, Broadcasting, Finance and Customs, Legal, Personnel, Police, Prisons, Public Relations, Statistics, and other/miscellaneous posts.

Source: Anthony Kirk-Greene, *On Crown Service: A History of HM Colonial and Overseas Civil Services 1837–1997* (London: I. B. Tauris, 1999), 52, 77.

hopes of an imperial renaissance, with the boundaries of the overseas empire simply shifted further to the east in Asia and into the Middle East and Africa as well.

Averting bankruptcy and extending the empire's lease on life, however, depended on American financial and military backing, which in the context of the Cold War and the growing threat posed by the Soviet Union was given, although not without conditions. With the end of the Lend-Lease system in 1945, Britain turned to the United States for a massive $3.5 billion loan, which it agreed to, but only if dollars would no longer be impounded in Britain's Foreign Exchange Pool, thus allowing private citizens and companies to purchase U.S. exports with dollars passing through Britain. The result was a frantic run on the Bank of England's reserves and the rapid depreciation of the pound against the dollar. In response to the postwar sterling crisis, the Labour government looked to empire for salvation, impressing on individual administrations the need to intensify the exploitation of imperial resources in an effort to use the colonies to earn dollars by exporting to the United States, and save dollars through substitution of imports from the dollar area to Britain. The pot was sweetened by the Overseas Resources Development Bill of 1947, which created two new public corporations: the Colonial Development Corporation (CDC), which was given authority to borrow up to £100 million, and the Overseas Food Corporation (OFC), with up to £50 million. The new state corporations were established ostensibly with the goals of both improving the general standard of living and welfare of colonial peoples and increasing the supply of colonial products abroad. The 1947–51 period was the only time, as Cowen and Shenton observe, that a fully blown Chamberlainite "colonial development offensive" was mounted by the imperial government to serve the direct interests of the British national economy.[3]

Throughout what remained of Britain's overseas empire, individual colonial administrations emerged from the war, under pressure from above, prepared to penetrate local agrarian society more deeply than ever before. Over the course of the next several years the colonial state would move on an unprecedented scale to act as the prime agent of agrarian change and rural development in Africa and elsewhere. Over seventy major agricultural development initiatives were listed by the Colonial Office as active in 1955, including pilot projects for water and soil conservation and food production; numerous land improvement and resettlement schemes; various mechanized cultivation projects for cotton, rice, and *padi* cultivation; tractor plowing and hiring units; drainage and irrigation schemes; and cooperative and group farming ventures, among many others.[4] Augmenting these efforts were numerous infrastructure, health, education, and welfare programs, not to mention such megaprojects undertaken by the CDC and OFC as the massive East Africa Groundnut Scheme discussed below. The magnitude and intensity of government activity in nearly every aspect of rural colonial peoples' lives from the late 1940s on amounted in all but name to what D. A. Low and John Lonsdale have strikingly termed "the second colonial occupation."[5] The "new imperialism" extended beyond the African colonies, encompassing Southeast Asia as well, where, as A. J. Stockwell notes, "Prestige, strategy, and economic considerations meant that, notwithstanding promises of eventual self-government, Britain was committed to the reimposition of control over former dependencies, at least in the short term, and to the defence and economic development of South-East Asia as a whole."[6] Nor was Britain alone in its response to the postwar crisis. In 1946 France followed Britain's lead, establishing its own Fonds pour l'Investissement en Développement Économique et Social, which provided metropolitan funds for development initiatives overseas on a much larger scale than previously. Technical personnel, research facilities, and new organizational and administrative abilities were also significantly increased, enabling French bureaucratic power to be extended into rural colonial areas as never before.[7]

And yet, despite the vast increase in imperial financing, bureaucratic services, and specialist personnel, and despite the increasingly ambitious tone and confidence of colonial planners and technical advisers, many of the new development ventures of the early postwar years proved to be disappointing, and, at times, downright disastrous. By 1954 it was clear to authorities in Whitehall that the metropolitan development drive had failed, and initiative was once more handed back to local colonial governments. Such an anticlimactic ending to the late colonial mission demands explanation. What happened? What went wrong and what are the lessons to be gained?

From Groundnuts to Administrative Impasse: The Antinomies of Late Colonial Development Practice

Unquestionably the most spectacular failure of the postwar "colonial development offensive" was the East Africa Groundnut Scheme, due, in large measure, to the ironic fact—ironic, given all the effort and resources that were poured into the project—that it was ill-conceived, hastily put into practice, and poorly managed. It was, in the words of one prominent colonial agricultural officer, Roger Swynnerton, "a vast and unproven project without a pilot scheme" formulated by, as he sarcastically dubbed them, "the three wisemen."[8] In the mind of one of those wise men, A. J. Wakefield, the scheme was envisioned as an agricultural revolution that not only would respond to the world food shortage but also would save East Africa from what he described as the early warning signs of a future Malthusian environmental crisis.[9] After sixteen years as director of agriculture in Tanganyika and another five as inspector general of agriculture for the West Indies, Wakefield had come to the conclusion that earlier efforts to reform peasant agriculture were simply not enough. If production was going to be raised to a level capable of feeding the region's growing population and at the same time provide surpluses to satiate world demand, then the productivity of already heavily populated lands would have to be substantially increased and previously uninhabited and submarginal lands would need to be opened up on a grand scale. Science would have to be fully harnessed through the engagement of advisory panels of eminent authorities, and through cooperation with leading research institutions like the East African Agricultural and Forestry Research Organization and Rothamsted.

The proposal which Wakefield and the other members of the planning mission eventually produced in February 1947 was truly colossal: 3,210,000 acres were to be cleared by 1952 and transformed into more than a hundred farming "units" of 30,000 acres each, which it was projected would be capable of producing between 600,000 and 800,000 tons of groundnuts annually once optimal yields (of 850 to 1,120 pounds per acre) had been attained.[10] The total capital expenditure for the scheme spread over the six years was estimated at £24 million, plus £1.25 million for the construction of a new railway and deepwater port. What is more, every aspect of the plan was to be fully mechanized (something unheard of before the war), from the initial bush-clearing to the plowing and planting to the lifting and harvesting of the nuts, involving a literal invasion of bulldozers and tractors and an army of African wage labor, together with all the logistical services needed to support such a campaign: from heavy repair workshops to roads, railways, and ports for supply, to housing, health services, and training for the workers. Perhaps the most radical administrative departure was that control of the scheme was not given to the Colonial Office, which was seen by many in the new Labour government to be inherently conservative and slow to act, but to the Ministry of Food, which initially called on the United African Company to manage the operation before the newly created Overseas Food Corporation could take it over in April 1948. It was thus to be, in every sense, a showcase of what science, technology, and the state could achieve, and it all had to happen in record-breaking time in order to produce export food surpluses to help support Britain's postwar economic recovery.

Despite Wakefield's propaganda and the high hopes of the new Labour government, however, the scheme was a white elephant from the very beginning, due, in large measure, to an insufficient examination of soil and climatic conditions at Kongwa, the area initially chosen for the scheme, and to underestimation of the difficulties involved in supplying and operating heavy tractors and planting machinery under tropical conditions. Those responsible for the initial plan did not anticipate, for example, the high attrition rate of the second-hand tractors, or the challenges involved in removing the long, sinewy roots of the Kongwa scrub, which clogged the ill-suited rooters, causing extended delays. As a result, only 7,500 acres were ready for planting by the end of 1947, a far cry from the 150,000 targeted by Wakefield. By 1949, after every imaginable effort had been made, only 50,000 acres were planted in all three areas of operation. Nor did yields meet expectations, with the average of the first harvest working out to less than half the original estimates. Numerous other difficulties arose as well, reflecting an overall lack of forethought and inept planning: the problem of how to get supplies and equipment in and groundnuts out of the selected areas was never fully considered; the cost of building the infrastructure and providing support services was significantly underestimated; the central board of directors sitting in London found it difficult to stay on top of what was really happening on the ground, and as a result grew increasingly aloof and out of touch with field staff; the use of contractors for land-clearing produced endless conflicts and divisions between the OFC's area managers and the contractors' agents; the agricultural machinery, designed as it was for use in North America, proved unsuited for East African conditions and had to be scrapped or redesigned, while mechanized plowing by heavy tractors also caused rapid soil erosion; and, finally, the soil in the selected areas proved too hard and compact during the dry months to be effectively worked, which meant that all mechanical operations had to be confined to the wet season.[11]

By the end of 1948 the scheme was in serious trouble. Field department heads submitted a joint memorandum to the chairman of the board, Sir

Leslie Plummer, airing their concerns and grievances, while authorities in London responded with a campaign of retrenchment and economy. To make matters worse, Kongwa was gripped with drought in 1949, revealing what should have been clear from the beginning: that this was an area of marginal and highly localized rainfall with difficult soil conditions that should never have been developed in the first place. Exasperated, OFC staff members began writing publicly to demand the resignations of Plummer and the minister of food, John Strachey, prompting the latter to make a secret visit in December to quell the "revolt." None was more enraged than the OFC's chief agricultural adviser, the renowned South African ecologist John Phillips, who complained that the "true facts" of the case were being misrepresented. Phillips considered writing the prime minister personally to urge an official probe, but the governor of Tanganyika, Sir Edward Twinning, discouraged the idea, suggesting that British colonial prestige was too much at stake.[12]

By mid-1951, however, Phillips had had enough, and wrote to the British scientific journal *Nature* to set the record straight. He freely acknowledged the lack of ecological planning in the early stages of the scheme, explaining that time pressures and a lack of trained personnel did not permit carrying out adequate soil and ecological surveys. He also admitted that much of the hundred thousand acres of land cleared at Kongwa had turned out to be unsuitable for arable farming, or else was still waiting in fallow to be developed. But he vigorously denied charges that things had been done haphazardly, noting that considerable preliminary work was undertaken in areas developed later: soil reconnaissance and topographical maps, in the case of Urambo, and aerial photo and ground surveys, in Nachingwea, had formed the basis for more refined land utilization and development plans. "I do not know of any equivalent area in Africa," Phillips asserted, "which has been as intensively surveyed for land-use, demarcated and mapped as that covered by the Corporation's areas at Kongwa, Urambo and the Southern Provinces."[13] Despite Phillips' elucidations, criticism continued, with the general secretary of the Association of Scientific Workers, Ted Ainley, writing to the secretary of state for the colonies to inform him that "the Council regrets the failure of the East African groundnut scheme and attributes it in part to an approach which was faulty and to the detriment of proper scientific effort and planning and a proper relationship with the local population."[14]

In the end, the groundnut fiasco was a painful reminder, given the careful investigations that had been made before the war by scientists like

Geoffrey Milne, Colin Graham Trapnell, and William Allan, and the value accorded to such investigations by the CO's specialist advisers, of the need for knowledge of locality to be gathered in advance of practice, and of the complexity and challenges of the African environment which no amount of capital or technology could undo. "May it be abundantly clear," Phillips would later advise, "that failure to study and understand Nature inevitably ends in disaster."[15] One wonders how such fundamental mistakes could ever have been allowed to happen. It seems that the crux of the problem was that the project was premised on an entirely new approach that had never before been tried in British tropical Africa: large-scale, mechanized production of crops.[16] There was, therefore, little actual information available, or past experience to guide those who either planned or carried out the scheme, since all previous research, as we have seen, was based on small-scale, organic, biological farming methods. Moreover, given the urgency to produce, there was no time for preliminary surveys and pilot projects to see if the new methods would actually work. To a great extent this was the fault of the Ministry of Food and its managing agents, the United Africa Company and the OFC. In any case, responsibility had been taken away from the CO, which, as we have seen, was in the meantime becoming more mindful of the need for careful examination of local conditions and constraints. Yet, beyond the inadequate planning and bureaucratic blundering, there lies the fact that the men who dreamed up and inexplicably pushed on with this colossus among misguided projects in the face of mounting evidence of failure, were, like Goethe's Faust, driven by an insatiable and illusory vision of what was possible which blinded them to the basic economic and ecological facts. As one OFC employee later remarked, "Implicit in the outlook of most of us at Kongwa, and of most Europeans that one met, was the idea that without any question one had a useful contribution to make to Africa; in fact that one was there to carry on the white man's burden."[17] But there is more to this story than a simple tale of modernization gone awry. For despite all the talk of agrarian revolution, the planners were limited, as a guiding principle, to the selection of submarginal areas with relatively poor climatic and soil conditions and few or no inhabitants, so as to interfere as little as possible with existing communities and land rights.

The enormous scale and stunning collapse of the Groundnut Scheme make it something of an aberration in British colonial Africa. So, too, it must be said, does the glaring lack of ecological examination and the generally perfunctory study that preceded it. Yet even the best-made development plans and interventions of the period, formed on the basis of years of preliminary trials and preparatory research, often fell victim to many of the same impulses, tensions, unexpected difficulties, and constraints plaguing the OFC. The rehabilitation of the Sukuma region of Tanganyika's Lake Province—one of the earliest, most extensively investigated, and ambitious integrated rural development projects in all of colonial Africa—is a case in point.[18] The main problems, of increasing erosion, overgrazing, losses in soil fertility, and declining crop yields due to densely concentrated human and livestock populations, had long been recognized by field officers in the area. Another great challenge in Sukumaland was the encroachment of tsetse fly, which forced the retreat of human and stock populations into increasingly concentrated areas. Most colonial observers before the war saw this movement, mistakenly, as it turned out, as part of a cyclical process of depopulation, triggered either by incessant ethnic conflict or the abandonment of land due to soil exhaustion, which they feared would eventually lead to the elimination of all people and beasts from the region.[19] In response, the government embarked on a tsetse eradication campaign in the 1920s, establishing a Department of Tsetse Research under C. F. M. Swynnerton, whose main research station at Shinyanga began carrying out experimental trials in controlled burning, bush isolation, and fly extermination. These investigations persuaded tsetse experts that bush clearing along a broad front, followed closely by planned settlements of sufficiently minimal density, was the most effective method of reopening infested areas.

In addition to tsetse research, with the aid of the Empire Cotton Growing Corporation agricultural experimental stations were set up in the 1930s at Ukiriguru and Lubaga, where experimental field trials were carried out on soil conservation measures, the selection of better strains of cotton, rice, and other food and cash crops, and the use of organic manures and grass leys. An agricultural training center was also established at Lubaga to demonstrate the principles of mixed farming, crop rotation, soil conservation, and extension methods to African agricultural instructors, and careful surveying and pilot projects were conducted by agricultural officers such as J. G. M. King, Denis Thornton, Norman Rounce, and others. In the mid-1930s, district agricultural officers began meeting with administrative officials and researchers in the Tsetse Research and Survey Department to discuss common problems, and in 1936 Donald Malcolm was asked to conduct an intensive study of land utilization in the region. Malcolm recommended a new integrated team approach that would go beyond district boundaries and departmental divisions to formulate a regional rehabilitation plan. At the same time, the regional assistant director of agriculture, Norman Rounce, and his officers began compiling a detailed map of the area, building on Geoffrey Milne's soil reconnaissance survey by recording bush areas, population and cattle densities, roads, rivers, elevations, and administrative boundaries. From this data, various "land-usage areas" were identified and an "ideal" land use pattern for the region derived, which would later serve as the basis for postwar planning.[20]

Although the ground had been well prepared before the war, the launch of the Sukumaland Development Scheme in 1947 was nonetheless a novel departure in terms of both the amount of financial resources made available and the new forms of administrative organization it employed. Following CO directives, the Tanganyikan government, like most colonies, prepared a ten-year development plan in 1946, and the following year introduced a separate Agricultural Development Fund, which included funding for eight agricultural schemes, the most important of which was the Sukumaland Scheme. The aim of the project, according to Rounce, was to improve living standards and preserve soil fertility in the Sukuma region by a more "orderly settlement of people and stock on the land at reasonable densities." This was to be achieved first by reducing pressures on the overcrowded and over-cultivated central areas through the opening up of previously uninhabited or under-inhabited parts to the west, east, and south, and subsequently through the adoption of improved methods of husbandry and soil conservation over the whole region.[21] The most immediate measures involved putting in new, permanent water supplies, constructing roads for access, eradicating the tsetse fly through bush-clearing; and controlled movement of people into areas selected for resettlement. The more difficult and longer-term challenge was rehabilitating the central areas while maintaining the new settlements through a program of controlled de-stocking and use of village grazing reserves; the introduction of improved cultivation methods such as early planting and weeding, crop rotation, grass leys, use of manure, and the introduction of plows and mechanical cultivation on heavier soils; and the implementation of soil conservation measures such as contour cultivation and tie-ridging. In both new and existing areas, farmers were to practice mixed farming on family holdings derived from what was known as the "Sukumaland Equation," which was used to determine the "ideal" human and stock population density as well as the volume of water that was needed at each watering point (see figure 7.1). According

In ONE HOMESTEAD there are on average TWO TAXPAYERS or a total of SEVENTEEN PEOPLE with an average of FOURTEEN CATTLE and TEN SMALL STOCK UNITS (at 5 small stock equalling one stock unit) equals SIXTEEN STOCK UNITS which produce altogether SIXTEEN TONS OF MANURE PER ANNUM. This manure is enough to manure EIGHT ACRES EVERY OTHER YEAR which is one acre more than the average acreage of arable for Sukumaland but the stock require TWO ACRES EACH OF PASTURE (the average for Sukumaland is ½ acres) equals THIRTY-TWO ACRES PLUS EIGHT OF ARABLE equals FORTY ACRES equals SIXTEEN HOMESTEADS PER SQUARE MILE equals 112 PEOPLE PER SQUARE MILE, SAY ONE HUNDRED. Three miles to walk to water is about THIRTY SQUARE MILES equals FIVE HUNDRED HOMESTEADS equals EIGHT THOUSAND STOCK x FIVE GALLONS OF WATER x 120 DAYS (August to November) equals FIVE MILLION GALLONS = 10,000,000 GALLONS (to allow for evaporation) PER 500 HOMES AND THIRTY SQUARE MILES.

Figure 7.1. The Sukumaland Equation indicates how the optimum density of one hundred people per square mile and optimum water supply for thirty square miles were arrived at. N. V. Rounce, *The Agriculture of the Cultivation Steppe of the Lake, Western and Central Provinces* (Cape Town: Longmans, Green, 1949), 105.

to the Equation, it was thought that a maximum density of one hundred people per square mile could be permitted, which worked out to approximately twenty acres for each family homestead. The total capital expenditure for the project over the ten years amounted to roughly £2 million in addition to normal and recurrent costs for staff, equipment, and services.[22] Malcolm's original proposal for a "flying squad" traveling from one district to another was taken up and reworked into an interdepartmental supra-district team, known as the Sukumaland Development Team (SDT), made up of experienced senior officers from the Agriculture, Veterinary, Forestry, and Water Development Departments, plus the deputy provincial commissioner for Sukumaland, who acted as the coordinating officer. The team was centrally located at Malya, along with all services and staff except field officers, to ensure full coordination of operations, and operated independently of other district or provincial offices and staff so as to avoid interference and divided allegiances. Each district also established an interdepartmental team made up of technical field staff to prepare district plans and carry out routine development work. Most crucial of all was the founding in 1947 of the Sukuma Federation of Chiefs, an organization of fifty Native Authorities which acted as the key intermediating agent between the team and local communities. Although initiatives came from the SDT, the support and cooperation of the Federation Council was critical, since all plans were subject to their approval and, most important of all, they were responsible for passing Native Authority Ordinances, which came to play an increasingly crucial part as the scheme wore on.

Despite years of preliminary investigation, new revenue sources, and the new integrated team approach, the results of the Sukumaland Development Scheme were less than satisfactory. Operations were already running down in some areas by 1953 and in the following year the team at Malya was disbanded. By 1955 the project had been completely abandoned. A number of problems can be identified in hindsight which help account for its premature closure. Although the Sukumaland Team carried out the surveying, mapping, and investigational work for the scheme, as well as operations of a special nature such as dam construction or tsetse bush-clearance, it did not have sufficient staff to execute and supervise the more routine development work, which was handled, as a result, by administrative and technical officers in the districts, who found themselves, especially early on, stretched extremely thin. In the new districts opened up for land settlement, for example, the shortage of officials made it impossible to seal off areas once the "optimum" density of people and stock had been reached, or to ensure the new land use patterns were adopted. Staffing problems aggravated more fundamental tensions between the team and district administrators, especially the older and more senior district commissioners, who resented the duplication of responsibility and interference both from Malya and the district teams. District commissioners complained there wasn't enough department staff to go around and that the team in Malya was living in an "ivory tower," divorced from realities on the ground. There was also a high turnover of field staff seconded to the team from the technical departments, who often continued to be treated as members of departmental staff first and of the scheme only second. The chronic shortage of personnel and bureaucratic wrangling at the district level delayed operations and made it impossible to effectively enforce orders passed by the Federation Council.

But the most intractable barrier to the scheme's success was the increasing lack of cooperation of the people for whom it was intended. Many of the measures proved unpopular with area cultivators and herders, and it was only after a great deal of time and persuasion that district commissioners and their staff were able to convince district chiefs and Native Authorities of the necessity of introducing the proposed changes. Farmers were reluctant, for example, to construct ridges along the contour, because unless this was done with absolute accuracy or graded in such a way as to allow drainage channels, water would build up and break through at the lowest point in the ridge, causing immeasurable damage to crops. Ties were encouraged as a solution to the problem, but farmers disliked the extra work involved, which had to be done when they were already fully occupied ridging and planting.[23] The measure that proved most contentious was the government's drive against overstocking by limiting livestock numbers in accordance with "grazing capacity." For two years such proposals were met with flat-out refusal and noncompliance by the Federation Council, and only with great reluctance were they implemented at all. Frustrated by the slow rate of progress, the development team came to rely increasingly on a set of orders enacted through the Council. Land Settlement Rules and Sukuma Livestock Restriction Rules, for example, were introduced in 1950 to ensure control over human and stock densities in the Land Usage Areas. Other rules and regulations covered the cultivation of cotton, marketing of cattle, soil conservation measures, ox plowing, the protection of forests, hilltop reserves, and cultivation near streams (see figure 7.2). At its height in 1952 and 1953, the scheme came to resemble a police operation more than an agricultural assistance program. Resentment over the restrictions and interference caused by the Native Authority Ordinances, especially the passage and implementation of the de-stocking rules, led to strong antipathy toward chiefs and the Sukumaland Scheme among many cultivators and herders, who openly disregarded Council orders even at the risk of heavy fines. By the mid-1950s, with the government virtually powerless to enforce natural resource legislation of any kind, the scheme was brought to a standstill. Team officials blamed the failure on the shortage of field staff, which made enforcement and supervision of regulations and settlement controls next to impossible, as well as the apparently innate conservatism of peasant farmers, which made them deeply suspicious of any government interference.

Despite official excuses, it is clear that many of the "failures to cooperate" were based on rational economic judgments. Simply put, the government's expectations were in many cases unrealistic. The promotion of mixed farming as a means of intensifying cropping in the more densely settled areas—one of the key components of the scheme—met with only limited success largely because it was either unprofitable or infeasible. It

1. All cultivated land not liable to waterlogging on slopes shall be ridged on the contour with ridges not less than 5' wide from crest to crest. An area per taxpayer (to be prescribed from time to time) of this ridge land shall be tie-ridged on the following soil and the ties shall be at intervals of 3 yards in the furrows and shall be at least 6" lower than the crests of the ridges. The soils are as follows:

   Ikarusi, Isanga, Luseni, Matongo, Kikungu, Nduha, Ibushi

   Any sections of the above soils which are subject to waterlogging are exempted from this order.
2. No cultivation shall take place within a specified distance of any drainage channel, stream or river bed or main road.
3. Where serious erosion is developing all people in the areas may be called upon to carry out communal work to repair and prevent further damage
4. It shall be an offence to damage or destroy any soil conservation works or any herb, tree or shrub, planted to prevent soil erosion.
5. Any person, the land in whose occupation borders upon an area declared as a reserve in accordance with the provisions of orders 6 and 8 shall plant a hedge or erect beacons to demarcate the boundary of the declared area where it touches the land in his occupation.
6. A Native Authority may declare reserves, which shall be demarcated, of the following types:
   a) Complete reserves, being areas which are so eroded or overgrazed as to be unsuitable for utilization.
   b) Special grazing reserves, being areas within which all cultivation shall be prohibited; and
   c) Ordinary grazing reserves, being areas within which all cultivation shall be allowed.
7. Grazing of stock is prohibited within both special and ordinary grazing reserves during the months March to July inclusive, unless otherwise prescribed by a Native Authority.
8. A Native Authority may declare woodland reserves which shall be demarcated, of the following types:
   a) Hilltop: being areas within which cultivation and grazing shall be prohibited and firewood, poles and thatching grass may be obtained by written permission of the Native Authority.
   b) Woodland: being areas where cultivation is prohibited but grazing and the cutting of firewood, poles and thatching grass may be allowed with the written permission of the Native Authority.
   c) Plantation: where cultivation, grazing and the cutting of firewood, poles and thatching grass may be allowed with the written permission of the Native Authority.
9. It shall be an offence for any person to burn grass or bush outside the limits of his cultivated area or to permit fire to escape from his cultivated area into any surrounding land.
10. Any person who becomes aware of a forest fire shall report at once to his headman.
11. Every person on being so ordered by his Native Authority shall assist in extinguishing any fires which shall break out on any lands within his jurisdiction.
12. A Native Authority may issue orders to burn any grass or bush for the purpose of securing early burning of vulnerable areas.

Figure 7.2. Soil conservation orders, Sukumaland. N. V. Rounce, *The Agriculture of the Cultivation Steppe of the Lake, Western and Central Provinces* (Cape Town: Longmans, Green, 1949), 104.

took considerable effort for farmers to pen their livestock every night and to prepare the manure for use, while the cost of purchasing an oxcart to transport the necessary quantities to their fields was beyond the means of all but the larger farmers. More than this, somewhere between 30 and 50 percent of homesteads outside of Mwanza District possessed no cattle at all. Even for many larger farmers, preparing and applying manure in sufficient quantities to have an impact on yields was simply not worth the added work and expense involved. Similarly, in many of the new areas opened up for cultivation, such as Geita District, settlers found that the diversity and poor quality of the soils made the government's ideal of each family working a twenty-acre compact, mixed farm completely untenable. The traditional system, where each family worked several widely scattered plots in order to take advantage of different types of soils for growing different crops, proved far better suited.[24]

The gulf separating official and farmer went deeper, however. Whereas the Sukumaland Scheme, like all colonial land settlement and improvement schemes at the time, was premised on the belief that increasing population pressures necessitated the development of stable farming systems that would permit more intensive production and land utilization, for the Sukuma themselves the most crucial limiting factor was not land, but labor, which led them to favor labor-saving, extensive cultivation strategies.[25] In other words, they determined that the same increase in production could be achieved with far less labor by extending the area under cultivation or settling new and more fertile lands, rather than by working the same land more intensively in order to produce a higher yield. Moreover, steady population growth leading to the continuous expansion of cultivation on the peripheries, as John Ford has shown, was the Sukuma's traditional and highly effective means not only of pushing back the tsetse fly belt but of regulating population densities in established areas as well.[26] In contrast, many colonial technical advisers denigrated extensive methods, agreeing with Rounce that "when the African opens up new land there is always the danger that he will return to his old 'soil mining' practices instead of farming . . . To gain a livelihood from the soil the African will have to adopt more intensive methods of cultivation. He will have to think how much more he can get from taking care of that one acre and tending it more conscientiously, rather than growing another acre in addition."[27] These two radically divergent approaches to land use in the region put Sukuma farmers on a collision course with colonial agriculture, veterinary, and livestock officers, who were convinced of the dangers incurred by overgrazing, overpopulation, and soil erosion, as well as of the technical soundness of their own prescribed solutions.

The steady enlargement of land under cultivation reflected something else as well. Sukuma farmers had been continually expanding cotton cultivation ever since the prewar "plant more crops" campaigns, but, beginning in the mid-1950s, several years of favorable growing conditions along with increasing market prices and the introduction of better strains of seed set off a prolonged period of even more accelerated cotton growing. The total number of bales of cotton ginned in the Lake Region in 1953 stood at 38,412, whereas by 1959 that figure had risen to 183,333. The majority of this increase in output did not come about from better yields per acre, however, but from the rapid increase in acreage put under cultivation by extensive methods, as more people began growing cotton and more substantial farmers adopted oxen plows and even tractors to cultivate larger holdings, especially in the drier areas of eastern, western, and southern Sukumaland.[28] Thus, far from reflecting the "innate" conservatism of African cultivators, resistance to the Sukumaland Scheme, with its comprehensive set of rules placing numerous restrictions on cultivation, was more a manifestation of the increasing commercialization of Sukuma agriculture and of the perception among farmers that such measures were designed to hinder rather than augment their ability to increase cash income and thus raise their living standards. It is significant, for example, that some of the most vocal opponents of the scheme were the cotton growers' cooperative societies that began to form in the early 1950s.[29]

The alienation of the Sukuma and their general refusal to comply with the various conservation and land use regulations that stood at the heart of the scheme marked the effective beginning of a broader cumulative growth of antigovernment sentiment that would spread throughout the territory in the mid-1950s—and enable incipient nationalist leaders to gather mass rural support for independence.[30] It was among the Sukuma, as Low and Lonsdale observe, that the decisive strains in the colonial relationship first appeared, helping to precipitate Tanganyika's early achievement of independence.[31] Cooperative organizations like Paul Bomani's Lake Province Growers' Association and early political associations such as the Tanganyika African Association were heavily involved in Sukumaland, helping to organize protests against de-stocking and anti-erosion regulations. This activity reached a peak with the formation of the Tanganyika African National Union (TANU) in 1954. The government banned TANU from registering in the area until 1958, but union members remained secretly active, holding meetings, selling membership cards, and organizing disturbances. Indeed, the growth of an anticolonial nationalist movement, which destabilized administration and development plans throughout the territory in the 1950s, was undoubtedly the biggest factor in the breakdown and abandonment not only of the Sukumaland Development Scheme but of a number of related initiatives in other regions, including the Uluguru Land Usage Scheme in the Uluguru Mountains, where riots broke out in 1955 in opposition to compulsory terracing, and the Maasai Development Plan in the Northern Province, where, as Dorothy Hodgson has recently shown, anger and frustration led many elders to lend their support to anticolonial political groups like the Kilimanjaro Union.[32] By the late 1950s, contact between district field officers and African farmers and herders had decisively broken down in many areas, making any formal attempts at development by the administrative and natural resource departments out of the question. This was especially true of the Sukuma, who refused even to discuss the subject of development, adopting a noncooperative attitude that would continue to shape their interactions with the state even after the end of colonial rule.[33]

The aborted development mission in Tanganyika was by no means unique in the late colonial world, in part because the Sukumaland Scheme served as an important early model for rural development thinking, influencing, for example, Swynnerton's ideas about land utilization and the organization of field services, which, as we will see, were put into practice in Kenya in the mid-1950s.[34] The influence of the scheme would also carry over to West Africa where J. G. M. King, who had worked at the Lubaga Experimental Station in the 1930s and helped Malcolm prepare the original Sukuma Land Utilization Report, was seconded to investigate the mixed farming scheme in Northern Nigeria.[35] Moreover, as the memoirs of colonial technical officers involved on the front lines of the "second colonial occupation" attest, similar land use and soil conservation measures were introduced throughout the British colonies after the war—and were met by resistance from farmers, partly because they could see no additional return for their labor, but also because of deep-seated mistrust and suspicion over government intentions. In Nyasaland, for example, where the director of agriculture, Richard Kettlewell, argued that postwar problems such as erosion and overpopulation were too urgent to wait for education and voluntary effort, the government resorted to legislating soil conservation measures. In 1946 a Natural Resources Ordinance was enacted, supplemented by further acts in 1949 and 1952, which gave the government extensive powers of enforcement and required African villagers to construct contour planting ridges and bunds, plant only at certain times, clear out old crops by certain dates, and refrain from cultivating on steep hillsides and riverbanks or cutting down trees without a permit. As in Sukumaland, the new agricultural rules proved highly unpopular and were often opposed, which again led to compulsion, with the agricultural field staff operating more as a quasi-police force, handing out fines and imprisoning offenders, than as extension workers. "It is tempting to think," Kettlewell reflected years later, "that, given more time and even a modest fraction of the post-independence aid Malawi has enjoyed, we could have made remarkable progress with the organization we had perfected in colonial times. But this is an illusion: without the goodwill and cooperation of the people and their leaders it would have come to nought."[36]

Such deep hostility to the postwar state's agricultural policies among peasant cultivators must be placed within the broader framework of African resistance to the formation of the Central African Federation. The National Party victory in South Africa in 1948 spurred whites in the two Rhodesias, with support from many colonial officials in Nyasaland, to resume their campaign for closer association of the three Central African territories as a way of counterpoising the Union, and in 1953 the Conservative government in London assented to the creation of a new federal state. At the same time, as part of the "second colonial occupation," new soldier-settlement schemes were being favorably discussed by authorities in both Northern Rhodesia and Nyasaland, triggering African fears of further losses of land to European settlers. Thus, for many rural Africans, postwar colonial development regulations and schemes were closely tied to the wider issue of increasing European encroachment on their land. In the early 1950s, for example, Harry Nkumbula, who had recently returned from studying abroad to form the rurally based African National Congress (ANC) in Northern Rhodesia, launched a determined campaign to undermine the African Farming Improvement Scheme (AFIS), begun in 1946, asserting that the real object of the scheme was to improve the land for white settlement.[37] A minority of more substantial farmers, usually in mission areas, had registered as members, some even forming Farmers' Associations that worked closely with the government, but the AFIS never held much appeal for most farmers of the Tonga Plateau, who either did not see the need for contour ridges and grass strips or felt the gains to be had from the scheme did not warrant the level of regulation and extra work involved. Tensions elevated after the AFIS was restructured in 1949 with tighter conservation rules and greater restrictions on acreage. The ANC capitalized on this anger in the run-up to federation, attacking government agricultural policies and conservation programs and spreading rumors that the scheme was a cover for the planned takeover of improved lands by whites. The strategy enjoy some success, especially in districts with a history of heavy land alienation, persuading some "improved farmers" to defy government orders and even de-register from the scheme.

The ANC was temporarily set back by the failure of the antifederation campaign, however, as many farmers began to have second thoughts about opposing the AFIS, which was enhanced with the offer of loans and extended to other areas in the mid-1950s. But the Congress continued with its antigovernment campaign, causing, for example, plans for a land usage survey of the plateau to be abandoned, and encouraging farmers and even some chiefs to boycott a cattle inoculation campaign in the Choma and Mazabuka Districts. Antagonism toward the federation campaign provoked similar reactions in Nyasaland, where, as Roger Tangri has shown, many local branches of the Nyasaland African Congress voiced hostility to the government's contour-bunding campaign in the late 1940s, encouraging defiance and promoting disturbances and unrest in rural areas.[38] Tensions came to a head in 1953 when African tenants and laborers rioted in the Cholo District of the Shire Highlands, staging short strikes, felling trees, burning chiefs' court houses and mission stations, and damaging white estates. Thus, opposition to the government's rigorous enforcement of better land husbandry and conservation practices and its greater interference generally in African agrarian life after the war created a broad undercurrent of rural discontent that helped to fuel the rapid politicization of both the Malawian and Zambian countryside in the 1950s, and, in turn, to ignite anticolonial nationalist sympathies.

A similar dynamic unfolded in the Gold Coast in response to the government's campaign to eradicate swollen shoot disease on cocoa farms. The identification in 1939 of swollen shoot as a viral disease of cocoa by Adrian Frank Posnette, a botanist and plant breeder at the West African Cocoa Research Institute at Tafo, led to plans for the control of the disease during the war.[39] A detailed survey of all cocoa areas was undertaken to determine the extent of the disease, which found that the infection was far more widespread than originally thought, with as many as fifty million trees (or roughly 12 percent of the total crop) in need of treatment. Dr. Harold A. Tempany, the CO's agricultural adviser, visited the colony during the war and concurred that the best line of action was to remove all sources of infection by cutting out diseased trees from farmers' orchards, by force if necessary.[40] The swollen shoot campaign got under way in 1945, when the government authorized the Agricultural Department to employ gangs of laborers to cut out diseased trees with or without farmers' permission, and by the end of 1947 some two million trees had been destroyed. Such heavy-handed tactics met with alarm and protest from cocoa farmers, who complained that government workers were indiscriminately cutting down healthy as well as diseased trees. "We farmers," as one memorandum declared to the Watson Commission in 1948, "do not deny that the swollen shoot must be fought. An anti-septic for killing of the germs must be sought instead of this primitive method of cutting out which must bring no good results. THE CUTTING OUT OF COCOA TREES should stop . . . we have been informed that the disease affects not only cocoa trees but some forest trees. The order is that for every affected cocoa tree, some others should be felled . . . How much of the whole will be left?"[41] Again, as in the case of Central Africa, there were deep suspicions, even among otherwise loyal Ashanti chiefs, that the government was deliberately trying to sabotage the cocoa industry and acquire the land for some secret purpose. As opposition intensified, leaders of the newly formed United Gold Coast Convention Party (UGCC) took up the farmers' cause, recognizing that this was an opportunity to broaden their support in the countryside and to link rural protests in the cocoa districts with the growing discontent and economic hardships in the main towns and urban centers, where wages had fallen behind the skyrocketing prices of imported and consumer goods in the 1940s. Tensions boiled over in February 1948 when a police detachment opened fire on a procession of aggrieved ex-servicemen, setting off riots in Accra, Kumasi, and other towns, which ended with a state of emergency being declared and the arrest of several prominent UGCC leaders. The Accra Riots, as historians have long recognized, were the key turning point in the lead-up to independence in Ghana, sparking the fateful schism between Kwame Nkrumah and the Convention.[42]

The well-tempered designs of postwar developmental imperialism, as the above examples illustrate, quickly came unstuck in the face of local noncompliance and the ferment of mass anticolonial political mobilization in the late 1940s and 1950s, forcing colonial officials, planners, and technical advisers to acknowledge the gaping fissures that were appearing in the aging edifice of colonial rule. Given the pressures coming both from the CO above and at the district and community level below, the late colonial state simply did not have the political capacity or the moral authority to carry through the dramatic changes envisioned by metropolitan politicians, officials, and experts. The grand project of colonial development may have been revitalized to some extent by the influx of specialist personnel and the new emphasis on integrative planning and the coordination of government services, but the incongruities and limits to colonial state power remained. Colonial governments, as countless studies have shown, lacked legitimacy and support in local communities and, as a result, resorted to greater state coercion and compulsory legislation to solve rural problems and meet demands for increased production, in the process inadvertently deepening social discontent and anticolonial resistance in many territories.[43] The foundations of the colonial state, as Joanna Lewis observes in her study of Kenya, were simply too tenuous and fragile. Democracy and respect for the rule of law were never a conspicuous part of British rule in colonial Africa, and in the absence of a functioning civil society to give moral legitimacy, and with governing structures and habits that were deeply patriarchal, racialist, and authoritarian in nature, the capacity of the state to deliver a metropolitan-style development and welfare program was critically and fatally circumscribed.[44] What is more, beyond or perhaps because of the opposition of local communities and leaders to authoritarian government policies and practices, critical tensions and ongoing policy debates developed *within* the late colonial state. Colonial power, as recent scholarship reminds us, cannot be taken for granted as such, but, like the concept of development itself, needs to be complicated, problematized, and historicized.[45] The postwar development mission was constrained as much by bureaucratic confusion, rivalry, and opposition from officials and experts operating within the colonial state as it was by local resistance and anticolonial opposition. These stresses and strains within the state were, in large measure, an expression of the tenuous grip of colonial power, and of fissures, both at the local and imperial level, which produced confusion of purpose and administrative impasse. Ultimately, it would cripple and impede the postwar drive altogether.

Debate and disagreement existed at every level of administration, from the lofty colonial secretary's room in Whitehall, with its impressive burr walnut map case, down to the most remote district boma at the farthest reaches of empire. At the Colonial Office, for example, discord surfaced between senior administrative staff and specialist experts over the increasing influence and often unrealistic expectations of many of the new advisory bodies. Many sympathized with G. F. Seel, assistant secretary of the East Africa Department in the early 1940s, who sarcastically noted that "[t]he technique of the Advisory Committee, as I see it, is, having made an exhaustive survey of the ground, to get out a blue-print for an elaborate bridge on a scale often quite out of proportion to the traffic in view." And, he added, "[W]hen the blue-print is accompanied—as has been allowed to happen—by an intimation that no assistance from the C.D.W. Fund can be expected unless it is accepted, deplorable misunderstandings are inevitable."[46] Such tensions reflected the widening gap between the geographical and subject departments, and the growing tendency of some advisory committees and subcommittees to take on a quasi-executive character in technical matters, advising the secretary of state directly without consulting officers on the "political side," as the geographical departments were often described. As a result, as Arthur Dawe critically diagnosed, "authority is divided, action is confused, and administrative officers are discouraged and hampered by the fact that there is no clear definition of their responsibilities . . . The unfortunate Governors are scourged with theories and crowned with memoranda which too frequently have no relation to the practical problems to which their activities should be directed."[47] Many colonial governors concurred, describing the policy directives of experts from London as, to use the governor of Kenya, Phillip Mitchell's words, wrapped in a "mist of unreality."[48] The greater undertaking of responsibility by specialist departments and councils at the CO also generated frictions and dissension in the professional services in the field. Staunch resistance was shown, for example, by agricultural officials to the CAC's proposal for the creation of a network of regional advisory councils that would be responsible for coordinating and planning development: they pointed out that this would blur the distinction between the executive functions of colonial technical departments and the advisory functions of experts in London.[49]

But the most serious clashes took place within the colonial state itself. It is clear, for example, that, like the eroding gullies that kept them awake at night, deep rifts emerged between colonial technical officers and their administrative cousins. Antagonisms also surfaced between field staff generally (both administrative and technical), whose direct contact with rural societies made them more cognizant of local conditions and constraints, and their superiors at provincial or central government headquarters. In Nigeria, for example, one of the most respected prewar agricultural directors, James Mackie, resigned in a cloud of controversy in 1945, apparently due to irreconcilable differences between himself and the governor over the relationship between the agricultural and administrative departments. Before the war, Mackie and the Department of Agriculture had initiated an extensive program of research, whose main efforts were directed at developing improved varieties of food crops, the study of intercropping practices and traditional bush fallow systems, and experiments with greening manuring and mixed farming. During the war Mackie urged the government, in keeping with the CAC's principles of colonial agricultural policy, to step up its efforts to improve the colony's self-sufficiency in foodstuffs and maintain soil fertility through mixed farming schemes, but the administration seems to have been less than enthusiastic, responding to many of Mackie's recommendations with "a penchant for delay." In the correspondence explaining his resignation, Mackie suggested that achieving agricultural development in a country with the size and population of Nigeria was not a question of miracles but of continuous research and sound extension work, which he felt had been compromised by the subordination of the department to the rest of the administration.[50] The main proposals of the department, Mackie claimed in frustration, were obstructed or sidelined because agricultural staff had lost executive authority and had to take orders from administrative officials. Even criticisms by outside technical advisers, such as those made by the CO's agricultural adviser, Dr. Harold A. Tempany, after his visit in 1943, were suppressed.

When the mixed farming scheme was resumed after the war, there was still almost no progress, with the new director of agriculture, Donald Brown, reporting only 8,800 mixed farmers in the northern region by 1950, out of a total population of some three million peasant farmers. The slow rate of introduction stemmed partly from the traditional separation of animal and crop husbandry in the region, with pastoralism and arable farming practiced by different ethnic groups: the Fulani herders and Hausa farmers, respectively. Moreover, as had been the case with the Sukuma, many farmers found that the effort and cost involved in enclosing their stock, finding, cutting, and carrying sufficient fodder for them, and carrying and spreading the manure from the pen was simply not worth the return in yields. Many opted instead to have their work-bulls cared for by local Fulani herdsmen or as part of communal herds when they weren't using them as draught animals. The agricultural department here also imposed many rules and restrictions, warning that those who failed to abide by them would have their oxen and equipment confiscated without compensation of government loans. Such requirements, not surprisingly, made the scheme unpopular among area farmers.[51] But the deeper explanation for the failure of mixed farming in Northern Nigeria was political. The experts' plans for agrarian innovation conflicted with the priorities of administrative officials, whose main concern was to uphold the authority of the Emirate aristocracy on which the legitimacy of the colonial state rested, and this meant blocking land use reforms that threatened to dislodge the existing social order. "Mixed farmers," as Abdul Raufu Mustapha explains, "would have had to be given enough land to guarantee their pasturage, with their cattle kraaled in rotating fields. To do so, however, would have struck at the very heart of the agrarian social structure developed with deep British involvement in the period 1900–30; peasants would have been dispossessed to make land available to capitalistic farmers."[52]

As this experience suggests, the concerns of and reports made by technical personnel were often criticized or even silenced by political and administrative authorities, who felt that tech officers were overstepping their bounds, not simply addressing the technical issues at hand but drawing in questions of wider policy that were none of their business. Yet technical officers did not always advocate revolutionary changes and transformative technology. Outside of areas of "vested interest," the tables were often turned. Again, in the case of Nigeria, some agricultural officers, like Tom Alan Phillips, who had been stationed in the eastern region since before the war, later complained that there was a general lack of understanding within the administration that adequate examination of local soil, water, and climatic conditions took two to three years, followed by interpretation of the data according to land management practices such as those pioneered by Trapnell and Allan. In the rush for development instigated by postwar population pressures, according to Phillips, many schemes went ahead in ignorance of existing information and without proper land use investigations. Any areas of apparently vacant land became targets for possible large-scale development schemes without anyone ever stopping to ask why the land was not already being used.[53] These examples point to an important but often overlooked dimension of late colonial rule: state authorities were not united in their aims, but fractured and uncertain about how best to tackle the emerging agrarian crisis they perceived to be unfolding.

A. J. Wakefield, a man who in one way or another was at the center of much of the drama of the 1940s, diagnosed the problem poignantly in a random set of notes he recorded on the eve of the ill-fated Groundnut Mission.[54] According to Wakefield, the attitudes of "reactionary" administrative officers and Colonial Office staff, who desired complete control over technical officers and agencies, were the main cause of the feeling of exasperation and frustration that was rapidly bringing about the disintegration of the whole colonial service. All contact of the technical departments with local people was through the channel of the district commissioner. With no roots in the lives of the people, the technical agencies were unable to activate their "latent power" and communities had little say in planning or carrying out programs designed for their own amelioration. Bureaucratic routine, subjection to the whims of individual governors, and big flashy projects, rather than more appropriate and less costly measures directed at the simple problems affecting daily lives, were the result. Technical departments were often at odds with one another as well, with concern for professional prestige keeping the agricultural, health, and veterinary departments from working together to bring about rural development. In the end, high policy amounted to little in the reality of day-to-day departmental feuding and differences. The administrative structure was sorely out of date and verging on impasse—yet, introducing much-needed reforms too rapidly, especially shifting more authority to the technical branches and fostering a more bottom-up approach in which local people were consulted, would very likely bring about the system's collapse. It was, Wakefield prophesied, a recipe for political disaster.

Mission Impossible: Conflicting Imperatives and the Enigma of Late British Colonialism

Above and beyond the surges of popular resistance, structural limitations, and departmental infighting that threatened to incapacitate the postwar colonial state almost from the beginning, officials and technical experts also had to juggle a conflicting set of imperatives that translated into increasingly incoherent and problematic forms of policy intervention. In other words, it is my contention that one of the fundamental reasons for the failure of the postwar colonial development mission was the enigma of the mission itself. The "back to the land" vision of the future that underscored the 1940 *Statement of Policy on Colonial Development and Welfare* and other key directives, such as the 1943 "Principles of Agricultural Policy in the Colonial Empire," would prove highly contradictory in practice. The Depression and the Second World War, as the previous chapter detailed, brought to light the deteriorating living and working conditions and the ecological constraints that many colonial territories and peoples were facing, motivating the imperial government to adopt a new, more palliative development and welfare strategy. But the war and immediate postwar years also renewed pressures for the more intensive utilization of colonial resources for British imperial and national interests, in turn making the incongruities inherent in the new development paradigm even more evident and intractable. Local officials and technical experts vacillated between reasserting order and stability, on the one hand, and answering the demand for intensifying production and productivity, on the other; between raising colonial living standards and welfare, and responding to the pressures of metropolitan needs; between maintaining soil fertility and conservation, and exploiting colonial resources.

By 1940 it was clear to all who cared to notice that colonial living standards were appallingly low, and that much of the blame could be attributed to the mistakes of past colonial interventions and capitalist enterprises. It was also clear that if colonial living standards were to improve, then local resources would have to be more effectively managed and utilized and colonial consumer demand substantially raised, since Britain was in no position (despite the rhetoric of the CD&W Acts) to finance the social services and other nonproductive works such as erosion control that the empire so desperately needed.[55] The question of how best to do this, however, remained open to debate. The starting point for many CO advisers and committee experts, as sketched out in chapter six, was healthy lands and peoples. From the emerging perspective of ecology, the soil was "nature's capital," from which all else followed. The first priority, therefore, was the maintenance and enrichment of the fertility of the soil and other natural resources through the development of stable and sustainable farming systems and rural communities. The emphasis was on meeting the nutritional and social needs of local peoples and communities in regard to food as well as livestock, timber, fuel, and other resources. Only then was growing crops for export to be considered. The aim of colonial policy, therefore, was a more efficient reorganization of peasant agriculture in the hope of creating stable, prosperous, healthy, self-subsisting rural communities in the face of increasing strain due to land degradation, rapid population growth, and rural-urban migration.

The practice of shifting cultivation, along with other traditional bush fallow farming systems, though recognized in many cases as admirably well-suited to local conditions in the past, was increasingly seen as unsustainable. Most indigenous tropical farming systems, it was felt, were too inefficient both in terms of land and labor to meet the demands of growing population pressures and higher living standards. Shifting cultivation would have to give way to permanent farming based on more intensive methods of production through the integration of animal husbandry with the cultivation of the soil. Shifting cultivation, in one form or another, as Frank Engledow concluded in his *Report on Agriculture, Fisheries, Forestry and Veterinary Matters in the West Indies,* "has been immemorially used over large areas in the Tropics. Up to the present time there has been in these regions sufficient land to enlarge the area under shifting cultivation as the population increased. But there are now clear signs that population, fostered by external influences, is outgrowing available land—in fact for many tropical countries, especially Africa, a turning point has been reached. If growing populations are to be supported even at present levels of life, the land must be kept in permanent cultivation and at a higher level of productivity. That is, permanent mixed rotational farming must displace shifting cultivation."[56] Thus, policy turned on the introduction of mixed farming, in which crops and animals would be raised for subsistence first, and then for export, in order to maintain soil fertility and provide the basis for improved standards of living and nutrition.

But rectifying the problem of soil fertility and developing stable, highly productive farming systems in the face of growing population pressures proved more elusive than originally imagined. Efforts to develop various organic and self-sufficient methods of maintaining soil fertility and increasing yields in order to allow a more continuous and intensive form of cropping remained, by the end of the war, inconclusive. The use of green manure crops to enrich the soil and cover crops to prevent wash had, for example, been successfully applied on tea and rubber estates in Ceylon before the war, but a long series of experimental trials in Nigeria had shown that such methods in no way compared with traditional bush fallow systems unless they were consistently employed for extended periods of ten years or more.[57] "One must call into question," one research officer exclaimed in exasperation, "whether soils of this nature, under the climatic conditions that exist, are in fact capable of carrying continuous intensive arable cultivation under population pressures to which they are being increasingly subjected."[58] Much hope had also been invested in the use of grass leys, which through the influence of Sir George Stapledon and others before the war had come to be seen as something of a golden key to erosion problems. But experiments carried out by the Research and Specialist Division of the Department of Agriculture in Kenya using both British grasses and local varieties such as Kenya Wild White as well as indigenous legumes were generally disappointing. The experiment aimed at the introduction of mixed farming based on a rotation of arable cropping and grass leys. Land was placed under grass for three to four years; livestock were penned at night to provide farmyard manure. Much effort was spent in the selection of grasses and methods of ley management, yet the experiments appeared to show no better results than tumble-down or bush fallow practices. It was discovered that grass roots soon disappeared under the effect of tropical microflora, microfauna, and insects, and after a few years the nitrogen content in the soil declined rapidly.[59] The benefits of manuring were also being questioned since decomposition due to termites was so rapid that its effect on soil structure, unless large amounts were applied, was often negligible. Annual dressings of three to six tons per acre did bring marked improvements in yields, but such quantities required larger numbers of livestock and larger areas for grazing than most African farmers had access to, especially in areas with increasing population pressures. In regions of the "wet tropics" such as the West Indies, trials with farm manure found that, to be effective, rates exceeding the old English standard of twenty tons per acre every four years would be needed, which raised the question of whether mixed farming was even practical there.[60]

Such mixed results led to a reassessment of indigenous methods. Research by W. S. Martin, an agricultural research officer in Uganda, for example, drew attention to the effect of grass leys in binding soil particles into aggregates, lending scientific approval to shifting cultivation and bush fallow practices as an effective method of restoring soil fertility.[61] Indigenous practices such as intercropping also received renewed attention and were found to promote higher yields than planting in pure stands. Experiments with mixed planting of crops also found the practice to be highly effective in protecting against insect pests and countering soil erosion.[62] Attention was given to other techniques as well, such as mound cultivation, which was found to be a simple and efficient method of conserving the organic matter in the soil, one more beneficial than plowing in grass leys because the mound acted as a compost heap, heating and fermenting the grass underneath.[63] But despite more serious consideration given to customary indigenous practices on the part of some colonial researchers, the most immediate reaction to the practical difficulties was to press on with the introduction of new inorganic and mechanical methods of cultivation and soil conservation, such as the use of artificial fertilizers, mechanized plowing, and mechanically constructed, broad-based terraces, which entailed radical changes in local farming systems. This reorientation of policy did not go unchallenged within colonial agricultural departments in the field. In Nigeria, for example, a protracted and heated debate surfaced in the late 1940s and early 1950s between those who faithfully adhered to the old organic or "Humus School" of thought championed by Faulkner and Mackie, and those anxious to introduce chemical fertilizers.[64] The battle raged for three years, with divisions largely along generational lines. Older agricultural officers, who had been there since before the war, were more aware of the complexity of many indigenous farming systems and practices and the diversity of local climatic and environmental conditions, while younger officers, many of them hired in the postwar recruitment drive, were confident in the power of state planning and a technocratic approach, involving mechanization, artificial fertilizers, chemical pesticides, economies of scale, and land consolidation.

By the late 1940s, CO advisers in London were also coming to the realization that higher standards of living could be attained only by substantially raising production and yields. Given higher population densities and a declining resource base, this meant that chemical fertilizers, mechanization, and new forms of cooperative farming might well be necessary. The CAC called for surveys to determine existing supplies of phosphate, potash, lime, and nitrogenous fertilizers within the empire and for properly designed fertilizer trials to be carried out on a whole range of crops and soil types within the territories.[65] By 1948 the council had come to the conclusion that the maintenance of tropical soil fertility at a level efficient enough for production of export crops in competitive world markets would be impossible without the use of artificial fertilizers.[66] In the same vein, it called for properly designed field trials covering the social and material aspects of mechanized farming, to be coupled with general research carried out at experimental stations in all colonies with the aim of understanding the wider impact mechanization would have not only on production but also on existing systems of land use and tenure.[67]

The shift toward fertilizers and mechanization corresponded with the productionist view of development spearheaded by the new financial adviser, Sidney Caine, who as noted earlier was put in charge of development and welfare policy in the spring of 1943. In his first important statement on colonial development, Caine argued that the problems of "natural poverty" and overpopulation were too formidable in many colonies to expect any significant improvement in the standard of living unless there was a dramatic boost in local productivity and output.[68] Caine's assessment of colonial poverty as "aboriginal," rather than as the consequence of previous colonial intervention, marked a significant break from the palliative views of colonial critics of the 1930s, and paralleled similar trends in Fabian thinking on the empire.[69] Caine was also critical of the policy adopted by the CAC and its obsession with improving local nutrition and peasant subsistence farming at the expense of export crops, which he suggested would only reinforce colonial "backwardness." The Caine memorandum of August 1943 is often singled out as the key document denoting a policy primarily directed toward increasing output and economic growth, rather than simply the provision of social welfare services and the improvement of subsistence farming.[70] Too much can be read into this, however. As was the case in the mid-1930s, Caine's prodding initiated a debate within the CO over the meaning and purpose of development, one centered on the question of how productivity could be rapidly intensified, and how this was to be reconciled with the commitment to colonial welfare.[71] By the end of the war these questions remained unanswered. It was agreed that the long-term goal was to make the colonies economically self-supporting, which implied increasing their real productive power, but in the meantime substantial spending was still needed on social services and raising consumption.

Plans were further complicated, however, by the postwar economic crisis. Delegates at the Food and Agriculture Organization's Preparatory Commission on World Food Policy warned in 1947 that if nothing was done to expand production and consumption both in the "advanced" countries and the "undeveloped" regions, the world might find itself in a greater depression than 1929, resulting in millions out of work and unparalleled social unrest.[72] In Britain there were great shortages of food and other wage goods which threatened to lead to poor nutrition and lower productivity of the workforce. Shortages of materials also placed a brake on the drive to increase production, and thus augured a return to serious unemployment—something that would have been a political disaster for the new Labour government, which promised full employment, increased consumption, and the swift implementation of the Beveridge Plan for welfare expenditure.[73] To these motivating factors must be added the Labour government's resolve to counter increasing U.S. domination of the British national economy. Thus, following the war, the British government impressed upon governors in the tropical colonies the imperative of increasing production of food and raw materials *to meet British need.* The real turning point came in response to the postwar sterling crisis, which, as noted above, sparked the Fabian-inspired colonial development offensive of 1947. The decision was made to brush aside Colonial Office dithering and attempt to resolve things directly through the creation of new public development corporations such as the OFC and the CDC, circumventing private capital by implementing fully mechanized, state-managed schemes for agricultural production on large-scale estates. At the same time, control over colonial commodity trade was maintained through state marketing boards in order to, in effect, subsidize British consumers.

In introducing the Overseas Resources Development Bill, the parliamentary undersecretary of state, D. R. Rees-Williams, explained, "There is a vast plan in our minds both for improving the living standards and welfare of colonial peoples and increasing the supply of foodstuffs and raw materials to the world."[74] No stone was to be left unturned: research and surveys of the empire were needed to locate new resources; the efficiency of existing cultivated areas were to be improved and new agricultural areas and other natural resources such as forestry and fisheries brought into production; and secondary industries would be established to process products for local consumption. The irony of the new endeavor was not lost on the secretary to the cabinet, Norman Brook, who warned that such a policy could also, "I suppose, be said to fall within the ordinary definition of 'Imperialism.' And, at the level of a political broadcast, it might be represented as a policy of exploiting native peoples in order to support the standards of living of the workers in this country."[75]

The new imperatives reinforced trends already set in motion by Caine for raising the efficiency and boosting the output of peasant agriculture, and could be felt in the tone set by the agricultural adviser, Geoffrey Clay, who declared at the opening of the second Colonial Service Summer School at Cambridge in September 1948 that the stage was set for a large expansion of colonial production. In the past, Clay noted, colonial efforts had aimed at the slow evolution of peasant agriculture, respecting established native practices and encouraging local initiative through community development and mass education, but such methods were slow and could not keep pace with the growth of population or the demand for higher standards of living. In the name of higher yields, the time had come to introduce "revolutionary changes" in the organization, methods, and systems of agriculture.[76] Drawing on the lessons of Britain's own history of enclosures, Clay called on colonial service officers to be bold in their conceptions and courageous in their experiments. Given current scientific knowledge of crop rotations, use of artificial fertilizers and grass leys, modern methods of control of pests and diseases, and mechanization, and given sufficient capital and leadership, Clay felt it would not be difficult to effect an agricultural revolution akin to the that of the eighteenth century, but in a fraction of the time. Soviet collectivization and the as yet still highly touted East Africa Groundnut Scheme were held out as examples of what could be done. Clay ended by citing Phillip Mitchell's caveat that little progress would be made by trying to persuade existing rural communities to adopt the new methods and practices voluntarily. It would, therefore, be necessary to set up new communities in areas at present uninhabited or sparsely populated in order to demonstrate the merits of the new development patterns, and, though he did not say it explicitly, a great deal more government compulsion would also be needed.

We have seen something of the tensions ignited by the postwar schemes for rural settlement and improvement, as well as the intractable constraints that proved the undoing of many of them. Similar challenges awaited many of the early trials of the new "modern package" of artificial fertilizers and mechanized farming. During the 1940s, and even before, extensive fertilizer field trials were begun in West and East Africa, which in some cases produced remarkable increases in yields, while in others the results were less striking and often inconclusive, depending on the fertilizers, soil types, and rotation of crops involved. In general, it was found that East African soils responded well to phosphates and superphosphates; in West Africa, sulphate of ammonia proved more valuable.[77] But the real problem with the new fertilizers was their cost: they were prohibitively expensive for most peasant farmers until heavy state subsidies were introduced in the 1960s. Experiments with mechanization fared little better. In Nigeria, for example, a policy of mechanization was introduced after the war, including large projects planned and funded by the Colonial Development Corporation near Iseyin and Abuja that pioneered the use of cable plows. Many of the older agricultural officers, according to Donald Brown, were stunned by the proposals, which made the earlier measures of contour plowing and ridging for the prevention of soil erosion, so forcefully advocated by Faulkner and Mackie, virtually impossible.[78] Nigerian agricultural staff were not alone in their hostility.

authorities. Experience in India, as noted in chapter one, as well as other parts of South and Southeast Asia, including Ceylon and Malaya, had shown that individualization could lead to land transfers, insecurity of tenure, the subdivision of holdings, peasant indebtedness, and landlessness, which threatened to tear apart the whole social framework of the village community—the very thing the new development initiatives were intended to counteract. Moreover, the experience of small farmers in the American Midwest, South Africa, and Australia suggested that individual freehold tenure in combination with growing indebtedness could also give rise to a highly exploitative mining of the soil.

This was a dilemma not lost on experts in London, who were exposed to an increasingly global perspective. As CO agricultural advisers concluded, "Efficiency in agriculture depends to an important extent on a satisfactory system of land tenure. Reforms are clearly necessary if the prevailing laws or customs prevent or discourage such things as the cultivation of land in units of economic size, long-term improvements and the best possible use of water resources. The same is true if there is nothing to prevent individual owners or occupiers from farming their land in such a manner as to damage or imperil the interests of their neighbours or their successors, or to prevent the transfer of the ownership of land from working farmers to land owners who are interested in cash profits."[87] Nor was the conundrum lost on colonial agricultural officers like Colin Maher, who had seen the plight of small farmers in the American Midwest during his visit in 1938. With this in mind, Maher urged in the case of Kenya the introduction of a system of secure leasehold in which tenants remained subject to "corporate tribal responsibility," "rather than a system of individual ownership which results inevitably in the breakdown of the family-group organisation with its social responsibilities."[88] In this way he hoped to avoid the disabling effect of peasant indebtedness and protect the African from the horrors of individual tenure. So did Donald Malcolm, who in defending the Sukuma system of village group tenure asserted, "In the middle of the nineteenth century, when Maine's *Ancient Law* was written, it was axiomatic that individualization of land holding was the goal toward which civilization was moving. The same view could not pass unchallenged today. There is, indeed, no reason to suppose that the social and land organization evolved in Africa is inferior, in its own environment, to its European counterpart."[89]

Similar reasoning was applied in the case of the West Indies, where "uncontrolled" peasant ownership was said to be the root cause of a general decline in the level of soil fertility and the spread of "crude" and inefficient agricultural practices.[90] The size of many peasant holdings was plainly inadequate, often less than two acres of poor-quality land, and the purchase price high, which burdened many smallholders from the start with heavy debts.[91] Stockdale, who had visited the United States in the 1930s and must have been struck by the similarities, explained the situation forcefully:

> One of the great problems in the West Indies is land tenure. In the past there has been a universal desire for freehold tenure and as a result the available resources of small cultivators has [*sic*] been absorbed in the purchase of land and there is nothing left for the building of houses or for development . . . The evils of the freehold system are becoming increasingly realised and many Dependencies are now accepting the alternative idea of a leasehold tenure, which provides for the proper maintenance and cultivation of the land on penalty for failure to do so of relinquishing the lease . . . the desire for freehold, which was previously very strong, is now weakening and there is growing recognition of the desirability of a system of land tenure designed to take care not only of the present generation but of the generations to come. There is no doubt that freehold, except under very careful management, is adverse to the interests of the land, the owners and the community as a whole.[92]

What disturbed Stockdale most were the government-sponsored land settlement schemes, the ideal of which was the individual peasant holding five acres on a freehold basis. Efforts to turn ex-slaves into contented peasant proprietors had begun in the West Indies at the turn of century, when, as we have seen, the depressed state of the sugar industry together with the mounting pressures of surplus population prompted Sir Henry Norman and Sydney Olivier to propose land settlement as a means of relieving unemployment.[93] By the Depression, most West Indian governments had pinned their faith on land settlement schemes and peasant ownership, although, as William Macmillan was quick to point out, it was "still only an aim, not an achievement."[94] The problem, as the experts professed, was that most of the land in the islands that was suitable for agriculture was already in use, which meant that no significant addition to cultivated land could be expected from new government schemes. The only solution, then, was to make better use of the land already under cultivation through a more intensive development of both peasant and estate agriculture. Such an "objective" argument, of course, carefully avoided the politically volatile question of the disproportionate distribution of land in favor of large proprietors.[95]

But the call for a more efficient reorganization of peasant agriculture went deeper: Olivier's favorite "peasant solution" had turned out not to be the panacea for all social and economic ills. Peasant cultivation, as several important surveys concluded, had proven less productive and efficient than cultivation on well-managed, large-scale plantations and estates.[96] Even sympathetic observers like Macmillan, much to the chagrin of Olivier, came to the conclusion that a free and independent peasantry as an end in and of itself would not undo the region's backwardness.[97] The problem, as Sir Alan Pim explained, was that in most cases peasant agriculturalists could not work adequate areas, command sufficient capital resources, or otherwise benefit from the economies of scale necessary to enable them to compete against the efficiency of larger estates and other forms of capitalist agriculture.[98] There was a dilemma, too, as officials in West Africa and South and Southeast Asia were also discovering, in that smallholders failed to remain simple peasants. The very independence of peasant farmers prevented officials from introducing measures to control disease or to prevent the partition and fragmentation of peasant holdings through inheritance or the sale of communal lands. By the 1930s, as Anne Phillips explains in the case of the Gold Coast, "the success of the [cocoa] industry had dissolved the 'peasants' into a complex of moneylenders, absentee landlords, independent smallholders and migrant labourers."[99] In the West Indies, settlement schemes based on the ideal of peasant proprietorship were intended to avoid the problems of inefficient production and class differentiation, but again, there was no way of averting the "undue" concentration of ownership as a result of indebtedness, or the "excessive and uneconomic subdivision of holdings."[100] Moreover, there was no way of ensuring the desired measures of rehabilitation and soil conservation would be carried through.[101] Yet, for reasons that were, as we have seen, largely social and political, the quest for stable rural communities of thriving peasant producers remained sacrosanct.

In the immediate postwar years, however, colonial officials were uncertain just how to make indigenous agricultural practices more sustainable and yet more productive at the same time. They remained loath to grant individual title, hoping instead to effect the desired agricultural reforms and conservation measures by the introduction of a number of alternative strategies—more effective agricultural education and extension work, leasehold tenure, government marketing boards, and even experiments in cooperative and collective farming—as ways around the problem. By far the most common trend was the introduction of systems of mixed farming on the basis of secure leasehold in new settlement areas, where peasant smallholders were granted consolidated plots on long-term leases (usually ninety-nine years), under which title was restricted or protected in an effort to prevent the fragmentation or alienation of land by leasing, mortgaging, sale, or partible inheritance. Such protected tenures were also designed to empower the state with the right to eject those who did not introduce specified improvements in cultivation or anti-erosion measures.[102] The challenges were much more formidable, however, in areas of established cultivation with complex patterns of land tenure, in which each family owned several scattered plots, often less than an acre each in size, on which they practiced extensive forms of shifting or bush fallow cultivation. Here it was hoped that government extension and demonstration work would convince the mass of cultivators to adopt the new recommended practices and that the success of the new land settlement schemes would lead more "progressive communities" to seek government help in reconditioning and consolidating existing areas along the same lines.

Attempts to introduce secure leasehold tenure in land settlement schemes provoked heated debate, especially in the West Indies, where an Agricultural Policy Committee set up in 1942 under the comptroller's agricultural adviser, A. J. Wakefield, called for a new land policy based on self-contained systems of mixed farming and leasehold tenancy. In keeping with the principles of the West Indian Royal Commission, their report urged the expansion of peasant agriculture and local food production in an effort to improve rural welfare, health, and nutrition. Modeling their recommendations on lessons learned from the Land Settlement Association's projects for unemployment relief in Britain, the committee proposed two kinds of schemes, both under a leasehold system of tenure.[103] First, there would be schemes for settlers who would combine casual employment on sugar estates and large-scale enterprises with growing at least part of their food on their own land. It was hoped that, in time, as settlers gained adequate knowledge and experience of mixed farming, they would be able to make the jump to full-time holdings. In the second type of community settlement, families would live and work on their own land, growing their own ground provisions rather than relying on imported foodstuffs, while at the same time growing permanent export crops.[104]

The local press heralded the policy as a "new life movement" and the West Indies version of the Beveridge Plan.[105] And, indeed, it was: rural settlement and betterment was as much a social welfare as a strict agricultural proposition. Much would depend on the active cooperation and support of local people, which meant, above all, an emphasis on education in the widest sense to instill new attitudes to rural life. As Stockdale explained to Christopher Cox, the CO's educational adviser, the committee's aim was better living for a rural environment, which meant practical educational efforts that contributed to the general stability, welfare, and happiness of the community.[106] Better nutrition, housing, and sanitation were the goals, as much as better husbandry practices. Settlement tenants were also to be instilled with a new "spirit of fellowship" through the pooling of equipment and resources on a communal basis and the cooperative organization of purchasing and marketing to take advantage of economies of scale.

The most significant scheme established along these lines was the Lucky Hill Community Project in the Parish of St. Mary, Jamaica, which the district Pioneer Club initiated in 1940. The project sought to combine the advantages of large-scale enterprise with the individuality of small family holdings by working an 873-acre former cattle estate on a communal basis. After two years of financial losses, however, the Department of Agriculture took over the scheme and substantially revised it in light of recommendations by Wakefield. It received a £20,000 grant from the CD&W Fund as well as assistance from the government of Jamaica, but continued to operate at a loss until 1945, when it was reorganized again. Over the next few years things began to turn around as the project recorded small profits from the sale of food crops, milk, and cattle. The project involved around forty settler families who worked a central mixed farm collectively, receiving wages as well as dividends from profits in proportion to work done. A general overseer was responsible for supervising and assigning tasks, while a Managing Committee in consultation with a Settlers Council drew up the estimates of expenditure, oversaw finances, and prepared annual reports. All farm buildings, water tanks and piping, beef and dairy cattle, machinery and equipment, roads, fences, and so on were purchased collectively out of a Development Capital Grant, and cottages were provided for each settler family with half an acre of land for growing vegetables, rearing chickens, and perhaps keeping a cow.

The moderate success of the scheme led one senior agricultural officer, James Wright, to suggest that "[t]he Lucky Hill Experiment may provide a part answer to the very vexed and complex problem of land settlement with which Jamaica is faced at the present time."[107] But Lucky Hill was unique among settlements in the West Indies. Most other schemes continued granting land on an individual ownership or cash tenancy basis, where occupiers combined small market or cash crop farming with part-time agricultural wage employment or nonagricultural activities.[108] Nor was there much success convincing small farmers to grow local food crops for home consumption, largely because it remained cheaper to buy most foodstuffs from abroad with the income derived from the sale of higher-valued export crops.[109] Most of Wright's colleagues, moreover, remained opposed to the concept of leasehold tenure. In view of the historical background of slavery and post-1830 land policies, the commissioner of lands in Jamaica strongly advised against altering in any way the present practice of granting proprietary rights and full ownership to peasants. Jamaican peasants, he warned, viewed land ownership as a source of security and independence and would vehemently resist curtailment as a retrograde step.[110] F. C. C. Benham, the comptroller's economic adviser, concluded that the whole idea of land settlement was misplaced: there simply was not enough suitable land to grant every family a self-contained holding that would provide a moderately good standard of living. Instead, he advocated a policy of medium and large farms of between fifty and one hundred acres, increased mechanization of the sugar industry, and doing away with import duties of any kind as the best route to raising standards of nutrition and living.[111] At a conference held in Jamaica in 1950, the directors of agriculture for the British Caribbean concluded that the group approach to land settlement and land use had not been a success thus far and required review.[112] Much of the development was on poor land, with little agricultural planning and with holdings too small to be suitable for intensive farming. Despite the disappointing results, however, the CO's agricultural adviser, Geoffrey Clay, remained hopeful, suggesting that a policy of peasant settlement, if well managed and properly carried out, still held economic promise and should remain part of general policy in the region, as should the principle of leasehold tenure.[113]

Experiments with leasehold tenancy and cooperative farming systems began after the war in British Africa as well. In Northern Rhodesia, a Land Tenure Committee established in 1943 favored leasehold over freehold in granting land for settlement to smallholders. In 1948, for example, a Peasant Farming Scheme was introduced which granted small-scale farmers twenty-acre holdings on ninety-nine-year leases in unoccupied

Meanwhile, the colonial secretary, Oliver Lyttleton, announced in June 1954 that henceforth the men and women of the Colonial Service would be embarking on a new era as members of a new HM Oversea Civil Service, but he assured them that terms of service, pensions, and other benefits would be safeguarded, and that in the event of termination or premature retirement due to constitutional changes they would receive compensation.[128] On the face of things, the new service looked a lot like the old one, but the name change was an important psychological marker signifying that the colonial endgame was not too far off. The announcement reverberated throughout the Colonial Office, affirming what was already obvious: the postwar experiment in greater central management and direction had failed. As R. V. Vernon explained in a meeting with the CO's chief agricultural advisers, Geoffrey Clay, Geoffrey Nye, and Frank Engledow, the colonial territories in their present state of political development were no longer willing to accept advice rendered by the Colonial Office and its advisory committees. It was no longer appropriate for advisory bodies like the CAC to take the initiative in formulating agricultural policy or to pressure colonial governments to establish comprehensive soil conservation programs. From now on, as had been the case before the war, colonial governments would seek advice from metropolitan experts on their own initiative as problems arose. The council's work was to be cut back, but, in truth, as Engledow confided, it had already served out its purpose; it had in fact become so large and unwieldy that it had not met since 1951.[129] A similar waning of momentum and authority marked the postwar activities of other CO advisory committees, such as the ACEC, where it was increasingly realized that only those "on the spot" were really in a position to deal with the rapidly changing realities of colonial education in the 1950s.[130]

Back in the field, the disappointing track record and conflicting aims of postwar development policy were met by some colonial regimes with, as we have seen, one last resurgence of innovation and administrative power. But for many colonial technical officers, it seems, the last years of empire were ones of frustration and disillusionment, leaving some deeply cynical about the whole ordeal. As Anthony John Smyth, a soil scientist and agricultural research officer in Western Nigeria in the 1950s, later confided, "This experience has made me skeptical of situations in which a failure to persuade farmers is blamed on a lack of extension facilities—lack of a proven product is more likely the cause. People who live close to the bread line are right to be conservative." In a similar vein, the lesson

Donald V. Chambers, a senior research officer in Tanganyika, was left with was that "the introduction of unproven or untested 'improvements' carries a great responsibility as most farmers have little or no margin for error." Looking back on the postwar years, Charles Lynn, whose long career included stays in the Gold Coast, Northern Rhodesia, and the Bahamas as well as the ICTA in Trinidad, professed, "I have often thought that one of the attractions of independence must [have been] relief from expatriates breathing down one's neck extorting development." But then he quickly added, "without the development of its natural resources a country [can] make no progress and one wonders for how much longer the world can afford to tolerate such a state of affairs."[131] As the colonial era drew to a close, a solution to the old colonial conundrum of how to balance the ideals of stable rural communities and sustainable environments with the global imperatives for growth and accumulation remained as elusive as ever.

Conclusion

Postcolonial Consultants, Agrarian Doctrines of Development, and the Legacies of Late Colonialism

> We believe that the failure of a great many development projects to achieve even their most fundamental objectives is due to a reluctance on the part of development practitioners to appreciate the significance of history. Projects are frequently designed as if time began with the project implementation schedule. Past lessons are seldom examined and still fewer professionals bother to enquire into the historical circumstances of the people their interventions seek to assist . . . There is little doubt now that as development professionals they have failed . . . We have attempted to describe . . . how so many good intentions can yield so many poor outcomes.
>
> —*Doug Porter, Bryant Allen, and Gaye Thompson*[1]

IT IS IN THE AFTERMATH of colonial failure in the early 1950s that the postwar "development project" is said to have been born.[2] Reflecting back on the era from the vantage point of the mid-1990s, Colin Leys suggested that what most people understand by the term "development theory" began in the 1950s to deal with the pressing problem of transforming the "newly emergent" nations of the "third world" into productive modern economies.[3] Indeed, it is often suggested that the "war on world poverty" was a manifestation of the particular historical conjuncture of the immediate postwar years, and, in particular, the rise of the Cold War and the American-led campaign to fight the spread of communism in the third world. David Moore recounts a familiar tale when he notes that "[i]n the wake of the Marshall Plan, the Soviet-Stalinist model of industrialization, the Cold War, the Chinese Revolution, and the emergence of the politically independent 'third world,' western policy-makers—led by the United States—had quite a number of problems facing them as they considered how to deal with former colonial subjects. The word 'development' was the perfect hegemonic catch-all for capturing the goals and aspirations of all parties to the situation. Its basic elements were constructed in the years between 1945 and 1955."[4]

The prevailing logic now was that improving the living conditions of peoples in the "resurgent" nations would quell the sort of social unrest that provided fertile ground for communist propaganda and influence. In this context, the steady stream of neo-Malthusian and other arguments of the late 1940s and 1950s drawing attention to the world "population problem" and its geopolitical and ecological consequences took on added significance.[5] In Africa, the censuses of the late 1940s indicated that population was increasing much faster than had previously been thought, and that birthrates were among the highest in the world.[6] Increasingly, the continent attracted the attention of international conservationists who worried that Africa's rapid population growth would endanger wildlife and habitat.[7] Nor was it lost on U.S. and other Western policymakers that the bulk of these surging populations lived not only in Africa but in other so-called underdeveloped areas, such as Asia, Latin America, and the Middle East, prompting fears of "swamping" and heightened international conflict along racial and class lines.[8] All this made development a pressing imperative, causing it to be couched in even more expansive terms: the "resources of the earth" had to mobilized in an effort to "put an end to hunger and poverty" and bring "freedom from want to all mankind."[9]

But although the late colonial development initiative may have failed by its own admission, its heavy bias in favor of state-centered ideologies and development structures would live on long after the formal transfer of colonial power. The new independent governments of the 1960s, as Fred Cooper relates, "sought to take over the interventionist aspect of the colonial state, and indeed intensify it, in the name of the national interest and (for a time) to demonstrate to voters that the state was improving their lives."[10] And though the new nationalist governments repudiated many of the former colonial policies, especially conservation measures, in favor of more extensive plans or high-profile projects, the principal effect was often the same: the expansion and entrenchment of bureaucratic state power in the name of development.[11]

But perhaps an even greater legacy in the long run has been the powerful set of assumptions and crisis narratives produced by late colonial experts, which have been considered part of the common knowledge of development professionals ever since. Recent scholarship has drawn attention to how conventional knowledge about development and environmental problems is historically constructed, reproduced, and translated into public policy paradigms that serve specific institutional and political purposes.[12] This study concludes with a look at the crucial role colonial agricultural and natural resource experts played in the growing institutionalization and globalization of colonial scientific knowledge and authority in the postcolonial epoch. Many of the men and women who sat on the Colonial Office's advisory bodies or who had their professional start as part of the colonial technical services would go on to enjoy prominent careers with the United Nations and other donor agencies. The agrarian doctrines of development laid out by them would, in turn, become deeply embedded in international programs and institutions in the decades following the end of colonial rule.

From Colonial Experts to Postcolonial Consultants: The Making of International Development Technical Assistance

Through the actions of a multitude of colonial scientific advisers and technical practitioners, the CO's earlier vision of integrated rural development would help set the pattern for much of the postcolonial epoch. The impetus behind the founding of the United Nations Food and Agricultural Organization (FAO), for example, owed much to the ideas and efforts of late colonial advisers. Gerard Clauson and Frank Engledow both attended the International Conference on Food and Agriculture held at Hot Springs, Virginia, in June 1943, where the principles of agricultural policy laid down in the 1943 CAC memo were expounded and endorsed. It was at this conference that plans for a new international organization to promote agricultural production and improve nutrition were launched, and in 1945 another prominent CO adviser, Sir (later Lord) John Boyd Orr, was appointed the new agency's first director-general.[13] Through the efforts of Lord Boyd Orr and others with strong imperial connections, the prewar movement for international scientific cooperation in devising a "world food plan" was taken up with renewed vigor. The initial surveys done by the FAO indicated that the war had caused an overall fall in the total world food supply. Conferences were held in Copenhagen in 1946 and Washington in 1947, and plans were drawn up for a World Food Board to unify the work of the FAO, the UN's Economic and Social Council, and the International Bank for Reconstruction and Development (the World Bank). Essentially, the aims of the proposed board were to stabilize world prices of the main foodstuffs through guaranteed markets and to double world food production in twenty years. This, it was felt, would provide sufficient food and other basic necessities to meet the needs of the two-thirds of the world's population dependent on agriculture for its livelihood, while also bringing about a great expansion in world trade, to the benefit of all. The plan was quashed in part because it conflicted with the Labour government's postwar colonial development offensive, but the FAO survived to become one of the leading agencies, along with the World Bank, responsible for the planning and oversight of rural development in the third world.[14]

Boyd Orr was not alone. Among the list of high-profile scientists with close connections to the Colonial Office who would go on to play a prominent role in the burgeoning web of international organizations after the war, none stands out more than the eminent British biologist and popularizer of science, Julian Huxley. Huxley had been an active adviser for the Advisory Committee on Education in the Colonies since 1930, when he surveyed East Africa to consider the feasibility of a new biological education curriculum for Africans.[15] Huxley also served along with Boyd Orr and Oldham on the African Research Survey Committee and made an extensive tour of West Africa in 1943 to advise on the creation of universities for the region. As a result of his interest in both education and conservation, Huxley was chosen to be the first director-general of the United Nations Educational, Scientific and Cultural Organization (UNESCO) in 1946. The new organization, as Huxley envisioned it, was to act as a nucleus for large-scale planning for world reconstruction, with its main energies directed to population planning, birth control, and wildlife and nature conservation. One of the projects that received support from Huxley's UNESCO was the new African Research Survey, carried out by E. B. Worthington in the 1950s, to promote, as Anker recently put it, "a plan for a scientifically engineered economy of nature for Africa that could foster economic growth and prosperity for its people, nations, and their leaders."[16]

While Boyd Orr and Huxley are among the more well-known examples, many other technical officers and researchers discussed in this book who began their careers in the colonial technical services and research networks would go on to become consultants and advisers for the UN's specialist agencies, especially the FAO and the World Health Organization (WHO), as well as the World Bank, British foreign aid agencies, and private consultancy companies, bringing with them the development models and strategies first devised during their formative years in the late empire. One of the more notable metamorphoses was that of Sandy Storrar, a senior agricultural adviser in Kenya from 1944 to 1965 and pioneer of farm planning, who became head of agriculture for the World Bank's Development Service for Eastern Africa in 1965, and in 1968 joined the bank's Washington project staff, where he remained until 1973. From there Storrar moved on to become the bank's agricultural adviser to the Indonesian Resident Mission, then chief of mission at the bank's office in Dacca, and, finally, senior adviser in Washington.[17] Archie Forbes, who had been a senior agricultural officer in Tanganyika, went on to a long career as a consultant for the United Nations, the World Bank, the FAO, and the Nordic Board, working in the Middle East, the Caribbean, and twenty-one African countries.[18] Anthony John Smyth, once a senior agricultural research officer in Western Nigeria in the 1950s, spent twelve years as a land classification specialist for the FAO from 1962 to 1974, before being hired as the director of the Overseas Development Administration's (ODA) Land Resources Development Center at Tolworth.[19] Nor should we overlook colonial researchers and practitioners in the fields of public health, nutrition, and epidemiology, such as Robert Hennessey, a former director of medical services and medical pathologist in Uganda between 1929 and 1944, who was appointed chairman of the WHO's African Regional Committee in 1953–54. Or R. L. Cheverton, the deputy director/director of medical services for several colonial territories between 1943 and 1953, who became an internationally recognized authority on malaria, organizing one of the first successful eradication campaigns while in Cyprus and serving as vice-president of the WHO's Conference on Malaria in Kampala, Uganda, in 1950.[20]

Beyond the more publicly renowned international agencies, many former colonial technocrats reincarnated themselves as policy or project advisers and consultants operating within the expanding network of British overseas aid agencies such as the ODA, the Ministry of Overseas Development (ODM), and the Commonwealth Development Corporation (CDC). C. E. Johnson, for example, a senior agricultural officer in Northern Rhodesia and later director of agriculture in Nyasaland, headed the ODA's technical cooperation mission for tropical agriculture to Bolivia in 1965, and then became an agricultural adviser in the Middle East Development Division from 1968 to 1974. He finished his career working as a consultant both for the World Bank and the United Nations Development Program (UNDP).[21] Anthony Blair Rains, who had been an animal husbandry and grassland research officer in Northern Nigeria in the 1950s, went on to serve as an agroecologist with the ODA in the late 1960s, 1970s, and 1980s, conducting ecological surveys in Botswana, Gambia, Nigeria, Tanzania, Swaziland, Lesotho, Fiji, and the Solomon Islands.[22] Countless others served as natural resource consultants, crop specialists, project leaders, and agricultural advisers for the ODA, ODM, and CDC in countries as far afield as Antigua, Botswana, Lesotho, Mauritius, St. Helena, and Thailand, as well as in central offices and on management boards in London.[23] Still others would find employment with private consultancy firms such as Hunting Technical Services Ltd (HTS), which formed in 1953 to provide technical services for natural resources development, especially in the emerging "underdeveloped areas."[24] HTS specialized, among other things, in aerial photography for detailed land resource surveys, and in the late 1960s and early 1970s the company pioneered the use of airborne radar imagery for the same purposes. Others would become independent tropical resource and development consultants. landing lucrative contracts with the World Bank, UN, ODM, the International Development Research Council, Ottawa, as well as various third world governments and private plantation companies.[25] Looking back on it, one former colonial research specialist judged that fifteen years spent in the Colonial Service had fitted him perfectly to become a development consultant.[26] And indeed, as Roger Swynnerton once rhetorically asked, "What would the World Bank, FAO, ODA and ODM and its association institutions, CDC and British consultancy firms have achieved in Third World development had they not had the highly professional cadres of the Colonial Agricultural service to draw on?"[27]

Beyond the vast network of professionals who shaped development practice in the field, many pioneering theoretical and conceptual contributions were also made by those who went on to occupy important academic and research appointments. T. W. Tinsley, for example, after years working on Swollen Shoot disease at the West African Cocoa Research Institute (WACRI) became head of the Insect Pathology Unit of the Department of Forestry, Oxford, and later director of the Natural Environment Research Council at the Institute of Virology, Oxford. Cyril Charles Webster, a research officer with years of experience in Burma, Nigeria, Nyasaland, Kenya, and Malaya, was appointed professor of agriculture at the ICTA in 1957, and later scientific adviser for the Agricultural Research Council in the UK from 1965 to 1975.[28] Webster's *Agriculture in the Tropics* was one of the leading textbooks in the field until well into the 1980s.[29]

Significant contributions to tropical ecology and tropical agricultural research were also made by J. W. Purseglove, lecturer in tropical agriculture at Cambridge, 1952–54, then professor of botany at the ICTA and University of the West Indies from 1957 to 1967, whose multivolume *Tropical Crops* was published by Longman as a companion to its definitive Tropical Agriculture Series on major crops; and by M. H. Arnold, who had worked at the Namulonge Cotton Research Station, Uganda, before going on to an important research career at the Cambridge Plant Breeding Institute, where he edited and contributed to a pioneering study on the application of agricultural science to national development.[30] Other seminal works were produced by William Allan, an agricultural research officer who worked closely with Colin Trapnell and J. Neil Clothier in Northern Rhodesia; John Ford, an entomologist in the Department of Tsetse Research in Tanganyika; Sir Charles Pereira, the deputy director of the East African Agricultural and Forestry Research Organization (EAAFRO) and later director of the Agricultural Research Council of Central Africa; John Phillips, who it will be recalled served as the OFC's chief agricultural adviser on the East Africa Groundnut Scheme; E. W. Russell, the second director of the EAAFRO; and Arthur Savile, the chief agriculturalist on the Swynnerton Plan in Kenya, among many others.[31]

Initially it was the colonial governments and technical services themselves that called upon the newly formed international organizations, especially the UN's specialist agencies, for technical assistance in the 1950s. Despite the great injection of development funds and the recruitment waves of the late 1940s and early 1950s, most colonial administrations remained chronically short of both trained technical staff and financing for the vastly enlarged scale of development work in the postwar years, and as a result looked to the new international organizations for additional aid. In Northern Rhodesia, for example, a nutritional survey which had been in the planning stages since 1949 finally got under way in the mid-1950s under the direction of Dean Smith, head of the Nutritional School at Makerere, with the assistance of two experts brought in from the WHO.[32] To take another example, a new and more extensive scheme for soil surveys in the West Indies was launched after World War Two under K. C. Vernon and F. C. Darcel. In 1952 arrangements were made with the FAO to have an outside soil expert visit to assist in training local surveyors in new soil-mapping techniques. M. Baldwin of the USDA Soil Survey Department was brought in to demonstrate the new Land Capability Survey methods that had been developed by the USDA using aerial photography.[33]

Increasingly, however, as international technical assistance expanded and colonial operations were wound up, it was former colonial experts and advisers who were called upon to provide their services to the burgeoning new international development infrastructure. As early as 1949, A. J. Wakefield was asked to prepare guidelines and lend advice on the application of technical assistance for the FAO under the UN's Expanded Programme of Technical Assistance.[34] The trend would continue, with the FAO actively seeking out the services of former colonial technical staff, especially after the advent of the UN's Special Fund in December 1958, which increased the scope of the organization substantially as it was called upon to carry out many more projects in various countries on a broader and more comprehensive scale than before. R. G. Heath, director of agriculture for the Federation of Malaya, was one of those asked to join the team of the first FAO Special Fund Project, an agriculture and irrigation feasibility survey of the Lower Volta River in Ghana, which concluded that a large-scale irrigation scheme to grow rice and sugarcane, modeled on the "cane farming" system of the West Indies or possibly the Gezira Irrigated Cotton Scheme in Sudan, would be expensive but feasible.[35] Heath added that "a 'national benefit' [would also derive] from the increased political stability engendered by a more settled and contented peasantry, with improved standards of living."[36] In the early postindependence period many expatriate colonial scientists and technical personnel also stayed on to assist the new governments in the transition and to help train replacement staff (in what was termed "indigenization" or "Africanization"), which often led to considerable friction between these former colonial officials and their new indigenous superiors.[37]

But over and above the remobilization of personnel, or the many investigative reports, international missions, project teams, and seminal texts that former colonial advisers worked on or contributed to, perhaps the most important and enduring legacy of the late colonial development initiative was the intellectual one. The developmental and environmental narratives and policy strategies laid out by the experts examined in this book would be passed on, becoming embedded and institutionalized within the expanding network of international aid agencies and organizations in the years and decades following the formal demise of colonial rule. For while it is clear that postwar conditions shifted the emphasis of developmentalist thinking toward a consideration of the exigencies of economic growth and the provision of international technical and monetary assistance, the central problem around which the practice of development had been framed since before the war remained: concern about the emergence of a surplus population, with all the potential political, social, and ecological ramifications that this implied. No matter how much planners and governmental elites both in the "developed" and the "underdeveloped" world proclaimed the dawning of a new age, the old colonial conundrum persisted. Like Sisyphus, the development expert would valiantly struggle on in the face of the absurd, searching for the elusive means of resolving an apparently irresolvable contradiction.

Paved With Good Intentions? Agrarian Bias and the Depoliticization of Poverty

From the ecologically and culturally sensitive viewpoint of the early twenty-first century, many of the principles behind the late colonial initiative are, perhaps surprisingly, worthy of praise. There was candid assessment in the corridors of Whitehall of the mistakes made in past colonial interventions and of the consequences of unfettered economic exploitation. Awareness was growing among advisory experts of the distinct climatic and environmental constraints of tropical development, and of the need for long-term planning and coordinated state action to tackle complex and interrelated problems like erosion and malnutrition. Frank Stockdale's advice was always that care should be taken not to generalize; each area had its own peculiar problems, and working plans needed to be prepared before definite schemes of reconditioning were carried out.[38] New agricultural methods and land use practices needed to be well-suited to local conditions. Along the same lines, it was becoming increasingly clear that, for solutions to be successful, they had to be based on the cooperation and participation of local communities and, whenever possible, built upon existing practices and cultural beliefs. The starting point, as previous chapters have shown, was healthy lands and peoples, not the externally driven demands of the world market. One might even say that the late colonial initiative as conceived in the early 1940s was "paved with good intentions."

Yet it would be hard to escape the overwhelmingly agrarian vision that was being projected onto the future of colonial, soon to be "third world," peoples. In the early 1940s colonial advisers constructed an intellectual framework that entrenched a deep rural bias in British colonial development thinking and practice that, as we have seen, continued to reverberate throughout the postwar era.[39] Colonial experts and advisers could not imagine that colonial peoples might live in any way other than a rural, agricultural life; therefore, the aim of colonial policy was a more efficient reorganization of peasant agriculture with the intent of forming or reforming stable, contented, and prosperous local communities which they believed were rapidly dissolving in the face of environmental devastation, burgeoning population pressures, and growing class divisions. As Lord Hailey's Advisory Panel on Native Land Tenure put it in 1945, "The main objective of land policy must be to conserve and develop land resources as one of the primary means of raising the general standards of living of the population of the territory . . . the aim is not merely to secure efficiency in production, it must be achieved by methods which will promote the ordered progress of the community and anything which causes undue disturbance to the existing social organization would prejudice the chances for success."[40] In other words, as T. A. M. Nash, a medical entomologist who worked on the Anchau Settlement Scheme in Nigeria in the 1940s, proclaimed, "Our policy is rural development as opposed to urban development. We want to keep people on the land and to encourage local industries."[41]

Development was imagined not as an attempt to remake colonial peoples in the image of the West by replicating the material conditions present in metropolitan Britain, but, rather, to prevent such an outcome, to manage and even forestall the process of transformation by mitigating the contradictions thrown up by private capitalist enterprise and colonial rule, namely the emergence of a surplus population of rural and urban unemployed and the loss of productive resources. Beneath the rhetoric of scientific triumph and state planning, one can see as much fear and uncertainty about the future as arrogance and confidence in Western progress. Moreover, by conceptualizing the social and ecological distress, strikes, and civil unrest plaguing the colonies as symptoms of endemic substandard living conditions and inadequate government services, colonial officials and advisers were confronting colonial poverty and disorder not as structural or political questions, but, rather, as technical problems that were remediable by large-scale governmental planning and state-directed welfare schemes. What James Ferguson has so vividly described as the "development apparatus" had its origins *here,* and for the very same reasons:

> "Development" institutions generate their own form of discourse, and this discourse simultaneously constructs [third world countries] as a particular kind of object of knowledge, and creates a structure of

> knowledge around that object. Interventions are then organized on the basis of the structure of knowledge, which, while "failing" on their own terms, nonetheless have regular effects, which include the expansion and entrenchment of bureaucratic state power, side by side with the projection of a representation of economic and social life which denies "politics" and, to the extent that it is successful, suspends its effects . . . it is an "anti-politics machine," depoliticizing everything it touches, everywhere whisking political realities out of sight, all the while performing, almost unnoticed, its own pre-eminently political operation of expanding bureaucratic state power.[42]

Depoliticization was certainly Joseph H. Oldham's intention in calling for a more "positive" policy of imperial trusteeship in Central and East Africa. The indigenous peoples, in his opinion, were simply too inexperienced and lacking in education to assume responsibility for their own affairs. Echoing E. D. Morel's earlier pronouncements, he judged democracy to be entirely ill-suited to tropical Africa.[43] Instead of introducing Western parliamentary institutions that would focus the ambitions of African leaders on political agitation, he hoped to funnel their energies into the promotion of their own economic and social advancement, while at the same time channeling greater responsibility for the region's development into the hands of the imperial government. Only London had the resources, organization, and "impartial" authority to implement the kind of farsighted policy of economic development and educational advance that was needed. The foundations of such a policy had already been laid with the expanding colonial scientific networks of the 1920s and 1930s. The next step was to bring these agencies into closer coordination through some central body of seasoned and respected experts in London, who would have the widest range of knowledge and experience placed at their disposal. The resources of modern science and the stabilizing influence of the "scientific attitude," rather than "sectional interests" or "arbitrary forces," would be brought to bear on the problems of tropical Africa.

Thus, when the upsurge of strikes and riots in the colonies first forced themselves onto the British political agenda in the late 1930s, they were confronted by the Colonial Office not as a labor problem, as such, but as a question of economic and social planning. "Centralisation," as Fred Cooper astutely observes, "was justified in the name of British scientific knowledge and the need for coordination rather than by the opposition of class interests."[44] Officials and experts were trying to assert control over the labor question by subsuming it within the framework of development and welfare. However, what Cooper does not pick up on is that this depoliticization of the labor question was closely tied to another salient feature of late colonial development thinking and practice: its deeply entrenched agrarian bias.[45] In the steady stream of reports and memoranda issued by the CO's development consultants and advisory bodies at the time, the problems of unemployment and growing rural-urban migration were directly linked to the perceived social breakdown of rural village life brought on by the penetration of Western influences, the growth of population, and the consequent desiccation of land and degradation of other natural resources. "Men of science," informed by knowledge of earlier experiences in both Britain and other parts of the empire, looked upon the specter of "detribalization," rural migration, and mass unemployment with grave foreboding.

Events in the West Indies provided, as Macmillan put it, a timely warning of the shape of things to come elsewhere in the empire. All the talk of African distinctiveness could not hide the fact that it was descendants of African slaves whose lineage-based systems and customs had been irretrievably broken who were now crowding into the tropical slums of Kingston, Port of Spain, and Bridgetown, swelling the ranks of the embittered and unemployed. It was in reference to the West Indies that these connections were most clearly evoked. "In the past," as Frank Stockdale noted in his Report as comptroller for development in the West Indies,

> the needs of the town dwellers have been given greater attention than those of the country, and it is not surprising therefore that the increase of the populations of the urban areas throughout the West Indies . . . is taking place. It is already becoming an embarrassment and its dangers should be publicly recognized. At present the life in the rural areas for the average worker is indeed drab and it cannot be a matter for surprise that workers on sugar estates in some areas are either young or past middle age. Those in their prime of life are either in the towns or otherwise engaged in occupations unconnected with agriculture . . . It is important that progressive measures should take full account of the present West Indian background, the drift to the towns, particularly of workers in the prime of their life, and the tendency to limit work to that which will provide a living at the present standards. To follow blindly western patterns of social security or of education is likely only to lead to disappointment. In schemes of development, the highest priority

> should be given to those which will help to improve conditions in the rural areas . . . The greatest asset of the West Indies is its land, and it is the land which must continue to be the major and primary basis of employment.[46]

If it was not possible to make the land support the burgeoning population, Stockdale warned, then other means of providing employment would have to be considered, including greater industrialization. "Much will depend," Stockdale noted in the case of Jamaica, "upon whether those who have transferred to the towns and have become urbanized can be induced to return to agricultural undertakings and employment in the country districts. If this retransfer to the country cannot be achieved, increased local industrial development must be contemplated and efforts made to encourage a greatly expanded tourist trade."[47]

This is exactly what W. Arthur Lewis, one of the pioneers of development economics, advocated while he was secretary of the Colonial Economic Advisory Committee (CEAC) during the war.[48] The problem for the West Indies, as Lewis diagnosed it, was rural unemployment and underemployment, which demanded a far-reaching "agrarian revolution" to raise capital for health programs and educational schemes, and for investment in new technologies, irrigation works, chemical fertilizers, and conservation methods to make peasant agriculture dramatically more efficient. Lewis criticized the Colonial Office for not having even "begun to put emphasis on increasing the yield of native production. In consequence the provision for agricultural education is meagre, the capital needed to expand production has not been set aside, and the necessary institutions are not being established . . . until it is accomplished, a potential source of wealth remains relatively underdeveloped."[49] But more than this, Lewis argued, there was an urgent need for planned industrialization through state intervention to absorb the surplus labor from the "overpopulated" agricultural sector. The islands could turn their large supply of cheap labor to their advantage by specializing in the production of labor-intensive manufactured goods for the export (meaning metropolitan) market. Such proposals were anathema to senior CO staff. They were willing to pay lip service to the idea of secondary "import replacement" industries producing mainly light consumer goods for the local domestic market—but the idea of export-promoting industrialization was sheer heresy. Not surprisingly, Lewis resigned in frustration from the CEAC in November 1944, putting on record the "touchy and uncooperative" attitude of the office as the main reason for his abrupt departure.[50] As far as CO officials and advisers were concerned, agriculture *was* and would necessarily remain the primary basis of economic life for the bulk of colonial peoples.

Lewis's annoyance with what he described as "the traditional emphasis of one of the oldest departments of the British Government" was indicative of the impatience which underscored so much development thinking and practice in the growth years of the 1950s and early 1960s. The self-assurance and the faith in science, technology, and the ability of state and international organizations to manage development and human progress were perhaps never greater. The thinking behind much of the new idealism, as Lewis argued, was that an "agricultural revolution" was needed to break through the resistant structure of agrarian societies; that subsistence farming had to be transformed in order to absorb inputs, such as fertilizers and mechanical equipment, that were essential for increased yields; that ancillary industries were necessary to provide inputs and processing and to draw off the "surplus population" on the land; that substantial increases were needed in credit provision, infrastructure, marketing facilities, and technical assistance. Calls for an agricultural revolution, as we have seen, were heard from colonial advisers such as Frank Engledow and Geoffrey Clay as well, and in the last years of empire some schemes were introduced which envisioned significant changes in land tenure and organization.

But for the generation of nationalist leaders who led the former colonies to independence in the late 1950s and early 1960s, such schemes did not go nearly far or fast enough. In capturing the state, they hoped to bring the results which in their judgment the old colonial regimes had failed to deliver. For many of them, the road to improvement lay not in reforming peasant agriculture, the seemingly obvious cause of their backwardness and impoverishment, but rather, as Lewis contended, in industrial development. Not surprisingly, Lewis and other development economists were eagerly sought after by the new nationalist governments. Eric Williams, whose People's National Movement came to power in Trinidad and Tobago in 1956, invited Lewis to help prepare the country's first five-year development program, in which industrial development figured prominently.[51] Professor Lewis was also invited by Kwame Nkrumah's interim government to write a report on industrialization and economic policy for the soon-to-be-independent nation of Ghana in 1953, and he remained Nkrumah's economic adviser until 1961. Lewis argued that surplus labor from the rural areas could be transferred at subsistence wages to the urban sector, where it would be absorbed by labor-intensive industries. In time, as the modern industrial sector was built up, it would provide the means to develop infrastructure and restructure the agricultural sector. In other words, the Lewis model favored structural change based on industry as the engine of growth. But because most newly independent nations lacked the necessary capital, technology, and managerial skills, he felt the best strategy was to try to attract foreign investment and companies by granting generous tax holidays and other incentives. In the case of Ghana, he recommended building infrastructure to attract foreign investment.[52] Lewis's ideas gave additional support to plans already afoot for the building of the massive Volta River Dam, which was to supply hydroelectricity for smelting Ghana's large bauxite deposits into aluminum for the world market, thus providing the catalyst for the country's rapid industrialization. Although the conventional wisdom of today is to censure the "industrialization by invitation" strategies of Lewis as well as the Faustian antics of Nkrumah for their uncritical acceptance of externally generated and unsustainable Eurocentric models, it is worth remembering that they were reacting against, as they saw it, years of colonial dithering and obstruction. In that context their call for dramatic solutions was indeed radical, even revolutionary. Nonetheless, the results were generally disappointing. Such strategies overlooked the problem of capital flight and undervalued agricultural investment, with the result that food imports increased and export earnings did not keep pace with the costs of industrial inputs. The new industries also failed to create effective rural-urban linkages, and, most crucial of all, they generated only limited employment opportunities.

While Nkrumah fantasized about transforming Ghana into an African industrial powerhouse, others dreamed of ending world hunger and feeding the earth's rapidly growing poor. By the mid-1960s researchers at the International Rice Research Institute at the University of the Philippines and the International Center for the Improvement of Corn and Wheat in Mexico, with funding from the Ford and Rockefeller foundations, were doing pioneering work on new high-yielding varieties (HYVs) of staple crops, particularly wheat, maize, and rice, which with proper applications of chemical fertilizer and irrigated water were producing yields up to eight times higher than existing strains. The so-called Green Revolution was kicked off with the release of the first HYVs—dwarf wheats in 1963 and dwarf rice varieties in 1966—which by the end of the decade had given a dramatic boost to agricultural production and exports in countries as varied as Colombia, China, India, Indonesia, Korea, Mexico, Pakistan, the Philippines, and Turkey.[53] The goal of food self-sufficiency seemed to be at hand, sparking some observers to speak of revolutionary "seeds for change."[54] But, beneath the euphoria, familiar problems awaited.[55] Informed by the same diffusionist assumptions held by earlier colonial scientists, agricultural experts thought that the new seed-fertilizer package would be adopted by the richer, more progressive farmers first, then spread to the poorer farmers, and that the HYVs would substantially raise labor requirements, thus providing employment for the growing numbers of land-poor and landless. But the high costs of fertilizer and irrigation proved to be a greater barrier to smaller, poorer farmers than expected, and the actual creation of employment for planting, cultivating, and harvesting much less. Sizeable profits were made by many adopters, especially larger commercial farmers, but they tended to invest those profits in labor-displacing mechanical inputs (often subsidized by aid agencies) and cheap herbicides. The overall effect in many cases was to heighten income inequality and increase rural unemployment.

By 1970 the specter of rural landlessness and rural-urban migration was provoking warnings of a "global employment crisis," with some economists demanding tariff and land reform.[56] Once again, practitioners and theorists of the new creed came face to face with the quandaries of what John Holt had so aptly described nearly three-quarters of a century earlier as the "craze for development." Eventually it would lead them back along the road to the late 1930s. For it was one thing to talk about rapidly increasing productivity. It was quite another to actually do so without effecting a complete social revolution potentially more volatile and dangerous than "rural underdevelopment" itself. Critics also began pointing to the many ecological consequences of the new "green" technology: the highly specialized seed varieties promoted monoculture and reduced biodiversity, while the heavy use of fertilizers and pesticides poisoned local fauna (including humans) and led to the eutrophication of lakes, streams, and rivers. Many of the inputs were also heavily dependent on gasoline or diesel fuel, while surging demands for irrigated water created new pressures for large dam-building projects.[57] Indeed, by the early 1970s, the social and ecological costs of the "decade for development" were becoming clear to all. One of the first conferences to examine the ecological consequences of international development was organized by the Conservation Foundation and the Center for the Biology of Natural

Systems at Washington University in 1968. Among those to present papers were some prominent former colonial experts now turned international consultants, including E. B. Worthington, John Phillips, and E. W. Russell.[58] Citing innumerable examples where the ecological costs of projects had been ignored or inadequately anticipated, the conference called for the restructuring of the criteria used for planning and project design and greater inclusion of ecological expertise in the decision-making process. The conference's findings were published in 1972, just as issues of rural poverty, widespread hunger, and environmental degradation were once again being brought to the fore by a series of droughts and famines that swept across the Sahel and South Asia, and by the 1973 oil embargo, which threatened to wipe out the gains in agricultural output of the previous two decades. Neo-Malthusian arguments and apocalyptic narratives were nothing new, but fears of a pending world crisis snowballed in the 1970s, fueling the growth of a new global environmentalism embodied in such influential studies as the Club of Rome's *The Limits of Growth*, and in important forums like the "Man and the Biosphere" program launched in 1971 and the UN's Conference on the Human Environment held in Stockholm in 1972. The fundamental message of the Stockholm conference was that the seemingly divergent imperatives of development and environmental protection could be reconciled through rational, integrated development planning.[59] The UN and its specialist agencies, along with other leading donor organizations such as USAID, the World Bank, and the private philanthropic foundations, began pouring money into research on the social and ecological impact of the seed-fertilizer revolution and a myriad of other problems, from mechanization to desertification to the maldistribution of resources and basic needs.

The socioecological critiques of postwar development that emerged in the late 1960s and 1970s were rapidly absorbed in development and environmental circles and co-opted by international aid agencies. By the early 1980s they had become official dogma. Forty years after Engledow's 1943 memo on "Principles of Agricultural Policy in the Colonial Empire," officials at the World Bank would take up the torch once again, reiterating many of the same agrarian doctrines. The ideas of and pioneering studies produced by colonial technical officers and advisory experts would find new life in the bank's Integrated Rural Development Strategies of the 1970s and 1980s. Once again, an array of conservation measures, land-use planning models, and resettlement programs were conjured up in an effort to increase the capacity of the land to absorb a growing surplus population. It was not without some irony that former colonial technical officers, many of whom, as we have seen, had become international consultants and advisers, noted the striking similarities between the colonial models of rural development devised in the late 1930s, 1940s, and early 1950s and the kinds of strategies that were in vogue in the 1970s and 1980s. Anthony Blair Rains, a grasslands research officer in Nigeria in the 1950s, for example, noted that the international community had once again turned its attention to the merits of the mixed farming, sown pasture, and fodder crops that had been pioneered in Northern Nigeria in the 1930s.[60] Similarly, Paul Tuley, who worked as a regional botanist in Eastern Nigeria in the late 1950s and early 1960s, noted that the "work today on mixed cropping, covers, search for more efficient legumes and mixing leguminous trees with intercropping . . . are merely combinations and repetitions of work that has been going on and considered since the thirties."[61] Robert Waddell, an agricultural officer in the 1950s, was also struck by the parallels, noting that intercropping, water conservation practices, crop response curves, shelter belt design, and on-farm trials were all being explored by colonial agricultural officers and researchers in the 1950s.[62] It seems the development debate had come full circle, but, sadly, few of its participants saw much value in examining the lessons of the past; what is worse, many of them remained unaware that there were any.

The new rural development strategies of the 1970s and 1980s aimed to create the conditions for the long-term social and economic viability of rural communities and to improve the standards of living of the rural poor by balancing the imperatives of growth with those of social welfare and the distribution of resources between rural and urban sectors.[63] In the 1980s the World Bank added important environmental criteria to the mix, encouraging governments in sub-Saharan Africa to be more aggressive in developing and implementing sustainable environmental policies, in addition to expanding basic health services and female education aimed at lowering family size as a way of tackling the problem of rapid population growth, which it felt was leading to serious environmental degradation and lower agricultural productivity.[64] The solution to the present population crisis, it felt, lay in raising agricultural productivity through the introduction of improved crop varieties and agricultural technologies, mixed farming systems, and more intensive use of chemical and organic inputs. In the name of sustainability, it also called for a shift from large-scale, high-input, mechanized agriculture to low-input, labor-intensive, and environmentally supportive techniques.

The failure of many of these later schemes has often been attributed to the mutual incompatibility of the objectives themselves.[65] The assumption that increasing peasant production for the market would also improve the welfare of rural communities as a whole proved not to be the case. Despite substantial increases in production and productivity, it seems the rural poor were not usually the beneficiaries of such "integrated rural development" projects; in many cases, rural poverty persisted and even increased. Resources continued to be channeled to relatively affluent farmers while poor peasants remained untouched or worse off. The old conundrums remained—and so, too, one might add, did the agrarian bias of development's practitioners. The former chief economist of USAID, John W. Mellor, concluded in 1990:

> The vast majority of the poor in developing countries are in rural areas. The less developed the country, the greater the proportion of the population that is poor and the greater proportion of the impoverished that live in rural areas . . . In all, over three quarters of the world's poor live in rural areas. The lesson that emerges is that abolition of hunger must be a rural process since this is where the majority of poor, hungry people are living and where the problem is most acute. The solution, too, must come from within that system. Even for the roughly 125 million of the poor in 1990 who will be urban based, vigorous rural development will reduce the competition for jobs in the major metropolitan areas, reduce rural-urban migratory flows, and thus help to reduce urban poverty as well.[66]

In other words, after nearly fifty years of accumulated wisdom and experience, development experts were still dreaming up plans for rural stabilization to ebb the flow of rural migration due to poverty—poverty which they continued to depoliticize under the compelling logic of the "laws of nature."

But no sooner had development practice unwittingly returned to its colonial roots than a shift in priorities began to be felt, one that would destabilize and, ultimately, undo the underlying assumptions and international framework that had sustained development theory and policy since the 1940s.[67] At the pith of the matter were fundamental changes in the "real world" of development, as Colin Leys puts it, that began in the early 1970s but whose full implications were not felt until the mid-1980s.[68] In response to its growing balance of payment problems, brought on by the cost of the Vietnam War and the end of the post–Second World War economic boom, the United States spearheaded the abandoning of the Bretton Woods system of fixed exchange rates in 1973. The shift to floating exchange rates and unregulated financial markets led to greater instability in the global economy, which was compounded by the related problem of rising oil prices in the wake of the 1973 OPEC oil embargo. These actions sparked widespread economic upheaval, leading to a general downturn in global economic activity coupled with a vast expansion of international borrowing. The result was a period of profound economic crisis marked by recession, inflation, and unprecedented levels of international debt. For most third world countries, the collapse of the Bretton Woods system and the subsequent recession of the late 1970s resulted in lower and even negative growth rates, especially in Latin America and the Caribbean, sub-Saharan Africa, the Middle East, and North Africa.[69] During this period the terms of trade moved sharply against primary product exporters, reflecting the relative decline in importance of tropical raw materials in world trade. This led to a dramatic drop in the developing countries' overall share of world exports and international investment. Faced with stagnating economies and declining per capita income levels, third world states increased their overseas borrowing until, by the mid-1980s, most were faced with serious debt servicing and balance of payment problems.

At the same time, a new orthodoxy began asserting itself in development economics, spearheaded by Peter Bauer, Deepak Lal, Bela Belassa, and others who blamed the retarding of development in the third world on inefficient and excessive state intervention in the economy, and called for a "new vision of growth" based on a return to free market principles.[70] The neoliberal "counter-revolution," as it has been called, underpinned the World Bank's and the International Monetary Fund's (IMF) new market-oriented lending policies and the Structural Adjustment Programs (SAPs) that were imposed on many developing countries as conditions of financing in the wake of the international debt crisis of the 1980s. In order to reestablish their creditworthiness, indebted countries were forced into drastic reductions of public spending and subsidies, privatization of state enterprises, and currency devaluations. At the same time, following the apparent market-friendly approach of the so-called East Asian Tigers, developing countries were encouraged to reorient their economies toward export intensification and open participation in the world market. This strategy of liberalization, according to Philip McMichael, turned the development state inside out and paved the way for what he terms the "globalization project."[71] And indeed, as Eric Helleiner observes, "by the early 1990s, an almost fully liberal order had been created . . . giving market actors a degree of freedom they had not held since the 1920s and completely overturning the Bretton Woods order."[72]

Yet the "New Global Economy" of the 1990s turned out to be as unstable and problematic as the old one. It did not lead to the promised restoration of growth rates to postwar levels.[73] The collapse of communism in Eastern Europe and the former Soviet Union and the increasing integration of these regions into the international market were accompanied by a marked rise of divisive nationalism, ethnic violence and social upheaval.[74] In sub-Saharan Africa and Latin America, SAPs had a devastating effect, causing even greater levels of inequality, declining per capita incomes, and increasing debt-service burdens. Even the once rapidly growing Asian markets teetered on the verge of collapse in the mid-1990s, sending shockwaves and inciting investor panic throughout the world, from Russia to Latin America. Market forces, it turned out, were not in reality sui generis, but had to be socially embedded in order to operate effectively. Disappointment with the results of liberalization and stabilization policies did not, however, produce a hasty retreat back to an "enlightened" state management. Following the precepts of rational-choice theory with its emphasis on "transaction costs" and "rule-setting institutions" as important mediators between price fluctuations and market actor choices, the "new developmentalism" of the 1990s sought a middle ground between market and state, in which development efforts were geared to support community action and civic engagement—or, in other words, civil society.[75] The World Bank and other international donor agencies for their part turned to a similar set of strategies, cobbling together an increasingly incoherent discourse of both market and state-led policy reforms. Commentators at the time described the bank's annual *World Development Reports* as eclectic mixtures of pragmatism and vacillation that left much of the neoliberal agenda still intact.[76] The results, for many countries, were stagnating or even declining per capita income levels, higher interest and unemployment rates, greater poverty and immiserization, and an increasing threat of social violence and political chaos.

As the curtain rises on the twenty-first century, development, it seems, is in deep trouble. "There is now overwhelming support," as the former chief economist and senior vice-president of the World Bank, Joseph

Stiglitz, has recently written, for the idea "that premature capital and financial market liberalization throughout the developing world, a central part of IMF reforms over the past two decades, was a central factor not only behind the most recent set of crises but also behind the instability that has characterized the global market over the past quarter century."[77] Some, like Jeffrey Sachs, the former director of the Harvard Institute for International Development, urge world leaders to devise a new basis for globalization, starting with supervisory controls on capital by domestic financial institutions, across-the-board debt write-downs, and the reorienting of aid programs toward international public goods like health and the mobilization of knowledge.[78] For others, rethinking and reinventing development simply doesn't cut it anymore. The earth is suffering from "planetary overload," they say, and simply cannot bear the environmental suffering and devastation that a fully industrialized and globalized world would entail.[79] Poststructuralists have gone furthest in this regard, calling upon the throng of experts, practitioners, members of grassroots movements, and students to liberate their minds from the faith in universal progress and to see development for what it is: a failed and impossible myth.[80] In its place, alternative gatherings like the World Social Forum and scholar-activists like Arturo Escobar, Majid Rahnema, and Vandana Shiva dream of the "post-development" world to come.[81] Drawing inspiration from a myriad of nongovernmental organizations and grassroots social movements, they reject both the supposedly ineluctable forces of the market and the rationalizing power of the nation-state, and look instead to preserve or reassert the primacy and autonomy of local communities. These communities, ideally, will be as self-reliant as possible, employing sustainable systems of production that are well adapted to local ecological conditions and geared primarily to meeting basic human needs.

As admirable as these ideals may be, it is hard not to notice the uncanny resemblance to earlier colonial doctrines, which were also intended to bring about a stable, sustainable world and to reconcile order and progress through cooperation and community. In many ways, it is the logical consequence of the agrarian biases that have been inherent in development practice from the start, but without the principle of trusteeship in the form of the state or the expert. Or is it? The scientists, planners, and project officers will be replaced, not, it seems, by "the people" directly, who are purported to have internalized the developers' perception of what they need or who have been torn apart by conflicting interests instilled by exposure to modernity, but, rather, by the wisest, most virtuous, and most authoritative persons of the group: those whose knowledge of both the old and the new is trusted and respected by everyone, and who through constant dialogue with other active members will be best able to determine "the direction, the quality and the content of changes desired by each community."[82]

Time will tell whether these global counter-movements will gain further momentum, but in the meantime uncertainty and instability have generated a growing sense of national and international insecurity. In the wake of the September 11, 2001, terrorist attacks on the World Trade Center and the Pentagon, there are renewed pressures on governments around the world to exercise greater "state effectiveness" in managing and stabilizing their populations.[83] Even more ominously, the world now faces the prospect of a return to empire. Like Joseph Chamberlain a century ago, the constructive imperialists of the twenty-first century are determined to break through the dithering and gridlock of debate, and to secure, by force if necessary, the resources, markets, and, most importantly, the order needed to sustain the United States as the world's dominant military-consumerist state.[84] "Now we are living," as Michael Ignatieff reminds us, "through the collapse of many of [the] former colonial states. Into the resulting vacuum of chaos and massacre a new imperialism has reluctantly stepped in—reluctantly because these places are dangerous and because they seemed, at least until Sept. 11, to be marginal to the interests of the powers concerned. But, gradually, this reluctance has been replaced by an understanding of why order needs to be brought to these places . . . Terror has collapsed distance, and with this collapse has come a sharpened American focus on the necessity of bringing order to the frontier zones."[85] Whether this new imperium will lead to a new reconfiguration and renaming of the "globalization project" remains unclear, but one thing is certain: the old colonial conundrums of how to reconcile order with progress, and how to balance the increasing challenge of environmental scarcity with the long overdue promise of better living standards and welfare for all, will continue to haunt development's practitioners and "trustees" for many years to come.

Notes

The following abbreviations are used throughout:

BLCAS	Bodleian Library of Commonwealth and African Studies
CO	Colonial Office
FCOL	Foreign and Commonwealth Office Library
GB	Great Britain
IMC/CBMS	International Missionary Council/Conference of British Missionary Societies Archives
PP	*Parliamentary Papers*
RA	Ross Archives

Introduction

1. Harold Wilson, *The War on World Poverty: An Appeal to the Conscience of Mankind* (London: Gollancz, 1953); the epigraph is from 22–23.

2. The term "Cold War lens" is taken from Matthew Connelly, "Taking Off the Cold War Lens: Visions of North-South Conflict during the Algerian War of Independence," *American Historical Review* 105, no. 3 (2000): 739–69.

3. See for example, Frans J. Schuurman, ed., *Beyond the Impasse: New Directions in Development Theory* (London: Zed Books, 1993); Richard Norgaard, *Development Betrayed: The End of Progress and a Coevolutionary Revisioning of the Future* (London: Routledge, 1994); Ronaldo Munck and Denis O'Hearn, eds., *Critical Development Theory: Contributions to a New Paradigm* (London: Zed Books, 1999).

4. Some notable contributions include David B. Moore, "Development Discourse as Hegemony: Towards an Ideological History, 1945–1995," in *Debating Development Discourse: Institutional and Popular Perspectives,* ed. David B. Moore and Gerald G. Schmitz (London: Macmillan, 1995), 1–53; Colin Leys, "The Rise and Fall of Development Theory," in *The Rise and Fall of Development Theory* (London: James Currey, 1996), 3–19; Philip McMichael, "Instituting the Development

Project," in *Development and Social Change: A Global Perspective*, 2nd ed. (Thousand Oaks, Calif.: Pine Forge Press, 2000), 3–42.

5. Wolfgang Sachs, ed., *The Development Dictionary: A Guide to Knowledge as Power* (London: Zed Books, 1992); Jonathan Crush, ed., *Power of Development* (London: Routledge, 1995); Arturo Escobar, *Encountering Development: The Making and Unmaking of the Third World* (Princeton, N.J.: Princeton University Press, 1995); Gilbert Rist, *The History of Development: From Western Origins to Global Faith* (London: Zed Books, 1997); Majid Rahnema and Victoria Bawtree, eds., *The Post-Development Reader* (London: Zed Books, 1997).

6. Escobar, *Encountering Development*, 33–34.

7. The most important influences here include David M. Anderson, "Organising Ideas: British Colonialism and African Rural Development" (unpublished paper, Social Science Research Council Workshop on Social Science and Development, 1993); M. P. Cowen and R. W. Shenton, *Doctrines of Development* (New York: Routledge, 1996); Frederick Cooper, *Decolonization and African Society: The Labor Question in French and British Africa* (Cambridge: Cambridge University Press, 1996); Frederick Cooper and Randall Packard, eds., *International Development and the Social Sciences: Essays on the History and Politics of Knowledge* (Berkeley: University of California Press, 1997).

8. Cooper and Packard, introduction to *International Development and the Social Sciences*, 7.

9. Ibid., 10.

10. In addition to the references cited separately below, see Monica van Beusekom and Dorothy Hodgson, "Lessons Learned? Development Experiences in the Late Colonial Period," *Journal of African History* 41, no. 1 (2000): 29–33; Joanna Lewis, *Empire State-Building: War and Welfare in Kenya, 1925–52* (Oxford: James Currey, 2000); David Anderson, *Eroding the Commons: The Politics of Ecology in Baringo, Kenya, 1890–1963* (Athens: Ohio University Press, 2002).

11. Helen Tilley, "Africa as a 'Living Laboratory': The African Research Survey and the British Colonial Empire: Consolidating Environmental, Medical, and Anthropological Debates, 1920–1940" (D.Phil. thesis, Oxford University, 2001); Tilley, "African Environments and Environmental Sciences: The African Research Survey, Ecological Paradigms and British Colonial Development, 1920–1940," in *Social History and African Environments*, ed. William Beinart and JoAnn McGregor (Oxford: James Currey, 2003), 109–30.

12. Suzanne Moon, "Empirical Knowledge, Scientific Authority, and Native Development: The Controversy over Sugar/Rice Ecology in the Netherlands East Indies, 1905–1914," *Environment and History* 10, no. 1 (2004): 59–81.

13. Monica M. van Beusekom, "Disjunctures in Theory and Practice: Making Sense of Change in Agricultural Development at the Office du Niger, 1920–1960," *Journal of African History* 41, no. 1 (2000): 79–99; van Beusekom, *Negotiating Development: African Farmers and Colonial Experts at the Office du Niger, 1920–1960* (Oxford: James Currey, 2002).

14. For the notion of the "development narrative," see Emery Roe, "'Development Narratives,' or Making the Best of Blueprint Development," *World Development* 19, no. 4 (1991): 287–300.

15. See for example, Howard Johnson, "The West Indies and the Conversion of the British Official Classes to the Development Idea," *Journal of Commonwealth and Comparative Politics* 15, no. 1 (1977): 55–83; D. J. Morgan, *The Official History of Colonial Development,* 5 vols. (London: Macmillan, 1980); J. M. Lee and Martin Petter, *The Colonial Office, War and Development Policy: Organization and the Planning of a Metropolitan Initiative, 1939–1945* (London: Institute of Commonwealth Studies, 1982); Stephen Constantine, *The Making of British Colonial Development Policy, 1914–1940* (London: Frank Cass, 1984); Michael Havinden and David Meredith, *Colonialism and Development: Britain and Its Tropical Colonies, 1850–1960* (London: Routledge, 1993); S. R. Ashton and S. E. Stockwell, eds., *Imperial Policy and Colonial Practice 1925–1945,* British Documents on the End of Empire, series A, vol. 1 (London: HMSO, 1996).

16. Ronald Robinson and John Gallagher with Alice Denny, *Africa and the Victorians: The Official Mind of Imperialism,* 2nd ed. (London: Macmillan, 1961); J. M. Lee, *Colonial Development and Good Government: A Study of the Ideas Expressed by the British Official Classes in Planning Decolonization, 1939–1964* (Oxford: Clarendon Press, 1967); A. D. Roberts, "The Imperial Mind," in *The Cambridge History of Africa,* vol. 7, ed. A. D. Roberts (Cambridge: Cambridge University Press, 1986), 24–76.

17. Edward Said, *Orientalism* (New York: Vintage Books, 1979); Ashis Nandy, *The Intimate Enemy: Loss and Recovery of Self under Colonialism* (Oxford: Oxford University Press, 1983); Bill Ashcroft, Gareth Griffiths, and Helen Tiffin, *The Empire Writes Back: Theory and Practice in Post-Colonial Literatures* (London: Routledge, 1989); Robert Young, *White Mythologies: Writing History and the West* (London: Routledge, 1990); David Spurr, *The Rhetoric of Empire: Colonial Discourse in Journalism, Travel Writing, and Imperial Administration* (Durham, N.C.: Duke University Press, 1993).

18. Dane Kennedy, "Imperial History and Post-Colonial Theory," *Journal of Imperial and Commonwealth History* 24, no. 3 (1996): 353. See also John MacKenzie, *Orientalism: History, Theory and the Arts* (Manchester: Manchester University Press, 1995).

19. Frederick Cooper, "Conflict and Connection: Rethinking Colonial African History," *American Historical Review* 99, no. 5 (1994): 1516–45; Nicholas Thomas, *Colonialism's Culture: Anthropology, Travel and Government* (Princeton, N.J.: Princeton University Press, 1994); Anne McClintock, *Imperial Leather: Race, Gender and Sexuality in the Colonial Conquest* (London: Routledge, 1995); Frederick Cooper and Ann Laura Stoler, eds., *Tensions of Empire: Colonial Cultures in a Bourgeois World* (Berkeley: University of California Press, 1997).

20. Richard Grove, *Green Imperialism: Colonial Expansion, Tropical Island Edens, and the Origins of Environmentalism, 1600–1860* (Cambridge: Cambridge

University Press, 1995); Grove, *Ecology, Climate and Empire: Colonialism and Global Environmental History, 1400–1940* (Cambridge: Cambridge University Press, 1997).

21. Tilley, introduction to "Africa as a 'Living Laboratory.'"

22. Cowen and Shenton, *Doctrines of Development,* 56–57. See also Michael P. Cowen and Robert W. Shenton, "The Invention of Development," in *Power of Development,* ed. Jonathan Crush (London: Routledge, 1995), 27–43.

23. For similar shifts in colonial policy among other leading European imperial powers see Alice Conklin, *A Mission to Civilize: The Republican Idea of Empire in France and West Africa, 1895–1930* (Berkeley: University of California Press, 1997); Suzanne Moon, "Constructing 'Native Development': Technological Change and the Politics of Colonization in the Netherlands East Indies, 1905–1930" (Ph.D. diss., Cornell University, 2000).

24. Daniel R. Headrick, *The Tools of Empire: Technology and European Imperialism in the Nineteenth Century* (New York: Oxford University Press, 1981); Michael Adas, *Machines as the Measure of Men: Science, Technology, and Ideologies of Western Dominance* (Ithaca, N.Y.: Cornell University Press, 1990).

25. Interest in the utility of science for empire was, of course, nothing new. There is now an extensive literature documenting how science in the eighteenth and early nineteenth centuries was used to identify and survey "new" lands, resources, and peoples, or develop new commercial crops and natural knowledge that helped lay the basis of plantation industries in far-reaching colonial outposts. What was new, however, was the belief that the state, in conjunction with scientific knowledge and expertise, was not only useful but central, and indeed *necessary,* for development to happen.

26. Joseph Chamberlain, Speech to the House of Commons, 22 August 1895, *Parliamentary Debates,* Commons, 4th ser., vol. 36, cols. 640–45.

27. Michael Worboys, "The Emergence of Tropical Medicine: A Study in the Establishment of a Scientific Specialty," in *Perspectives on the Emergence of Scientific Disciplines,* ed. Gerard Lemaine et al. (The Hague: Mouton, 1976), 75–98; Worboys, "Manson, Ross and Colonial Medical Policy: Tropical Medicine in London and Liverpool, 1899–1914," in *Disease, Medicine and Empire: Perspectives on Western Medicine and the Experience of European Expansion,* ed. Roy MacLeod and Milton Lewis (London: Routledge, 1988), 21–37; Paul Cranefield, "Joseph Chamberlain," in *Science and Empire: East Coast Fever in Rhodesia and the Transvaal* (Cambridge: Cambridge University Press, 1991), 121–36; Richard Drayton, "The Government of Nature," in *Nature's Government: Science, Imperial Britain, and the "Improvement" of the World* (New Haven, Conn.: Yale University Press, 2000), 221–68.

28. J. E. Lewis, "'Tropical East Ends' and the Second World War: Some Contradictions in Colonial Office Welfare Initiatives," *Journal of Imperial and Commonwealth History* 28, no. 2 (May 2000): 43.

29. Joanna Lewis, *Empire State-Building,* 5–7.

30. David M. Anderson, "Organising Ideas: British Colonialism and African Rural Development," 4.

31. Cooper and Packard, introduction to *International Development and the Social Sciences,* 13; van Beusekom and Hodgson, "Lessons Learned?" 31.

32. Anne Phillips, *The Enigma of Colonialism: British Policy in West Africa* (London: James Currey, 1989); Bruce Berman, *Control and Crisis in Colonial Kenya: The Dialectic of Domination* (London: James Currey, 1990); Bruce Berman and John Lonsdale, *Unhappy Valley: Conflict in Kenya and Africa,* 2 vols. (Athens: Ohio University Press, 1992); Cooper, *Decolonization and African Society.*

33. Mahmood Mamdani, *Citizen and Subject: Contemporary Africa and the Legacy of Late Colonialism* (Princeton, N.J.: Princeton University Press, 1996), 51–52.

34. Michael P. Cowen and Nicholas Westcott, "British Imperial Economic Policy during the War," in *Africa and the Second World War,* ed. David Killingray and Richard Rathbone (London: Macmillan, 1986), 20–67; A. N. Porter and A. J. Stockwell, *British Imperial Policy and Decolonization, 1938–64,* vol. 1, *1938–51* (London: Macmillan, 1987); Cooper, "Forced Labor, Strike Movements, and the Idea of Development, 1940–1945," in *Decolonization and African Society,* 110–70; Joanna Lewis, *Empire State-Building,* 5–7.

35. For origins of this term see D. A. Low and J. M. Lonsdale, "Introduction: Towards the New Order, 1945–1963," in *History of East Africa,* vol. 3, ed. D. A. Low and Alison Smith (Oxford: Clarendon Press, 1976), 1–64.

36. There is now an extensive historiography on local responses to the late imperial initiatives. See for example, Lionel Cliffe, "Nationalism and the Reaction to Enforced Agricultural Change in Tanganyika during the Colonial Period," in *Socialism in Tanzania: An Interdisciplinary Reader,* ed. Lionel Cliffe and John Saul (Nairobi: EAPH, 1972), 17–24; William Beinart, "Soil Erosion, Conservationism and Ideas about Development: A Southern African Exploration, 1900–1960," *Journal of Southern African Studies* 11, no. 1 (October 1984): 52–83; David Throup, *The Economic and Social Origins of Mau Mau, 1945–53* (London: James Currey, 1987); Steven Feierman, *Peasant Intellectuals: Anthropology and History in Tanzania* (Madison: University of Wisconsin Press, 1990); Leonard Leslie Bessant, "Coercive Development: Land Shortage, Forced Labour, and Colonial Development in the Chiweshe Reserve, Colonial Zimbabwe, 1938–1946," *International Journal of African Historical Studies* 25, no. 1 (1992): 39–65; Francis K. Danquah, "Rural Discontent and Decolonization in Ghana, 1945–1951," *Agricultural History* 68, no. 1 (1994): 1–19; A. Fiona D. Mackenzie, *Land, Ecology and Resistance in Kenya, 1880–1952* (Portsmouth, N.H.: Heinemann, 1998).

37. Cowen and Shenton, *Doctrines of Development,* 296–97; Michael P. Cowen and Robert W. Shenton, "The Origin and Course of Fabian Colonialism in Africa," *Journal of Historical Sociology* 4, no. 2 (1991): 143–74.

38. Joanna Lewis, *Empire State-Building,* 13–20.

39. Ashton and Stockwell, introduction to *Imperial Policy and Colonial Practice,* xxviii; Cosmos Parkinson, *The Colonial Office from Within, 1909–1945* (London: Faber and Faber, 1945), 55–56.

40. See David Gilmartin, "Scientific Empire and Imperial Science: Colonialism and Irrigation Technology in the Indus Basin," *Journal of Asian Studies* 53, no. 4 (November 1994): 1144; Joanna Lewis, *Empire State-Building,* 15.

41. See Gilmartin, "Scientific Empire and Imperial Science."

42. David Anderson, "Depression, Dust Bowl, Demography, and Drought: The Colonial State and Soil Conservation in East Africa during the 1930s," *African Affairs* 83, no. 332 (July 1984): 342–43.

43. Libby Robin, "Ecology: A Science of Empire?" in *Ecology and Empire: Environmental History of Settler Societies,* ed. Tom Griffiths and Libby Robin (Edinburgh: Keele University Press, 1997), 70.

44. Karl Ittmann, "The Colonial Office and the Population Question in the British Empire, 1918–62," *Journal of Imperial and Commonwealth History* 27, no. 3 (1999): 62.

45. John M. Lonsdale, "The Depression and the Second World War in the Transformation of Kenya," in *Africa and the Second World War,* ed. David Killingray and Richard Rathbone (London: Macmillan, 1986), 97–142.

46. David Anderson, "Depression, Dust Bowl, Demography, and Drought," 331–33; Grove, *Ecology, Climate and Empire,* 33–36; Thomas Dunlap, *Nature and the English Diaspora: Environment and History in the United States, Canada, Australia and New Zealand* (Cambridge: Cambridge University Press, 1999), 179–89.

47. Peder Anker, "The Oxford School of Imperial Ecology," in *Imperial Ecology: Environmental Order in the British Empire, 1895–1945* (Cambridge, Mass.: Harvard University Press, 2002), 76–117.

48. Ittmann, "Colonial Office and the Population Question," 62–63.

49. James Ferguson, *The Anti-Politics Machine: "Development," Depoliticization, and Bureaucratic Power in Lesotho* (Cambridge: Cambridge University Press, 1990), xiv; Joanna Lewis, *Empire State-Building,* 19.

50. Melissa Leach and Robin Mearns, "Environmental Change and Policy: Challenging Received Wisdom in Africa," in *The Lie of the Land: Challenging Received Wisdom on the African Environment,* ed. Melissa Leach and Robin Mearns (Portsmouth, N.H.: Heinemann, 1996), 1–33.

51. Connelly, "Taking Off the Cold War Lens," 741. See also Matthew Connelly, "The Failure of Progress: Algeria and the Crisis of the Colonial World," in *A Diplomatic Revolution: Algeria's Fight for Independence and the Origins of the Post–Cold War Era* (Oxford: Oxford University Press, 2002), 17–38.

Chapter 1: Setting the Terms of the Debate

1. Joseph Chamberlain, "Speech at Walsall," 15 July 1895, in *The Concept of Empire: Burke to Attlee, 1774–1947,* ed. George Bennett (London: A. and C. Black, 1953), 314.

2. Salisbury's coalition government was made up of Conservatives and Liberal Unionists, but was dominated by the Conservatives. To symbolize the alliance, however, the term *Unionist* was used.

3. Robert V. Kubicek, *The Administration of Imperialism: Joseph Chamberlain at the Colonial Office* (Durham, N.C.: Duke University Press, 1969), 69.

4. Paul Cranefield, "Joseph Chamberlain," in *Science and Empire: East Coast Fever in Rhodesia and the Transvaal* (Cambridge: Cambridge University Press, 1991), 121–36; Richard Drayton, "The Government of Nature," in *Nature's Government: Science, Imperial Britain and the "Improvement" of the World* (New Haven, Conn.: Yale University Press, 2000), 221–68; Michael Worboys, "The Emergence of Tropical Medicine: A Study in the Establishment of a Scientific Specialty," in *Perspectives on the Emergence of Scientific Disciplines,* ed. Gerard Lemaine et al. (The Hague: Mouton, 1976), 75–98; Worboys, "Manson, Ross and Colonial Medical Policy: Tropical Medicine in London and Liverpool, 1899–1914," in *Disease, Medicine and Empire: Perspectives on Western Medicine and the Experience of European Expansion,* ed. Roy MacLeod and Milton Lewis (London: Routledge, 1988), 21–37.

5. GB, *PP,* C. 8655, L (1897), *Report of the West India Royal Commission, with subsidiary report by D. Morris, Assistant Director of the Royal Botanic Gardens, Kew,* 1–176; Julian Amery, *The Life of Joseph Chamberlain,* vol. 4, *1901–1903: At the Height of His Power* (London: Macmillan, 1951), 234–55.

6. E. H. H. Green, *The Crisis of Conservatism: The Politics, Economics, and Ideology of the British Conservative Party, 1880–1914* (London: Routledge, 1995), 72.

7. C. A. Bayly, *Imperial Meridian: The British Empire and the World, 1780–1830* (London: Longman, 1989), 100; Drayton, *Nature's Government,* 227–29.

8. Michael Adas, *Machines as the Measure of Men: Science, Technology, and Ideologies of Western Dominance* (Ithaca, N.Y.: Cornell University Press, 1989); Robert H. MacDonald, *The Language of Empire: Myths and Metaphors of Popular Imperialism, 1880–1918* (Manchester: Manchester University Press, 1994); Thomas Metcalf, *Ideologies of the Raj,* New Cambridge History of India 3.4 (Cambridge: Cambridge University Press, 1994); Alice Conklin, *A Mission to Civilize: The Republican Idea of Empire in France and West Africa, 1895–1930* (Berkeley: University of California Press, 1997).

9. Frederick Cooper and Ann Laura Stoler, eds., *Tensions of Empire: Colonial Cultures in a Bourgeois World* (Berkeley: University of California Press, 1997).

10. H. W. Arndt, "Economic Development: A Semantic History," *Economic Development and Cultural Change* 29, no. 3 (April 1981): 461.

11. Roy Porter, "Progress," in *The Creation of the Modern World: The Untold Story of the British Enlightenment* (New York: W. W. Norton, 2000), 424–45; Neil McKendrick, John Brewer, and J. H. Plumb, *The Birth of a Consumer Society: The Commercialization of Eighteenth-Century England* (Bloomington: Indiana University Press, 1982), 332.

12. Drayton, *Nature's Government,* 85–128.

13. Patricia Seed, *American Pentimento: The Invention of Indians and the Pursuit of Riches* (Minneapolis: University of Minnesota Press, 2001), 31.

14. Karen Kupperman, *Settling with the Indians: The Meeting of English and Indian Cultures in America, 1580–1640* (Totowa, N.J.: Rowman and Littlefield, 1980); William Cronon, "Bounding the Land," in *Changes in the Land: Indians, Colonists, and the Ecology of New England* (New York: Hill and Wang, 1983), 54–81; Anthony Pagden, *Lords of All the World: Ideologies of Empire in Spain, Britain and France, c. 1500–c. 1800* (New Haven, Conn.: Yale University Press, 1995), 77.

15. Christopher Hill, *The English Bible and the Seventeenth-Century Revolution* (London: Allen Lane, 1993), 136.

16. John Locke, "Of Property," in *Second Treatise of Government* (Indianapolis, Ind.: Hackett, 1980 [1690]), 18–29.

17. Linda Colley, "Dominance" and "Majesty," in *Britons: Forging the Nation, 1707–1837* (New Haven, Conn.: Yale University Press, 1992), 147–94 and 195–236.

18. Bayly, *Imperial Meridian,* 116–21; Drayton, *Nature's Government,* 88–89.

19. See Bayly, *Imperial Meridian;* Michael Duffy, "World-Wide War and British Expansion, 1793–1815," and P. J. Marshall, "Britain without America—A Second Empire?" in *The Eighteenth Century,* ed. P. J. Marshall, vol. 2, *The Oxford History of the British Empire,* ed. William Roger Louis (Oxford: Oxford University Press, 1999), 184–207 and 576–95.

20. Donald Denoon, *Settler Capitalism: The Dynamics of Dependent Development in the Southern Hemisphere* (Oxford: Oxford University Press, 1983), 19–23.

21. Alfred W. Crosby, "Ills," in *Ecological Imperialism: The Biological Expansion of Europe, 900–1900* (Cambridge: Cambridge University Press, 1986), 195–216.

22. The best source on the construction of nature and debates over land use in the nineteenth- and twentieth-century Anglo settler world is Thomas Dunlap, *Nature and the English Diaspora: Environment and History in the United States, Canada, Australia and New Zealand* (Cambridge: Cambridge University Press, 1999). See also Tom Griffiths and Libby Robin, eds., *Ecology and Empire: Environmental History of Settler Societies* (Edinburgh: Keele University Press, 1997).

23. Edward Gibbon Wakefield, "The Art of Colonization," in *A Letter from Sydney and Other Writings,* ed. Ernest Rhys (London: J. M. Dent and Sons, 1929), 109–254; Richard Charles Mills, *The Colonization of Australia (1829–42): The Wakefield Experiment in Empire Building* (London: Sidgwick and Jackson, 1915), 90–139.

24. Arndt, "Economic Development," 462.

25. Michael P. Cowen and Robert W. Shenton, "The Invention of Development," in *Doctrines of Development* (New York: Routledge, 1996), 3–59.

26. David Ludden, "India's Development Regime," in *Colonialism and Culture,* ed. Nicholas B. Dirks (Ann Arbor: University of Michigan Press, 1992), 251–52; Metcalf, *Ideologies of the Raj,* 15–27.

27. Eric Stokes, *The English Utilitarians and India* (Oxford: Clarendon Press, 1959), 5.

28. Ibid., 9.

29. Louis Dumont, "The 'Village Community' from Munro to Maine," *Contributions to Indian Sociology* 9 (December 1966): 71.

30. Munro believed that giving the traditional *ryots* back their stake in the land would strengthen communal bonds through bypassing all intermediaries and settling directly with each peasant, who would then be considered the sole owner of his individual holding. Elphinstone and Metcalf took a more conservative stance, preferring to invest formal proprietary rights in the village community as a whole or in the coparcenary brotherhoods, but none of them, it must be noted, was hostile to reform. They were all sympathetic believers in an accurate land survey, which would clearly define the peasant cultivator's rights to the soil. See Stokes, *English Utilitarians and India,* 23; Dumont, "'Village Community' from Munro to Maine," 63–64; Ann B. Callender, *How Shall We Govern India? A Controversy among British Administrators, 1800–1882* (New York: Garland, 1987), 64–66.

31. Callender, *How Shall We Govern India?* 61.

32. David Ricardo, *The Works and Correspondence of David Ricardo,* vol. 1, *On the Principles of Political Economy and Taxation* (Cambridge: Cambridge University Press, 1951), 74–75; and "Of the Rent of Land," in vol. 2., *Notes on Malthus's Principles of Political Economy* (Cambridge: Cambridge University Press, 1951), 110–11.

33. Stokes, *English Utilitarians and India,* 91.

34. Metcalf, *Ideologies of the Raj,* 28–43.

35. Adas, *Machines as the Measure of Men,* 279.

36. John Rosselli, *Lord William Bentinck: The Making of a Liberal Imperialist, 1774–1839* (Delhi: Thomson Press, 1974), 277–97.

37. Zaheer Baber, *The Science of Empire: Scientific Knowledge, Civilization, and Colonial Rule in India* (Albany: State University of New York Press, 1996), 205.

38. J. A. Richey, ed., *Selections from Educational Records, Part II: 1840–1859* (Calcutta: Bureau of Education, 1922), 365.

39. Suresh Chandra Ghosh, *Dalhousie in India, 1848–56: A Study of His Social Policy as Governor-General* (New Delhi: Munshiram Manoharlal, 1975).

40. Stokes, *English Utilitarians and India,* 108.

41. C. A. Bayly, "The Consolidation and Failure of the East India Company's State, 1818–57," in *Indian Society and the Making of the British Empire,* vol. 2, *New Cambridge History of India* (Cambridge: Cambridge University Press, 1988), 106–35.

42. D. A. Washbrook, "Law, State and Agrarian Society in Colonial India," *Modern Asian Studies* 15, no. 3 (1981): 649–721.

43. Bayly, *Indian Society and the Making of the British Empire,* 132.

44. Thomas R. Metcalf, *The Aftermath of Revolt: India, 1857–1870* (Princeton, N.J.: Princeton University Press, 1964), 204–7; Stokes, *English Utilitarians and India,* 119–20.

45. Bayly, *Indian Society and the Making of the British Empire,* 189–91.

46. Metcalf, *Aftermath of Revolt,* 137.

47. Quoted ibid., 148.

48. John D. Hargreaves, *Prelude to the Partition of West Africa* (London: Macmillan, 1963), 64–78.

49. Metcalf, *Ideologies of the Raj,* 52–53.

50. Ibid., "Creation of Difference," 66–112, and "Ordering of Difference," 113–59.

51. See Henry Sumner Maine, *Village Communities in the East and West,* 2nd ed. (London: John Murray, 1872), 190–97.

52. A number of scholars have pointed to the important influence of Maine on the shift in official thinking. See especially Clive Dewey, "The Influence of Sir Henry Maine on Agrarian Policy in India," Gordon Johnson, "India and Henry Maine," and C. A. Bayly, "Maine and Change in Nineteenth-Century India," in *The Victorian Achievement of Sir Henry Maine: A Centennial Reappraisal,* ed. Alan Diamond (Cambridge: Cambridge University Press, 1991), 353–75, 376–88, and 389–97; Metcalf, "Creation of Difference" and "Ordering of Difference," *Ideologies of the Raj,* 66–112 and 113–59.

53. David Arnold, "India's Place in the Tropical World, 1770–1930," *Journal of Imperial and Commonwealth History* 26, no. 1 (1998): 2.

54. Richard Grove, "Edens, Islands and Early Empires," in *Green Imperialism: Colonial Expansion, Tropical Island Edens, and the Origins of Environmentalism, 1600–1860* (Cambridge: Cambridge University Press, 1995), 16–72.

55. Philip D. Curtin, *The Image of Africa: British Ideas and Action, 1780–1850* (Madison: University of Wisconsin Press, 1964), 71.

56. Mark Harrison, "'The Tender Frame of Man': Disease, Climate and Racial Difference in India and the West Indies, 1760–1860," *Bulletin of the History of Medicine* 70, no. 1 (Spring 1996): 74–75.

57. Curtin, *Image of Africa,* 67.

58. Mark Harrison, "'Tender Frame of Man,'" 82–83.

59. Nancy Stepan, *The Idea of Race in Science: Great Britain, 1800–1960* (London: Macmillan, 1982); Adas, "The Limits of Diffusion," in *Machines as the Measure of Men,* 271–341; Mark Harrison, "'Tender Frame of Man,'" 87–93.

60. Anne McClintock, "The Lay of the Land: Genealogies of Imperialism," in *Imperial Leather: Race, Gender and Sexuality in the Colonial Conquest* (London: Routledge, 1995), 21–74.

61. Gareth Stedman Jones, *Outcast London: A Study in the Relationship between Classes in Victorian Society* (London: Oxford University Press, 1971).

62. David Arnold, "White Colonization and Labour in Nineteenth-Century India," *Journal of Imperial and Commonwealth History* 11, no. 2 (1983): 133–58.

63. Saul Dubow, "The Idea of Race in Early 20th Century South Africa: Some Preliminary Thoughts" (African Studies Seminar Paper, University of the Witwatersrand, April 1989), 6.

64. Joseph Chamberlain, "Speech of 12 December, 1887, Toronto," in Willoughby Maycock, *With Mr. Chamberlain in the United States and Canada, 1887–88* (Toronto: Bell and Cockburn, 1914), 104–11.

65. Benjamin Kidd, *The Control of the Tropics* (New York: Macmillan, 1898), 56.

66. Arnold, "India's Place in the Tropical World," 10. See also Dubow, "Idea of Race in Early 20th Century South Africa," 18–21.

67. Charles Wentworth Dilke, *Problems of Greater Britain* (London: Macmillan, 1890), 2:443–44.

68. Dorothy O. Heely, "'Informed Opinion' on Tropical Africa in Great Britain 1860–1890," *African Affairs* 68, no. 272 (1969): 195–217.

69. George Baden-Powell, "The Development of Tropical Africa," in *Proceedings of the Royal Colonial Institute,* vol. 27 (London: Royal Colonial Institute, 1896), 240.

70. Cowen and Shenton, *Doctrines of Development,* 254–60.

71. Baden-Powell, "Development of Tropical Africa," 230–31.

72. L. Westenra Sambon, "Acclimatization of Europeans in Tropical Lands," *Geographical Journal* 12, no. 6 (1898): 589–99; Rubert Boyce, "The Colonization of Africa," *Journal of the African Society* 10, no. 40 (July 1911): 392–97.

73. Kidd, *Control of the Tropics,* 3, 46–47.

74. Charles Bruce, *The Broad Stone of Empire,* vol. 1 (London: Macmillan, 1910).

75. Daniel R. Headrick, *The Tools of Empire: Technology and European Imperialism in the Nineteenth Century* (New York: Oxford University Press, 1981); Adas, "Attributes of the Dominant: Scientific and Technological Foundations of the Civilizing Mission," in *Machines as the Measure of Men,* 199–270; Conklin, *Mission to Civilize,* 4–5.

76. Adas, *Machines as the Measure of Men,* 210–21.

77. Conklin, "The Setting: The Idea of the Civilizing Mission in 1895 and the Creation of the Government General," and "Public Works and Public Health: Civilization, Technology, and Science (1902–1914)," in *Mission to Civilize,* 11–37 and 38–72.

78. Suzanne Moon, "Empirical Knowledge, Scientific Authority, and Native Development: The Controversy over Sugar/Rice Ecology in the Netherlands East Indies, 1905–1914," *Environment and History* 10, no. 1 (2004): 64.

79. Juhani Koponen, *Development for Exploitation: German Colonial Policies in Mainland Tanzania, 1884–1914* (Helsinki: Finnish Historical Society, 1994); Andrew Zimmerman, "'What Do You Really Want in German East Africa, *Herr Professor?*' Counterinsurgency and the Science Effect in Colonial Tanzania," *Comparative Studies of Society and History* 48, no. 2 (April 2006): 419–61.

80. See Frederick Lugard, *The Rise of Our East African Empire* (London: Frank Cass, 1968 [1893]), 1:389–403; C. T. Hagberg Wright, "German Methods of Development in East Africa," *Journal of the Royal African Society* 1, no. 1 (1901): 23–28.

81. Bernard Semmel, "Joseph Chamberlain's 'Squalid Argument,'" in *Imperialism and Social Reform: English Social-Imperial Thought 1895–1914* (London: Allen

and Unwin, 1960), 83–98; Cowen and Shenton, "Development Doctrine in 'Underdeveloped' Britain," in *Doctrines of Development,* 254–93.

82. Michael Worboys, "Science and British Colonial Imperialism, 1895–1940" (Ph.D. thesis, University of Sussex, 1979), 28–81.

83. The five geographical departments in 1895 were West Indian, North American and Australasian, West African, South African, and Eastern. See George V. Fiddes, *The Dominions and Colonial Offices* (London: G. P. Putnam's Sons, 1926), 8–23; Kubicek, *Administration of Imperialism,* 15; Anne Thurston, *Sources for Colonial Studies in the Public Record Office,* vol. 1, *Records of the Colonial Office, Dominions Office, Commonwealth Relations Office and Commonwealth Office* (London: HMSO, 1995), 10, 159–61.

84. Kubicek, *Administration of Imperialism,* 15–16; Ralph Furse, *Aucuparius: Recollections of a Recruiting Officer* (Oxford: Clarendon Press, 1962).

85. Fiddes, *Dominions and Colonial Offices,* 24–30; Henry L. Hall, *The Colonial Office: A History* (London: Longmans, 1937), 42–43.

86. Ronald Hyam, *Elgin and Churchill at the Colonial Office, 1905–1908* (London: Macmillan, 1968), 430; Worboys, "Science and British Colonial Imperialism," 143–90.

87. Lucile H. Brockway, *Science and Colonial Expansion: The Role of the British Royal Botanic Gardens* (London: Academic Press, 1979); Drayton, *Nature's Government;* G. B. Masefield, *A History of the Colonial Agricultural Service* (Oxford: Clarendon Press, 1972); Worboys, "Science and British Colonial Imperialism," 28–81.

88. Drayton, *Nature's Government,* 221–29.

89. Roy MacLeod, ed., *Government and Expertise: Specialists, Administrators and Professionals, 1860–1919* (Cambridge: Cambridge University Press, 1988); Cranefield, *Science and Empire,* 133.

90. Peter Alter, *The Reluctant Patron: Science and the State in Britain, 1850–1920* (Oxford: Berg, 1987), 6.

91. Ibid., 92–93.

92. Bernard Semmel, "A Party of National Efficiency: The Liberal-Imperialists and the Fabians," in *Imperialism and Social Reform,* 53–82; G. R. Searle, *The Quest for National Efficiency: A Study in British Politics and Political Thought, 1899–1914* (Berkeley: University of California Press, 1971); Frank M. Turner, "Public Science in Britain, 1880–1919," *Isis* 71, no. 259 (1980): 589–608; Alter, *Reluctant Patron,* chap. 4; John Turner, "'Experts' and Interests: David Lloyd George and the Dilemmas of the Expanding State, 1906–19," in *Government and Expertise: Specialists, Administrators and Professionals, 1860–1919,* ed. Roy M. MacLeod (Cambridge: Cambridge University Press, 1988), 203–23.

93. Cranefield, *Science and Empire,* 133–34.

94. Kubicek, *Administration of Imperialism,* 175–76; S. B. Saul, "The Economic Significance of 'Constructive Imperialism,'" *Journal of Economic History* 17, no. 2 (June 1957): 173–92. For a view that emphasizes the importance of the policies initiated during the Chamberlain years in stimulating a major economic transformation in certain colonies, see Richard M. Kesner, *Economic Control and Colo-*

nial Development: Crown Colony Financial Management in the Age of Joseph Chamberlain (Oxford: Clio, 1981), 216–17.

95. K. K. D. Nworah, "Humanitarian Pressure-Groups and British Attitudes to West Africa, 1895–1915" (Ph.D. diss., University of London, 1966), 7.

96. Joseph Chamberlain, "British Trade and the Expansion of Empire," speech given before the Birmingham Chamber of Commerce, 13 November 1896, in *Foreign and Colonial Speeches* (London: George Routledge and Sons, 1897), 146.

97. Chamberlain, "The True Conception of Empire," speech made at the Royal Colonial Institute Dinner, 31 March 1897, in *Foreign and Colonial Speeches,* 244–45.

98. Hyam, *Elgin and Churchill at the Colonial Office,* 49.

99. Quoted in Bernard Porter, *Critics of Empire: British Radical Attitudes to Colonialism in Africa, 1895–1914* (London: Macmillan, 1968), 75.

100. Ibid., 292–93.

101. Mary H. Kingsley, *Travels in West Africa: Congo Français, Corisco and Cameroons,* 3rd ed. (London: Frank Cass, 1965 [1897]). See also Deborah Birkett, "West Africa's Mary Kingsley," *History Today* 37 (May 1987): 10–16; and Birkett, *Mary Kingsley: Imperial Adventuress* (London: Palgrave Macmillan, 1992).

102. Robert D. Pearce, "Missionary Education in Colonial Africa: The Critique of Mary Kingsley," *History of Education* 17, no. 4 (1988): 283–94.

103. For Kingsley's views on racial differences, see Adas, *Machines as the Measure of Men,* 316–17, 329–30, 341.

104. Birkett, *Mary Kingsley,* 132.

105. Birkett, "West Africa's Mary Kingsley," 13–14.

106. In her second book, Kingsley pleaded, "If you will try Science, all the evils of the clash between the two culture periods could be avoided, and you could assist these West Africans in their thirteenth century to rise into the nineteenth-century state without having the hard fight for it that you yourself had. This would be a grand humanitarian bit of work; by doing it you would raise a monument before God to the honour of England such as no nation has ever yet raised to Him on earth." See Mary Kingsley, *West African Studies* (London: Frank Cass, 1964 [1899]), 236.

107. Paul B. Rich, *Race and Empire in British Politics* (Cambridge: Cambridge University Press, 1986), 27–49; Birkett, *Mary Kingsley,* chap. 9.

108. Birkett, "West Africa's Mary Kingsley," 16.

109. See *Journal of the African Society* 1 (1901): xxxi. See also Kenneth Robinson, "Experts, Colonialists, and Africanists, 1895–1960," in *Experts in Africa: Proceedings of a Colloquium at the University of Aberdeen,* ed. J. C. Stone (Aberdeen: Aberdeen University African Studies Group, 1980), 56–57.

110. See J. D. Fage, "When the African Society Was Founded, Who Were the Africanists?" *African Affairs* 94, no. 376 (1995): 369–81.

111. Catherine Ann Cline, *E. D. Morel, 1873–1924: The Strategies of Protest* (Belfast: Blackstaff, 1980), chap. 2.

112. Nworah, "Humanitarian Pressure-Groups," 22.

13. Richard A. Lobdell, "British Officials and the West Indian Peasantry, 1842–1938," in *Labour in the Caribbean: From Emancipation to Independence*, ed. Malcolm Cross and Gad Heuman (London: Macmillan Caribbean, 1988), 200.

14. "Subsidiary Report by D. Morris on the Agricultural Resources and Requirements of British Guiana and the West Indies Islands," in *Report of the West India Royal Commission*, app. A.

15. In truth, a special government agricultural committee expressed great displeasure with the curator's work, charging him with negligence in transplanting, choosing worthless land for the site, and constructing poor roads. It wanted him placed under the supervision of a superior government officer, such as the colonial engineer, and terminated if his work did not improve by the end of the term. See CO 879/65, Confidential Print: West Africa, Correspondence: Botanical and Forestry Matters in West Africa, 1889–1901, no. 24, Administrator R. B. Llewelyn, Gambia, to Marquess of Ripon, S/S/C, 7 May 1895.

16. CO 879/65, Confidential Print: West Africa, no. 188, Thiselton-Dyer, Royal Gardens, Kew, to Colonial Office, 26 November 1900.

17. CO 879/69, Confidential Print: West Africa, Further Correspondence: Botanical and Forestry Matters in West Africa, 1901–1904, no. 37, Ralph Moor, High Commissioner for the Protectorate of Southern Nigeria, to Chamberlain, S/S/C, 15 March 1903.

18. CO 879/69, Confidential Print: West Africa, no. 12, Thiselton-Dyer, Kew, to CO, 4 November 1901.

19. CO 879/65, Confidential Print: West Africa, no. 7, 47, 49, 52, 251; CO 879/69, Confidential Print: West Africa, no. 37; CO 879/105, Confidential Print: Africa, Further Correspondence: Botanical and Forestry Matters in British Tropical Colonies and Protectorates in Africa, 1909–1911, no. 4, 6, 64, 170.

20. Worboys, "Science and British Colonial Imperialism, 1895–1940," 44–54; William Kelleher Storey, *Science and Power in Colonial Mauritius* (Rochester: University of Rochester Press, 1997), 97–123; Drayton, *Nature's Government*, 260–62.

21. BLCAS Mss. W. Ind. s. 42, Arthur Chapman Barnes papers, "Agriculture and the Agricultural Services in Jamaica," 1938.

22. F. A. Stockdale, T. Perch, and H. F. Macmillan, *The Royal Botanic Gardens, Peradeniya, Ceylon, 1822–1922* (Colombo: H. W. Cave, 1922).

23. See Storey, *Science and Power in Colonial Mauritius*, 97–108.

24. Under Johnson, new staff was added and several new agricultural experiment stations were opened at Tarquah, Kumasi, Ashanti, and Tamali. By 1911 the department consisted of twenty-seven members, including the director, an entomologist, an agriculturist, a traveling inspector, six curators, five native traveling instructors, three first- and second-class overseers, two garden assistants, and five clerks. See CO 879/88, Confidential Print: Africa, Correspondence: Botanical and Forestry Matters in British Tropical Colonies and Protectorates in Africa, 1905–1907,

no. 23, Acting Governor Bryan, Gold Coast, to Lyttleton, S/S/C, 9 June 1905; CO 879/105, Confidential Print: Africa, enclosure in no. 172, Governor Thornburn, Gold Coast, to S/S/C, 29 November 1911.

25. CO 879/105, Confidential Print: Africa, no. 4, Southern Nigeria: Imperial Institute to CO, 29 November 1909; no. 6, Northern Nigeria: Imperial Institute to CO, 8 December 1909; no. 64, Hesketh-Bell, Governor of Northern Nigeria, to S/S/C, 7 November 1910.

26. The list includes W. Maxwell-Lefroy and Albert Howard, who went on to serve in India; William Nowell, who became the first director of the East Africa Agricultural Research Station at Amani in Tanganyika; and Frank Stockdale, L. Lewton Brain, F. W. South, and Dr. Harold A. Tempany, who found important posts in Ceylon and Malaya. Stockdale and Tempany would emerge as the CO's chief agricultural advisers in the 1930s and 1940s. See BLCAS Mss. W. Ind. s. 61(1), Professor Frederick Hardy, "A History of Soil Science at the Imperial College of Tropical Agriculture, Trinidad, West Indies, 1922–1956," pt. 1.

27. See for example, RA, Rnum 51/352, "Mr. Churchill's Speech," in *Report of Proceedings at a Banquet Given by the Incorporated Chamber of Commerce of Liverpool to the Under-Secretary of State for the Colonies, Mr. Winston L. S. Churchill, 5 May 1906* (Liverpool, 1906), 5–13; Winston Churchill, "The Development of Africa," *Journal of the African Society* 6, no. 23 (April 1907): 291–96; RA, Rnum 32/094, "Speech by H.M. Parliamentary Under Secretary of State for the Colonies, Colonel J. E. B. Seeley, at the 58th Annual General Meeting of the Incorporated Chamber of Commerce of Liverpool, 8 May 1908," *Incorporated Chamber of Commerce of Liverpool Monthly Magazine Supplement*, 1–10.

28. CO 879/69, Confidential Print: West Africa, no. 134, Thiselton-Dyer, Royal Gardens, Kew, to CO, 18 July 1904.

29. See for example, CO 879/88, CO 879/99, CO 879/105, Confidential Print: Africa, which are filled with correspondence between the CO and the Imperial Institute, but almost none with Kew. This is in sharp contrast to the pre-1905 period when the reverse was the case.

30. Donal P. McCracken, *Gardens of Empire: Botanical Institutions of the Victorian British Empire* (London: Leicester University Press, 1997), 207.

31. Wyndham R. Dunstan, "Some Imperial Aspects of Applied Chemistry," *Bulletin of the Imperial Institute* 4 (1906): 317.

32. CO 879/88, Confidential Print: Africa, no. 4, Wyndham Dunstan, Imperial Institute, to the CO, 12 January 1905.

33. CO 879/88, Confidential Print: Africa, nos. 109 and 111, Dunstan, Imperial Institute, to the CO, 17 May 1907, 12 June 1907.

34. See CO 879/88, Confidential Print: Africa, nos. 65, 69, 70, 73, 90, 94, 98.

35. CO 879/99, Confidential Print: West Africa, no. 75, CO Memorandum, 31 March 1909, revised 23 April 1909; Worboys, "Science and British Colonial Imperialism," 59.

36. Worboys, "Science and British Colonial Imperialism," 63.

should have been created. For a good overview of the formation of the London School, see Philip Manson-Bahr, *History of the School of Tropical Medicine in London, 1899–1949* (London: H. K. Lewis, 1956), chap. 3; Farley, *Bilharzia,* 20–30.

61. The expedition to Central Africa was headed by Charles W. Daniels of the London School along with John W. W. Stephens and Samuel R. Christophers. See CO 885/7/119, no. 35, Chamberlain to Lord Lister, President of the Royal Society, 6 July 1898; Robert V. Kubicek, *The Administration of Imperialism: Joseph Chamberlain at the Colonial Office* (Durham, N.C.: Duke University Press, 1969), 146; Worboys, "Science and British Colonial Imperialism," 95; Cranfield, *Science and Empire,* 129–33.

62. Once again the initiative for a coordinating committee came from Read, and was set in motion by another key senior official, Sir Charles Lucas. See Kubicek, *Administration of Imperialism,* 144.

63. Ibid., 153.

64. CO 885/7/119, no. 150, "Proposal for the Establishment of a Liverpool School of Tropical Diseases," enclosed in Alfred L. Jones to Colonial Office, Liverpool, 6 January 1899. For the origins and history of the Liverpool School, see B. H. Maegraith, "History of the Liverpool School of Tropical Medicine," *Medical History* 16, no. 4 (1972): 354–68; Farley, *Bilharzia,* 20–30; Edwin R. Nye and Mary E. Gibson, "Ross and the Liverpool School of Tropical Medicine," in *Ronald Ross: Malariologist and Polymath: A Biography* (London: Macmillan, 1997), 91–192; Helen J. Power, *Tropical Medicine in the Twentieth Century: A History of the Liverpool School of Tropical Medicine, 1898–1990* (London: Kegan Paul International, 1999).

65. Power, *Tropical Medicine in the Twentieth Century,* 16; CO 885/7/129, no. 59, CO to A. L. Jones, 12 July 1900.

66. See Worboys, "Manson, Ross and Colonial Medical Policy," 26; Power, "Tropical Medicine in Temperate Liverpool," in *Tropical Medicine in the Twentieth Century,* 11–46.

67. Power, *Tropical Medicine in the Twentieth Century,* 24.

68. Ronald Ross, H. E. Annett, and E. E. Austen, *Report of the Malarial Expedition of the Liverpool School of Tropical Medicine and Medical Parasitology* (London: University Press of Liverpool, 1900), 2–3.

69. Ibid., 37–43.

70. CO 885/7/119, Miscellaneous, enclosure 2 in no. 267, Memorandum by W. T. Prout, 21 November 1899.

71. Mary P. Sutphen, "Not What, but Where: Bubonic Plague and the Reception of Germ Theories in Hong Kong and Calcutta, 1894–1897," *Journal of the History of Medicine and Allied Sciences* 52, no. 1 (January 1997): 112.

72. Kubicek, *Administration of Imperialism,* 146–47. For the correspondence between the Colonial Office and the Royal Society, see CO 885/7/119, Miscellaneous, Appointment of a Commission to Investigate Malaria, Correspondence, August 1900; CO 885/7/129, Miscellaneous, Investigation of Malaria, Further Correspondence, 1900.

73. Harrison, *Mosquitoes, Malaria and Man,* 126. See J. W. W. Stephens and S. R. Christophers, "Distribution of *Anopheles* in Sierra Leone, Parts I and II," in *Report to the Malaria Committee of the Royal Society,* 1st series (London: Harrison and Sons, 1900), 42–74.

74. J. W. W. Stephens and S. R. Christophers, "The Native as the Prime Agent in the Malarial Infection of Europeans," in *Further Reports to the Malaria Committee of the Royal Society,* 2nd series (London: Harrison and Sons, 1900), 3–19.

75. J. W. W. Stephens, "The Discussion on the Prophylaxis of Malaria," *British Medical Journal,* 17 September 1904, 630.

76. See J. W. W. Stephens and S. R. Christophers, "The Malarial Infection of Native Children," "On the Destruction of Anopheles in Lagos," and "The Segregation of Europeans," all 1 October 1900, in *Reports of the Royal Society Malaria Commission,* 3rd series (London: Harrison, 1900); Foreign and Commonwealth Office Library, Malaria Pamphlets 36, *Summary of Researches on Native Malaria and Malarial Prophylaxis: On Blackwater Fever, Its Nature and Prophylaxis,* by J. W. W. Stephens and S. R. Christophers, 4 February 1903.

77. See Ross, *Memoirs,* 437–39.

78. Ibid., 443–44.

79. Logan Taylor, *British Medical Journal,* 15 September 1902; Taylor, "Extermination of Mosquitoes in Sierra Leone," *British Medical Journal,* 17 January 1903; Dr. W. T. Prout, "The Extermination of Mosquitoes in Sierra Leone," *British Medical Journal,* 6 June 1903, 1349.

80. CO 885/7/132, Miscellaneous, "Measures to Be Taken for the Prevention of Malaria," Memorandum by Sir Michael Foster, 28 July 1900. See also CO 885/7/129, Miscellaneous, no. 11, The Malaria Investigation Committee to the Colonial Office, 24 February 1900.

81. CO 885/7/129, Miscellaneous, no. 82, Sir William MacGregor to Colonial Office, 6 December 1900; CO 885/7/139, Miscellaneous, no. 37, MacGregor to Chamberlain, Lagos, 8 July 1901.

82. "Notes on Antimalarial Measures Now Being Taken in Lagos," A Discussion on Malaria and its Prevention, Section of Tropical Diseases, British Medical Association, *British Medical Journal,* 14 September 1901, 682.

83. Henry Strachan, "Discussion on the Prophylaxis of Malaria," Proceedings of the Tropical Diseases Section of the British Medical Association, *British Medical Journal,* 17 September 1904, 637–39.

84. On the early career of William Simpson, see R. A. Baker and R. A. Bayliss, "William John Ritchie Simpson: Public Health and Tropical Medicine," *Medical History* 31, no. 4 (1987): 452–56; I. J. Catanach, "Plague and the Tensions of Empire: India, 1896–1918," in *Imperial Medicine and Indigenous Societies,* ed. David Arnold (Manchester: Manchester University Press, 1988), 149–71; Mark Harrison, *Public Health in British India,* 202–26; Sutphen, "Not What, but Where"; Mary P. Sutphen, "Striving to Be Separate? Civilian and Military Doctors in Cape Town during the Anglo-Boer War," in *War, Medicine and Modernity,* ed.

108. GB, *PP,* Cd. 5581, LII (March 1911), "Correspondence Relating to the Recent Outbreak of Yellow Fever in West Africa," no. 1, Sir Rubert Boyce to the Colonial Office, 30 August 1910; CO 879/105/960, African, "Report of Sir Rubert Boyce on Yellow Fever in West Africa," February 1911.

109. CO 879/102/940, African, Minutes of the 15th, 16th, and 17th Meetings of the AMSCTA, 4 October, 1 November, and 6 December 1910.

110. CO 879/102/940, African, Minutes of the 18th Meeting of the AMSCTA, 13 December 1910.

111. GB, *PP,* Cd. 5581, no. 4, Secretary of State Harcourt to the Governors of the West African Colonies, 7 February 1911, 743.

112. CO 879/108/975, African, no. 84, Governor of the Gold Coast to the Secretary of State, 23 August 1911.

113. CO 879/108/975, African, no. 121, Governor of the Gold Coast to the Secretary of State, 20 September 1911.

114. CO 879/107/966, African, no. 78, Governor of the Gambia to the Secretary of State, 31 March 1911.

115. CO 879/107/966, African, enclosure in no. 109, Memorandum by the Principal Medical Officer, H. Strachan, 10 March 1911.

116. CO 879/107/966, African, enclosure in no. 132, Senior Sanitary Officer to the Principal Medical Officer, Zungeru.

117. CO 879/107/966, African, no. 132, Acting Governor Temple to the Secretary of State, 29 April 1911.

118. CO 879/108/975, African, no. 33, Sir H. H. Bell to the Colonial Office, 25 July 1911.

119. CO 879/109/984, African, enclosure in no. 144, Sir Hesketh Bell's Observations on the Senior Sanitary Officer's Memorandum on the Question of Segregation, 20 February 1912.

120. CO 879/110/992, African, no. 17, Secretary of State to the Acting Governor of Northern Nigeria, 9 July 1912.

121. Isongesit S. Ibokette, "Contradictions in Colonial Rule: Urban Development in Northern Nigeria, 1900–1940" (Ph.D. thesis, Queen's University at Kingston, Canada, 1989), 358–59.

122. Ibid., 342–65; CO 879/120/1088, African, no. 13, Governor of Nigeria to the Secretary of State, 31 January 1921, enclosing a memo by H. B. S. Montgomery, Deputy PMO, and G. J. Pirie, Acting SSO, entitled, "The Principle of Segregation, and the Difficulties that Arise from the Present Method of Application, with Suggestions as to How these Difficulties May be Best Removed."

123. CO 879/117/1044, African, no. 107, Governor of the Gold Coast to the Secretary of State, 6 November 1916.

124. CO 879/117/1044, African, no. 89, Governor of the Gold Coast to the Secretary of State, 11 October 1916.

125. See for example, CO 879/112/999, African, no. 177, Memorandum on the Segregation Principle in West Africa, 10 June 1913; CO 879/119/1075, African, Memo-

randum by Professor Simpson on the Demand of Indians to Reside in the Areas Set Apart for Europeans in East Africa, 20 December 1919; no. 83, "Segregation and Town-Planning in Northern and Southern Nigeria," Memorandum by Professor Simpson, 11 August 1920; CO 879/120/1088, African, no. 35, Memorandum on European Residential Quarter at Lagos, by A. J. Horn, Medical Secretary of the AMSCTA, 2 June 1921.

126. See for example, CO 879/102/940, African no. 40, Governor of the Gold Coast, John Rodger, to the Secretary of State, 18 October 1909, in reference to the Gold Coast Annual Medical and Sanitary Report for 1908, 61.

127. Most older West African towns had a significant class of well-to-do Africans who opposed any plans to redesign or clear urban areas based on segregationist principles. Such was the case in Bathurst in the Gambia and in Accra on the Gold Coast. In Freetown, the most influential members of the Krio community at first gave their support to the Hill Station project. When they began to realize the implications it held for race relations in general, however, they reconsidered, seeing it instead as a sinister plot to transform Sierra Leone into another "white man's country." See Dumett, "Campaign against Malaria," 181; Leo Spitzer, *The Creoles of Sierra Leone : Responses to Colonialism, 1870–1945* (Madison: University of Wisconsin Press, 1974), 58–61; Philip D. Curtin, "Medical Knowledge and Urban Planning in Tropical Africa," *American Historical Review* 90, no. 3 (1985): 603.

128. See Alice Conklin, *A Mission to Civilize: The Republican Idea of Empire in France and West Africa, 1895–1930* (Berkeley: University of California Press, 1997), 65–72. For German East Africa, where state development interventions were the direct cause of one of the earliest and most significant anticolonial insurrections, the Maji Maji Rebellion of 1905, see Andrew Zimmerman, "'What Do You Really Want in German East Africa, *Herr Professor?*' Counterinsurgency and the Science Effect in Colonial Tanzania," *Comparative Studies of Society and History* 48, no. 2 (April 2006): 443–61.

Chapter 3: Science for Development

1. Niall Ferguson, "Empire for Sale," in *Empire: The Rise and Demise of the British World Order and the Lessons for Global Power* (NewYork: Basic Books, 2003), 245–302.

2. Robert Holland, "The British Empire and the Great War, 1914–1918," in *The Twentieth Century,* ed. Judith Brown and William Roger Louis, vol. 4, *The Oxford History of the British Empire,* ed. William Roger Louis (Oxford: Oxford University Press, 1999), 136.

3. Ibid.; Ferguson, *Empire,* 313–28.

4. Michael Adas, *Machines as the Measure of Men: Science, Technology, and Ideologies of Western Dominance* (Ithaca, N.Y.: Cornell University Press, 1989), 365–80.

5. The British textile industry was particularly hard hit. Its dependence on Germany for about 80 percent of its dyestuffs threatened an important part of

the national economy, thus impeding the government's ability to mobilize the war effort. See Frank M. Turner, "Public Science in Britain, 1880–1919," *Isis* 71, no. 259 (1980): 598–608; Ian Varcoe, "Scientists, Government and Organised Research in Great Britain 1914–16: The Early History of the DSIR," *Minerva* 8, no. 2 (April 1970): 192–93.

6. Peter Alter, *The Reluctant Patron: Science and the State in Britain, 1850–1920* (Oxford: Berg, 1987), chap. 4; Varcoe, "Scientists, Government and Organised Research in Great Britain," 214–16; Roy M. MacLeod and E. Kay Andrews, "The Origins of the DSIR: Reflections on Ideas and Men, 1915–1916," *Public Administration* 48, no. 1 (Spring 1970): 23–48.

7. James E. Cronin, *The Politics of State Expansion: War, State and Society in Twentieth-Century Britain* (London: Routledge, 1991), 65; Mary Langan and Bill Schwartz, eds., *Crises in the British State, 1880–1930* (London: Hutchinson, 1985).

8. John Turner, "'Experts' and Interests: David Lloyd George and the Dilemmas of the Expanding State, 1906–19," in *Government and Expertise: Specialists, Administrators and Professionals, 1860–1919,* ed. Roy MacLeod (Cambridge: Cambridge University Press, 1988), 203–23.

9. By the end of the war, twelve new departments and 160 new boards and commissions had been set up, and the number of central government employees mushroomed from 325,000 in 1914 to more than 850,000 in 1918. See Geoffrey Kingdon Fry, *The Growth of Government: The Development of Ideas about the Role of the State and the Machinery and Functions of Government in Britain since 1780* (London: Frank Cass, 1979), chap. 3; Cronin, *Politics of State Expansion,* 71–72.

10. GB, *PP,* Cd. 9230, XII (1918), Ministry of Reconstruction, "Report of the Machinery of Government Committee."

11. Roy Macleod and E. Kay Andrews, "The Committee of Civil Research: Scientific Advice for Economic Development, 1925–30," *Minerva* 7, no. 2 (1969): 681–82.

12. On the changing role of experts within the state in the interwar period, see Gail Savage, *The Social Construction of Expertise: The English Civil Service and Its Influence, 1919–1939* (Pittsburgh: University of Pittsburgh Press, 1996).

13. See Max Beloff, *Imperial Sunset,* vol. 1, *Britain's Liberal Empire, 1897–1921* (London: Macmillan, 1982), 210; A. D. Roberts, "The Imperial Mind," in *The Cambridge History of Africa,* vol. 7, ed. A. D. Roberts (Cambridge: Cambridge University Press, 1986), 43–47; Stephen Constantine, *The Making of British Colonial Development Policy, 1914–1940* (London: Frank Cass, 1984), 31–35; William Roger Louis, *In the Name of God, Go! Leo Amery and the British Empire in the Age of Churchill* (New York: Norton, 1992), 63–64.

14. Michael Havinden and David Meredith, *Colonialism and Development: Britain and Its Tropical Colonies, 1850–1960* (London: Routledge, 1993), 133–34.

15. Louis, *In the Name of God, Go!* 63. On Milner's support for the rationalization of government, see Walter Nimocks, *Milner's Young Men: The "Kindergarten" in Edwardian Imperial Affairs* (Durham, N.C.: Duke University Press, 1968), 132; Ian M. Drummond, *British Economic Policy and the Empire, 1919–1939* (London:

Allen and Unwin, 1972), 36–88; Eric Stokes, "Milnerism," *Historical Journal* 5, no. 1 (1962): 52.

16. Michael Worboys, "Science and British Colonial Imperialism, 1895–1940" (Ph.D. thesis, University of Sussex, 1979), 195.

17. Ibid., 199.

18. Amery to Milner, 13 November 1924, Milner Papers 209, as quoted in Constantine, *Making of British Colonial Development Policy,* 141.

19. Louis, *In the Name of God, Go!* 81.

20. Leopold Amery, speech of 30 July 1919, in *Parliamentary Debates,* Commons, 5th ser., vol. 118 (1919), cols. 2172–85, 2230–40.

21. Leopold S. Amery, *My Political Life,* vol. 1, *England Before the Storm, 1896–1914* (London: Hutchinson, 1953), 340.

22. GB, *PP,* Cmd. 2387 (April 1925), "Report of the East Africa Commission."

23. The commissioners justified the subvention of imperial funds by noting that "approximately half the capital sum would be spent in Great Britain on rails, bridging material, rolling stock, etc., which at this time would provide work for engineering industries of Great Britain and so lesson unemployment charges." See ibid., 182.

24. Ibid., 78.

25. Major Church was also the long-time general secretary of the Association of Scientific Workers. See Archibald G. Church, "Science and Administration in East Africa," *Nature* 115 (10 January 1925): 37–39; "Mr. Ormsby-Gore and Tropical Development," *Nature* 123 (21 January 1929), 27; "Colonial Development and the Scientific Worker," *Nature* 124 (21 September 1929): 433–34; and *East Africa, a New Dominion: A Crucial Experiment in Tropical Development and Its Significance to the British Empire* (Westport, Conn.: Negro Universities Press, 1970 [1927]).

26. Church, *East Africa, a New Dominion,* 62–63, 87. Church admitted, however, that only scant and unreliable quantitative information was available regarding density and distribution of population, births and deaths, length of life, agricultural returns, distribution and chemistry of soils, density and distribution of cattle, climate and distribution of rainfall, mineral resources, water power, forest resources, internal trade, existing and potential overseas markets, and conditions of labor. Given this, the question must be asked: on what basis did he form his opinion of the vast potential wealth and natural resources of tropical Africa?

27. Ibid., 108–9.

28. See for example, GB, *PP,* "Report by the Rt. Hon. W. G. A. Ormsby-Gore on His Visit to West Africa during the Year 1926," Cmd. 2744, IX (1927), 211; and "Report by the Rt. Hon. W. G. A. Ormsby-Gore on His Visit to Malaya, Ceylon and Java during the Year 1928," Cmd. 3235, V (1928–29), 791.

29. See Worboys, "Science and British Colonial Imperialism," 218.

30. MacLeod and Andrews, "Committee of Civil Research," 685.

31. "Report of the East African Commission," Cmd. 2387, 94.

32. The publication of the report was instrumental in the creation of the Committee of Civil Research (CCR), under which a number of expert subcommittees were set up to investigate a wide variety of imperial and domestic problems, including tsetse fly research in Tanganyika, desert locust control in North Africa, "dietetics" in Kenya, as well as general investigations on quinine production, rubber research, and the mineral content of natural pasturages. The CCR was transformed into the Economic Advisory Committee in May 1930, after the new Labour government decided it needed a central committee of expert economists as well as scientists to contend with the nation's worsening economic crisis. The CCR worked closely with the Empire Marketing Board (EMB), which was set up in May 1926 to administer an annual grant of £1 million for the purpose of furthering the marketing of empire products in Britain. The grant was given as a "non-tariff" preference to honor the government's commitments stemming from the Imperial Economic Conference of 1923. Even this amount continued to be disputed by Churchill and the Treasury, with the result that the 1926 grant was reduced to £500,000. Treasury squabbling over the EMB's funding continued up until its demise in 1933. In his memoirs, Amery, who was the board's first chairman, attached great importance to the research activities of the EMB, noting that "For those of us who played any part in all the outpouring of creative original work which characterized the Empire Marketing Board it was 'Heaven to be alive' during those years." See Amery, *My Political Life,* 1:354. For the significance of the East Africa Commission, see MacLeod and Andrews, "Committee of Civil Research," 700–703; Michael Worboys, "British Colonial Science Policy, 1918–1939," in *Les Sciences Coloniales: Figures et Institutions,* ed. Patrick Petitjean (Paris: Orstom, 1996), 99–111. For the EMB see Robert Self, "Treasury Control and the Empire Marketing Board: The Rise and Fall of Non-Tariff Preference in Britain, 1924–1933," *Twentieth Century British History* 5, no. 2 (1994): 153–82; J. M. Lee, "The Dissolution of the Empire Marketing Board, 1933: Reflections on a Diary," *Journal of Imperial and Commonwealth History* 1, no. 1 (1973): 49–58; Worboys, "Science and British Colonial Imperialism," 244–94.

33. Amery, *My Political Life,* 1:335.

34. S. R. Ashton and S. E. Stockwell, eds., *Imperial Policy and Colonial Practice, 1925–1945,* British Documents on the End of Empire, ser. A, vol. 1 (London: HMSO, 1996), xxiii; Max Beloff, *Imperial Sunset,* vol. 2, *Dream of Commonwealth, 1921–42* (London: Macmillan, 1989), 85; Paul B. Rich, *Race and Empire in British Politics* (Cambridge: Cambridge University Press, 1986), 64.

35. J. M. Lee and Martin Petter, *The Colonial Office, War and Development Policy: Organisation and the Planning of a Metropolitan Initiative, 1939–1945* (London: Institute of Commonwealth Studies, 1982), 31–46; J. M. Lee, *Colonial Development and Good Government: A Study of the Ideas Expressed by the British Official Classes in Planning Decolonization, 1939–1964* (Oxford: Clarendon Press, 1967), 41–46.

36. See for example, comments by Sir George Schuster, the CO's financial adviser, after his tour of Central and East Africa in 1928: "[I]t is astounding to find

each little Government in each of these detached countries working out, on its own, problems which are common to all, without any knowledge of what its neighbours are doing and without any direction on main lines of policy from the Colonial Office." See Schuster, *Private Work and Public Causes* (Cowbridge, Wales: Brown and Sons, 1979), 78.

37. Much of the new growth in technical advisers and advisory bodies came in response to recommendations made by the Colonial Office Organization Committee under the chairmanship of the permanent undersecretary, Samuel Wilson. Amery appointed the committee in March 1927 to examine ways of more effectively coordinating the work of the geographical departments to meet the office's new role in colonial development and to improve the provision of technical and specialist advice. The Wilson Committee recommended the improvement of existing machinery for obtaining expert advice through the creation of standing committees like the Colonial Advisory Medical and Sanitary Committee, and by the appointment of expert advisers similar to the legal and medical advisers to ensure greater coordination of general questions pertaining to the colonial empire as a whole. See CO 885/29/382, Miscellaneous, Report of the Committee on the Organisation of the Colonial Office, February 1927; CO 885/30/391, Miscellaneous, Colonial Office Organization Committee, Second Interim Report, 7 November 1927. On the increasing importance of technical advisers and bodies within the CO generally, see CO 967/2B, Private Office Papers, Proposed Committee on Appointment of Technical Advisers to Secretary of State Amery, 1926; A. D. Roberts, "The Imperial Mind," in *The Cambridge History of Africa,* vol. 7, ed. A. D. Roberts (Cambridge: Cambridge University Press, 1986), 41–49.

38. The General Department was split in two in 1928 as a first step toward the reorganization of duties along subject lines. One department dealt with personnel questions, while the other concentrated on specialist services such as communications, economic development, labor, health, and education. In 1934 the General Department was renamed the General Division, and a separate Economics Department was set up, followed by a Social Services Department in 1938. After 1940 the number of subject departments and specialist subject advisers mushroomed. By 1950 there were a total of sixteen separate departments under the General Division, dealing with everything from defense, international relations, communications, and information, to welfare, social services, research, finance, production, and marketing. See Cosmo Parkinson, *The Colonial Office from Within, 1909–1945* (London: Faber and Faber, 1945), 56–70; Charles Jeffries, *The Colonial Office* (London: Allen and Unwin, 1956), 110–11.

39. Michael Crowder, "The First World War and Its Consequences," in *UNESCO General History of Africa,* vol. 7, *Africa Under Colonial Domination, 1880–1935,* ed. A. Adu Boahen (Berkeley: University of California Press, 1985), 291.

40. GB, *PP,* Cmd. 730, XII (June 1920), 253, "Report of the Committee on the Staffing of the Agricultural Departments in the Colonies"; GB, *PP,* Cmd. 920, XII

(August 1920), 279, "Report of the Committee on the Staffing of the Veterinary Departments in the Colonies and Protectorates"; GB, *PP*, Cmd. 939, XII (September 1920), 267, "Report of the Departmental Committee appointed to Enquire into the Colonial Medical Service."

41. Worboys, "Science and British Colonial Imperialism," 65–70.

42. CO 885/28/371, Miscellaneous, Report of a Committee Appointed to Consider the Recruitment and Training of Officers for Agricultural Departments of the Non-Self-Governing Dependencies, March 1925.

43. BLCAS Mss. Brit. Emp. s. 476, Oxford Development Records Project—Food and Cash Crops Collection, Box 3 (21), C. E. Johnson; BLCAS Mss. W. Ind. s. 55, B. J. Silk, Colonial Service Training, 1940–42.

44. As Frank Stockdale noted, the Cambridge-Trinidad experience was designed to give "a certain uniformity of outlook to all scholars" to ensure that colonial governments were securing the services of technically qualified officers who were carefully and fully trained and more prepared than previous entrants to investigate the unique problems of tropical agriculture. See CO 323/1203/4, Report by Frank Stockdale, Agricultural Adviser to the S/S/C, on his Visit to the West Indies, Bermuda, British Guiana, and British Honduras, 1932.

45. BLCAS Mss. Brit. Emp. s. 476, Box 4 (32), Prof. J. W. Purseglove, 1912.

46. Joseph Hutchinson, "The Role of the ICTA in Tropical Agriculture," *Tropical Agriculture* 51, no. 4 (October 1974): 460; N. W. Simmonds, "The Earlier British Contribution to Tropical Agricultural Research," *Tropical Agricultural Association Newsletter* 11, no. 2 (June 1991): 4–5.

47. See CO 323/966/5, Organization of Agricultural Administration and Research in the Colonies, Protectorates and Mandated Territories, 1926; GB, *PP*, Cmd. 2825, VII (March 1927), 563, "Agricultural Research and Administration on the Non-Self-Governing Dependencies: Report of a Committee Appointed by the Secretary of State for the Colonies."

48. In addition to proposals for greater recruitment and training, the report suggested setting up a Central Clearing House of Information and Intelligence to be entrusted to the Imperial Institute. The Imperial Agricultural Research Conference took up the latter suggestion in October 1927, recommending the establishment of eight new imperial bureaus along the lines of the Imperial Bureaus of Mycology and Entomology. The new bureaus for soil science, animal nutrition, animal health, animal genetics, agricultural parasitology, plant genetics, fruit and plantation production, and pasturage crops were financed by the Dominion governments (with the colonial empire taken as one unit) through a common fund of £20,000 per annum, administered by an Executive Council that was responsible equally and directly to all governments. See Worboys, "Science and British Colonial Imperialism," 254–56.

49. On the movement for greater imperial scientific cooperation, see Roy M. MacLeod, "On Visiting the 'Moving Metropolis': Reflections on the Architecture of Imperial Science," in *Scientific Colonialism: A Cross-Cultural Comparison*, ed.

Nathan Reingold and Marc Rothenberg (Washington, D.C.: Smithsonian Institution Press, 1987), 242–43.

50. The idea to house the central organization at the CO seems to have come from F. B. Smith of the Cambridge School of Agriculture, who submitted a minority report criticizing the idea of a general council separate from the Colonial Office. Smith noted that the CO had no particular organization for dealing with agricultural work, nor did it employ any officers trained or with experience in tropical agriculture. As it stood, initiatives came from the colonies, while the various geographic departments dealt with the work separately. He felt there should be an agricultural department at the CO, with the Agricultural Research Council attached and staffed by tropical agricultural experts. Anticipating Treasury opposition, CO permanent staff rejected the idea of a central department of agriculture, but it did concur that the new council and adviser should be based at the CO rather than the Imperial Institute. See CO 323/966/5, Minority Report by F. B. Smith, School of Agriculture, Estate Management Branch, Cambridge, 19 October 1926; Minutes by E. B. Boyd, 27 October and 3 November 1926; and Draft Letter for Members of the Research Special Sub-Committee Attaching a Memo by Secretary of State Amery, 5 November 1926.

51. Leopold S. Amery, *My Political Life*, vol. 2, *War and Peace, 1914–1929* (London: Hutchinson, 1953), 344.

52. GB, *PP*, Cmd. 2884, VII (1927), 834, "Colonial Office Conference, 1927, Appendices." See also Amery, *My Political Life*, 2:344–45.

53. Gilks envisioned an Imperial Science Service with a director and research activities controlled by a council of the Privy Council, along the lines of the Medical Research Council. See GB, *PP*, Cmd. 2883 (1927), Colonial Office Conference, 1927, Summary of Proceedings, "Report of the Committee on Colonial Scientific and Research Services," sections 8 and 9; CO 323/972/8, Proposal for an Imperial Science Service by John Gilks, 23 January 1927.

54. See CO 323/984/3, Agricultural Research, 1927, Draft Despatch by Grindle to the Members for a Proposed Committee, 21 June 1927.

55. GB, *PP*, Cmd. 3049, VII (March 1928), 547, "Colonial Agricultural Service: Report of a Committee Appointed by the Secretary of State for the Colonies"; CO 323/999/2, Colonial Agricultural Service Organisation Committee (Lovat Committee), Minutes of Meetings Held between 30 June 1927 and 27 January 1928.

56. In making this recommendation, the committee took into account evidence heard from colonial delegates at the Imperial Agricultural Research Conference of 1927, who advised that no further central research stations should be created until the staff of the EAARS was brought up to full strength. The establishment of further links in the chain would follow when sufficient funds were available, and would be carried out on the advice of the proposed Colonial Advisory Council on Agriculture and Animal Health. For the moment, it recommended that the scheme include funding for one complete central research station at Amani. See CO 323/1006/12, Imperial Agricultural Research Conference, Report and Recommendations, 1928.

57. Of the £127,000 total, £93,000 was to be for the service, £20,000 for the central research station, and £14,000 for the central headquarters council. See CO 323/1046/8, Colonial Advisory Council: Treasury Sanctions; CO 323/999/1, Report of Committee on the Colonial Agricultural Services, 1928.

58. CO 323/1001/6, Colonial Agricultural Service—Lord Lovat's Report, 1928, Circular by Amery to all Colonies and Protectorates, 16 May 1928.

59. CO 323/1001/6, Minutes Concerning Proposed Scheme, by E. B. Boyd, 25 April 1928; CO 323/1048/8, Colonial Agricultural Service, Proposal for a United Service (CAC 3), May 1929.

60. CO 323/1017/3, Colonial Agricultural Service, Extract from Address by Governor of Ceylon to Legislative Council, 21 June 1928.

61. CO 323/1048/8, Colonial Agricultural Service, Proposal for a Unified Service (CAC 3), May 1929, Annex—Summary of Replies, Malaya.

62. T. J. Barron, "Science and the Nineteenth-Century Ceylon Coffee Planters," *Journal of Imperial and Commonwealth History* 16, no. 1 (1987): 15–23; L. A. Wickremeratne, "Economic Development in the Plantation Sector, 1900 to 1947," and Swarna Jayaweera, "Land Policy and Peasant Colonization, 1914–1948," in *History of Ceylon*, vol. 3, *From the Beginning of the 19th Century to 1948*, ed. K. M. De Silva (Peradeniya: University of Ceylon Press Board, 1973), 428–45, 446–60.

63. Algernon Aspinall, "The Imperial College of Tropical Agriculture," *Tropical Agriculture* 11, no. 2 (1934): 40–43.

64. H. M. Burkill, "Murray Ross Henderson, 1899–1983, and Some Notes on the Administration of Botanical Research in Malaya," *Journal of the Malaysian Branch of the Royal Asiatic Society* 56, no. 2 (1983): 87–101.

65. CO 323/1048/8, Colonial Agricultural Service, 1929, Annex—Summary of Replies, Malaya.

66. CO 323/1017/3, Colonial Agricultural Service, To Secretary of the Treasury, from William Ormsby-Gore, Parliamentary Under-Secretary of State, 18 June 1928.

67. CO 323/1017/3, Colonial Agricultural Service, Extract from Minutes of the 26th Meeting of the EMB, 20 June 1928.

68. CO 323/1017/3, Colonial Agricultural Service, Letter to Ormsby-Gore, Parliamentary Under-Secretary of State, from Stephen Tallents, Secretary of the EMB, 27 November 1928.

69. See CO 323/1528/5, CAC Appointments, 1937; CO 323/1022/9, Colonial Agricultural Service, Establishment of Central Fund, Draft Circular from Secretary of State Amery to all Colonies and Protectorates, 4 January 1929.

70. The first meeting of the CAC was held on 26 March 1929. Ormsby-Gore presided as chairman. Its members formed an impressive list of authoritative experts from the British scientific establishment, including Frank Stockdale as vice-chairman; Lt. Gen. Sir William Furse, director of the Imperial Institute; Dr. A. W. Hill, director of the Royal Botanic Gardens, Kew; Dr. G. K. Hill; Dr. G. K. Marshall, director of the Imperial Bureau of Entomology; Dr. E. J. Butler, director of the Imperial Bureau of Mycology; T. B. Wood, Drapers' Professor of Agriculture,

Cambridge; Dr. W. H. Andrews, director of the Veterinary Lab, Ministry of Agriculture and Fisheries; Sir John Russell, director of the Rothamsted Experimental Station; Dr. (later Lord) John Boyd Orr, director of the Rowett Research Institute, Aberdeen; and Dr. A. T. Stanton, the chief medical adviser to the Colonial Office. See CO 323/1046/4, Appointment of Advisory Council on Agriculture, 1929, Minute by Creasy, 6 March 1929; FCOL CAC, Minutes of the Meetings, 1929–1943, Minutes of the 1st Meeting of CAC, Tuesday, 26 March 1929.

71. FCOL CAC, Minutes of the 1st Meeting of CAC, 26 March 1929. See also CO 323/1046/4, Appointment of Advisory Council of Agriculture, 1929, Opening Statement by Chairman at First Meeting.

72. FCOL CAC, Minutes of the 2nd meeting of CAC, 19 June 1929; CO 323/1048/8, Colonial Agricultural Service, Proposal for a Unified Service, 1929. The subcommittee included Stockdale, A. W. Hill, R. Vernon, Professor John B. Farmer, Ralph Furse, and Creasy.

73. A Personnel Division was set up as part of the General Department in October 1930 to oversee the unification, with Ralph Furse and his staff incorporated as an Appointments Department, assisted by an independent Colonial Services Appointments Board. See CO 885/33/426, Miscellaneous, Report of the Administrative Service Sub-Committee of the Colonial Service Committee, 1931.

74. It amounted to a single method of entry and the acceptance, although not the enforcement, of the principle that members would be liable to compulsory transfer by the secretary of state, with or without promotion, to any listed posts. See Charles Jeffries, *The Colonial Empire and Its Civil Service* (Cambridge: Cambridge University Press, 1938), 63–89, 166–70; Anthony Kirk-Greene, *On Crown Service: A History of HM Colonial and Overseas Civil Services, 1837–1997* (London: I. B. Tauris, 1999), 30–37.

75. For the demise of the EMB see Beloff, *Imperial Sunset*, vol. 2, 189–93; Lee, "Dissolution of the Empire Marketing Board," 49–58.

76. FCOL CAC, Minutes of the 2nd Meeting of CAC, 19 June 1929; 8th Meeting, 16 March 1931; 9th Meeting, 2 July 1931. Unfortunately, the soil scheme had to be scrapped after the West African governments failed to contribute toward the salaries and expenses of the team of four soil experts. Once again the financial difficulties of the Depression were cited as the reason. See FCOL CAC, Minutes of the 12th Meeting of CAC, 20 April 1932.

77. The best source on the Colonial Development Act of 1929 and its results is Constantine, *Making of British Colonial Development Policy*, 164–226. See also Havinden and Meredith, *Colonialism and Development*, 160–86; D. J. Morgan, *The Official History of Colonial Development*, vol. 1, *The Origins of British Aid Policy, 1924–1945* (London: Macmillan, 1980), 37–42; David Meredith, "The British Government and Colonial Economic Policy, 1919–1939," *Economic History Review* 28, no. 3 (1975): 484–99.

78. The £1 million a year was to be given primarily for defraying the interest on loans raised by colonial governments for a maximum of ten years. It was

103. David Anderson hits it on the head when he remarks, "But it would be wrong to see Frank Stockdale as the orchestrator of a campaign to draw attention to the erosion issue. He did not create the problem, it landed on his desk in the form of reports and memoranda from the various colonies and, as the official with overall responsibility for colonial agricultural policy, he set about trying to make sense of it and devising policies that would tackle it." See Anderson, "Depression, Dust Bowl, Demography and Drought: The Colonial State and Soil Conservation in East Africa during the 1930s," *African Affairs* 83, no. 332 (July 1984): 342.

104. FCOL CAC, Minutes of the 9th, 12th, and 23rd Meetings of CAC, 2 July 1931, 20 April 1932, 9 October 1934.

105. CO 323/1088/7, Report by F. A. Stockdale, Agricultural Adviser to the Secretary of State for the Colonies, on his visit to West Africa, 1929, 16 January 1930 (CAC 19).

106. FCOL CAC, Minutes of the 17th Meeting of CAC, 11 April 1933; 30th Meeting, 21 July 1936.

107. E. E. Cheesman, "The Economic Botany of Cocoa: A Critical Survey of the Literature to the End of 1930," *Tropical Agriculture,* supplement, June 1932; D. B. Murray, "Cocoa Research at ICTA," *Tropical Agriculture* 51, no. 4 (1974): 477–79.

108. Geoffrey Evans, "Research and Training in Tropical Agriculture," *Journal of the Royal Society of Arts,* 87, no. 4499 (1939): 341; G. G. Gianetti, "The Gold Coast Cacao Industry," *Tropical Agriculture* 12, no. 2 (1935): 311–12.

109. CO 323/1410/3, Report of the Agricultural Adviser on His Recent Visit to the West African Dependencies, October 1935–February 1936, 31 March 1936 (CAC 270); FCOL CAC, Minutes of the 30th Meeting of CAC, 21 July 1936.

110. Even after 1937, the Department of Agriculture and the WACRI found themselves hamstrung by shortages of staff and resources, which led to an inadequacy of research and experimental stations and a lack of contact between staff and farmers. See Seth La-Anyane, *Ghana Agriculture: Its Economic Development from Early Times to the Middle of the Twentieth Century* (London: Oxford University Press, 1963), 95–96, 109–10; R. H. Green and S. H. Hymer, "Cocoa in the Gold Coast: A Study in the Relations between African Farmers and Agricultural Experts," *Journal of Economic History* 26, no. 3 (1966): 309, 318; Geoffrey B. Kay, *The Political Economy of Colonialism in Ghana: A Collection of Documents and Statistics, 1900–1960* (Cambridge: Cambridge University Press, 1972), 227–36, 272–76.

111. GB, CO, Report and Proceedings of the Conference of Colonial Directors of Agriculture, July 1931, Colonial no. 67 (London, 1931).

112. FCOL CAC, Minutes of the 25th Meeting of CAC, 12 March 1935; 56th Meeting, 6 July 1943.

113. Leopold Amery, *My Political Life,* 2:341.

114. See Worboys, "Science and British Colonial Imperialism," 43.

115. Self, "Treasury Control and the Empire Marketing Board," 155; Worboys, "Science and British Colonial Imperialism," 244–94.

116. Worboys, "Science and British Colonial Imperialism," 246.
117. Ibid., 221–32.

Chapter 4: The "Human Side" of Development

1. Penelope Hetherington, *British Paternalism and Africa, 1920–1940* (London: Frank Cass, 1978), 45–60; Gilbert Rist, *The History of Development: From Western Origins to Global Faith* (London: Zed Books, 1997), 58–68.

2. Frederick Lugard, *The Dual Mandate in British Tropical Africa* (London: Frank Cass, 1965 [1922]).

3. Ibid., 617.

4. Frederick Lugard, "Education in Tropical Africa," *Edinburgh Review* 242, no. 493 (July 1925): 3.

5. Ibid., 4.

6. J. M. Lee, *Colonial Development and Good Government: A Study of the Ideas Expressed by the British Official Classes in Planning Decolonization, 1939–1964* (Oxford: Clarendon Press, 1967), 1–32; Hetherington, *British Paternalism and Africa,* 49–50.

7. Alice Conklin, "Civilization through Coercion: Human *Mise en Valeur* in the 1920s," in *A Mission to Civilize: The Republican Idea of Empire in France and West Africa, 1895–1930* (Berkeley: University of California Press, 1997), 212–45.

8. Monica M. van Beusekom, "Colonisation Indigène: French Rural Development Ideology at the Office du Niger, 1920–1940," *International Journal of African Historical Studies* 30, no. 2 (1997): 299–323.

9. See CO 885/21/248, Memorandum by Herbert Read, 12 April 1910; CO 323/894, Tropical African Advisory Medical and Sanitary Committee: Minutes Regarding the Extension of the Scope of the Committee so as to Cover All Colonies, 4 August 1922.

10. Stanton was one of Manson's research protégés at the London School of Tropical Medicine, where he worked as a demonstrator before going on to become assistant to Henry Fraser, the director of the Institute of Medical Research at Kuala Lumpur in the Federated Malay States, in 1907. Stanton succeeded Fraser as director of the institute in 1916. See *s.v.* "Sir (Ambrose) Thomas Stanton," *Who Was Who,* vol. 3, *1929–1940,* 2nd ed. (London: Adam and Charles Black, 1967).

11. CO 323/979/11, Proposal for a Colonial Medical Research Committee, 1927. The list of tropical medical science agencies connected with the CO by the late 1920s included medical research institutes in Malaya, Nigeria, the Gold Coast, and East Africa, as well as bacteriological laboratories in Ceylon, Hong Kong, Mauritius, and British Guiana, among others. There were also important private laboratories that worked closely with local administrations, such as the Wellcome Tropical Research Laboratories in Khartoum and the Liverpool School's Alfred Lewis Jones Research Laboratory in Sierra Leone under the direction of Dr. Donald B. Blacklock. In Britain, the list of institutions and agencies included the two Schools of Tropical Medicine in London and Liverpool; the Department

28. For the Pacific, see Alfred W. Crosby, *Ecological Imperialism: The Biological Expansion of Europe, 900–1900* (Cambridge: Cambridge University Press, 1985), 256–57, 267–68; Donald Denoon et al., eds., *The Cambridge History of the Pacific Islanders* (Cambridge: Cambridge University Press, 1997), 243–49; Karl Ittmann, "The Colonial Office and the Population Question in the British Empire, 1918–62," *Journal of Imperial and Commonwealth History* 27, no. 3 (1999): 62. For Africa, see Norman Leys, *Kenya*, 3rd ed. (London: Hogarth, 1926), 298; Joseph H. Oldham, "Population and Health in Africa," *International Review of Missions* 15 (1926): 402–17; CO 885/33/418, Papers Relating to the Health and Progress of Native Population in Certain Parts of the Empire, 1929–31.

29. CO 879/118/1061, Africa: Further Correspondence Relating to Medical and Sanitary Matters in Tropical Africa, 1918–19, no. 138, Acting Governor of Sierra Leone, E. Evelyn, to the Secretary of State, 21 May 1919; CO 885/28/378, Miscellaneous, Part II: Correspondence Relating to Medical and Sanitary Matters, 1926, no. 42, J. C. Maxwell to the Secretary of State, 1 May 1926.

30. See Stephen H. Roberts, *Population Problems of the Pacific* (New York: AMS, 1969 [1927]), 58–85; Megan Vaughan, *Curing Their Ills: Colonial Power and African Illness* (Cambridge: Polity Press, 1991), 143.

31. See Ittmann, "Colonial Office and the Population Question," 62; Manderson, *Sickness and the State*, 201; CO 885/67/472, Miscellaneous, "Certain Aspects of the Welfare of Women and Children in the Colonies," Memorandum by Dr. Mary G. Blacklock, Liverpool School of Tropical Medicine, 19 June 1936; Vaughan, *Curing Their Ills*, 143.

32. See, for example, CO 879/119/1075, African: Medical and Sanitary Matters in Tropical Africa, Further Correspondence, 1919–1920, no. 2, Special Meeting of the AM&SCTA, 15 July 1919.

33. GB, *PP*, Cmd. 2387 (April 1925), "Report of the East Africa Commission," 169–70.

34. See Keith Williams, "'A Way Out of Our Troubles': The Politics of Empire Settlement, 1900–1922," in *Emigrants and Empire: British Settlement in the Dominions between the Wars*, ed. Stephen Constantine (Manchester: Manchester University Press, 1990), 22–44; Dane Kennedy, "Empire Migration in Post-War Reconstruction: The Role of the Overseas Settlement Committee, 1919–1922," *Albion* 20, no. 2 (1988): 403–19.

35. L. S. Amery, *My Political Life*, 2:360–61; Edward W. M. Grigg (Lord Altrincham), *Kenya's Opportunity: Memories, Hopes and Ideas* (London: Faber and Faber, 1955), 71.

36. William Roger Louis, *In the Name of God, Go! Leo Amery and the British Empire in the Age of Churchill* (New York: Norton, 1992), 94–99; Max Beloff, *Imperial Sunset*, vol. 2, *Dream of Commonwealth, 1921–1942* (London: Macmillan, 1989), 215–19; Martin Chanock, *Unconsummated Union: Britain, Rhodesia and South Africa, 1900–45* (Manchester: Manchester University Press, 1977), 190–94.

37. Amery and Ormsby-Gore, "Problems and Development in Africa," 329 (italics added).

38. Significantly, the commission refused to publish the figures on population on the grounds that such estimates were unscientific and unreliable. Indeed, acknowledging that there was a crisis of *depopulation* would have brought into question the commission's favorable assessment of the benefits to the population of further contact and Western influence.

39. "Letter of Dr. Norman Leys, Medical Officer, Nyasaland, to the S/S/C, 7 February 1918," in John W. Cell, *By Kenya Possessed: The Correspondence of Norman Leys and J. H. Oldham, 1918–1926* (Chicago: University of Chicago Press, 1976), 91–136. On Leys's role within anticolonial humanitarian politics in the 1920s and 1930s, see Diana Wylie, "Confrontation over Kenya: The Colonial Office and Its Critics, 1918–1940," *Journal of African History* 18, no. 3 (1977): 428–29; and "Norman Leys and McGregor Ross: A Case Study in the Conscience of African Empire, 1900–39," *Journal of Imperial and Commonwealth History* 5, no. 3 (1977): 301.

40. See Leys, *Kenya,* 298.

41. Wylie, "Confrontation over Kenya," 430–31.

42. See Ian R. G. Spencer, "The First World War and the Origins of the Dual Policy of Development in Kenya, 1914–1922," *World Development* 9, no. 8 (1981): 740–42.

43. J. H. O. to Hon. E. F. L. Wood, 17 May 1921, quoted in Roland Oliver, *The Missionary Factor in East Africa* (London: Longmans, Green, 1952), 256.

44. GB, *PP,* "Memorandum on Indians in Kenya," Cmd. 1922 (July 1923).

45. See C. J. D. Duder, "The Settler Response to the Indian Crisis of 1923 in Kenya: Brigadier General Philip Wheatley and 'Direct Action,'" *Journal of Imperial and Commonwealth History* 17, no. 3 (1989): 349–73; Christopher P. Youe, "The Threat of Settler Rebellion and the Imperial Predicament: The Denial of Indian Rights in Kenya, 1923," *Canadian Journal of History* 12, no. 3 (1978): 347–60.

46. William Ormsby-Gore, speech of 10 April 1923, in *Parliamentary Debates,* Commons, 5th ser., vol. 162 (1923), col. 1139.

47. Joseph H. Oldham, "Christian Missions and the Education of the Negro," *International Review of Missions* 7 (1918): 242–47; Kenneth James King, "Africa and the Southern States of the USA: Notes on J. H. Oldham and American Negro Education for Africans," *Journal of African History* 10, no. 4 (1969): 663; Edward H. Berman, "American Influence on African Education: The Role of the Phelps-Stokes Fund's Education Commissions," *Comparative Education Review* 15, no. 2 (1971): 132–45; Charles W. Weber, "The Influence of the Hampton-Tuskegee Model on the Educational Policy of the Permanent Mandates Commission and British Colonial Policy," *Africana Journal* 16 (1994): 73–76.

48. Kenneth James King, *Pan-Africanism and Education: A Study of Race Philanthropy and Education in the Southern States of America and East Africa* (Oxford: Clarendon Press, 1971), 101.

72. Sybille Kuster notes that very similar reasoning was used in Southern Rhodesia in the 1920s by H. S. Keigwin, the director of native development. See Kuster, *Neither Cultural Imperialism nor Precious Gift of Civilization: African Education in Colonial Zimbabwe, 1890–1962* (Munster: Lit Verlag, 1994), 106–7.

73. Constantine, *Making of British Colonial Development Policy,* 157–58.

74. CO 885/31/397, Miscellaneous, Minutes of the 18th Meeting of the Advisory Committee on Education in the Colonies (ACEC), 25 September 1930.

75. IMC/CBMS Archives, Box 93, Oldham to the Archbishop of Canterbury, 21 May 1925 and 27 May 1925.

76. See Deborah Lavin, "Margery Perham's Initiation into African Affairs," in *Margery Perham and British Rule in Africa,* ed. Alison Smith and Mary Bull (London: Frank Cass, 1991), 52.

77. BLCAS Mss. Lugard 13/1, Letter to Lugard from Oldham, Nairobi, 5 April 1926.

78. Moreover, Oldham came to have second thoughts about Grigg and Kenya, moving from being a warm supporter of the governor to one of the main opponents of his plans for closer union in East Africa. See BLCAS Mss. Lugard 13/1, Sir Edward Grigg, Governor of Kenya, to Oldham, 12 June 1926; Cell, *By Kenya Possessed,* 81–82.

79. See Jan Christian Smuts, *Africa and Some World Problems* (Oxford: Clarendon Press, 1930).

80. Raymond L. Buell, *The Native Problem in Africa,* 2 vols. (New York: Macmillan, 1928).

81. John Flint, "Macmillan as a Critic of Empire: The Impact of an Historian on Colonial Policy," in *Africa and Empire: W. M. Macmillan, Historian and Social Critic,* ed. Hugh Macmillan and Shula Marks (London: Gower, 1989), 220–21.

82. Cell, *By Kenya Possessed,* 82.

83. Joseph H. Oldham, *White and Black in Africa: A Critical Examination of the Rhodes Lectures of General Smuts* (London: Longmans, Green, 1930).

84. Ibid., 44.

85. Lavin, "Margery Perham's Initiation into African Affairs," 52.

86. Oldham, *White and Black in Africa,* 4.

87. Seppo Sivonen, *White Collar or Hoe Handle: African Education under British Colonial Policy, 1920–1945* (Helsinki: Suomen Historiallinen Seura, 1995), 178–88.

88. See IMC/CBMS Archives, Africa: General, Box 204, no. 44–47; Frederick D. Lugard, "The International Institute of African Languages and Cultures," *Africa* 1, no. 1 (January 1928): 3–12; Kenneth Robinson, "Experts, Colonialists, and Africanists, 1895–1960," in *Experts in Africa: Proceedings of a Colloquium at the University of Aberdeen,* ed. J. C. Stone (Aberdeen: Aberdeen University African Studies Group, 1980), 59–62.

89. CO 533/618, E. Africa: International Institute for Study of African Languages and Cultures, 1926, Minutes of Meeting at SOAS, University of London, 21 and 22 September 1925.

90. IMC/CBMS Archives, Box 204, no. 44, "Memorandum on the Place of the Vernacular in African Education and on the Establishment of a Bureau of African Languages," 31 December 1924.

91. CO 533/618, Letter from Lugard, Chairman of the Executive Council of the IIAL&C, to all African Colonies and Protectorates, 25 January 1927; Memorandum by Prof. D. Westermann, IIAL&C.

92. CO 822/35/7, IIAL&C: Programme of Civil Research 1931, IIAL&C, "A Five-Year Plan of Research."

93. Ibid.

94. Henrika Kuklick, *The Savage Within: The Social History of British Anthropology, 1885–1945* (Cambridge: Cambridge University Press, 1991), 208.

95. Among the participants were E. E. Evans-Pritchard, Audrey Richards, Hilda Beemer (later Kuper), Isaac Schapera, Phyllis Kaberry, Meyer Fortes, S. F. Nadel, Max Gluckman, Godrey Wilson, and Monica Hunter (later Wilson). See Sally Falk Moore, "Changing Perspectives on a Changing Africa: The Work of Anthropology," in *Africa and the Disciplines: The Contributions of Research in Africa to the Social Sciences and Humanities,* ed. Robert Bates (Chicago: University of Chicago Press, 1993), 10n8.

96. Kuklick, *Savage Within,* 210; Bronislaw Malinowski, "Practical Anthropology," *Africa* 2, no. 1 (1929): 22–38; and "The Rationalization of Anthropology and Administration," *Africa* 3, no. 4 (October 1930): 405–29.

97. Sivonen, *White Collar or Hoe Handle,* 183.

98. Investigators sent out by the IIALC in the 1930s who also studied under Malinowski at the LSE included Audrey Richards and Margaret Read in Northern Rhodesia; Mr. and Mrs. Gordon Brown in Tanganyika; S. F. Nadel to study the Nupe in Nigeria; Meyer Fortes in the Gold Coast; Godrey Wilson and Monica Hunter (later Wilson) in Tanganyika and Nyasaland; Kalervo Oberg to study the Banyankole in Uganda; and Gunter Wagner among the Kavirondo in Kenya. See CO 822/35/7, Africa, IIAL&C, Programme of Civil Research, 1931, Minute by Tom Tomlinson to Cecil Bottomley, 3 October 1931; CO 847/2/8, Africa, IIAL&C, Programme of Civil Research, 1932, Minute by Tomlinson to Fiddian, 9 March 1933; CO 847/3/7, Africa, IIAL&C; Programme of Civil Research, 1934, Letter from Oldham to Bottomley, 4 June 1934, Minute by S. Peel, 30 January 1934.

99. CO 847/3/7, Africa, IAL&C, Programme of Civil Research, 1934, Minute by S. Peel, 30 January 1934.

100. The most prominent London lobby groups included the Anti-Slavery and Aborigines Protection Society, the International Institute of African Languages and Cultures, Chatham House, the Royal African Society, the Conference of British Missionary Societies at Edinburgh House, the Imperial Institute's Africa Circle, the Labour Party's Advisory Committee on Imperial Questions, the London Group on African Affairs, the Friends of Africa, the League for Coloured Peoples, and the International African Service Bureau. See Lavin, "Margery Perham's Initiation into African Affairs," 45–61. See also Mona Macmillan, *Champion*

Research," *Tropical Agricultural Association Newsletter* 11, no. 2 (June 1991): 2–7; William Kelleher Storey, *Science and Power in Colonial Mauritius* (Rochester: University of Rochester Press, 1997), 108–23; A. Fiona D. MacKenzie, "The Construction of Colonial Agricultural Knowledge, Kenya, 1914–1952," in *Land, Ecology and Resistance in Kenya, 1880–1952* (Portsmouth, N.H.: Heinemann, 1998), 98–124.

14. R. H. Green and S. H. Hymer, "Cocoa in the Gold Coast: A Study in the Relations between African Farmers and Agricultural Experts," *Journal of Economic History* 26, no. 3 (1966): 299–319; MacKenzie, "Construction of Colonial Agricultural Knowledge, Kenya, 1914–1952," in *Land, Ecology and Resistance in Kenya.*

15. For the importance of studying local conditions and conducting field experiments and trials, see BLCAS 100.47 s. 1, British Empire Command Papers, 1920–21, Cmd. 720 (1920), "Report of the Committee on the Staffing of the Agricultural Departments in the Colonies, June 1920; BLCAS Mss. W. Ind. s. 28, Duncan Stevenson, "The Value of Forestry," Lecture delivered to the Belize Literary and Debating Club, 6 February 1925; O. T. Faulkner and J. R. Mackie, *West African Agriculture* (Cambridge: Cambridge University Press, 1933).

16. Alan Pim, *Colonial Agricultural Production: The Contribution Made by Native Peasants and by Foreign Enterprise* (London: Oxford University Press, 1946), 3; G. B. Masefield, *A Short History of Agriculture in the British Colonies* (Oxford: Clarendon Press, 1950), 70–71, 86–87; Anne Phillips, *The Enigma of Colonialism: British Policy in West Africa* (London: James Currey, 1989).

17. See BLCAS Mss. Afr. s. 823, James Mackie Papers (1), Papers on Nigerian Agriculture, 1939–45; Faulkner and Mackie, *West African Agriculture.*

18. For the early mixed farming trials and extension work, see BLCAS Mss. Afr. s. 1739, "Mixed Farming in Northern Nigeria," Report by J. G. M. King, 1939; J. G. M. King, "Mixed Farming in Northern Nigeria, Part I: Origins and Present Conditions," *Empire Journal of Experimental Agriculture* 7, no. 27 (1939): 271–85; BLCAS Mss. Brit. Emp. s. 476, Oxford Development Records Project—Food and Cash Crops Collection, Box 1 (2), Anthony Blair Rains; Masefield, *History of the Colonial Agricultural Service,* 76–89. In addition, the research program concentrated on the development of improved varieties of yams, maize, cassava, and other food crops, an extensive series of field trials on the use of green manures, mixed cropping, *boma* or farmyard manures and fertilizers, careful recording of palm oil yields, and the introduction of different breeds of West African cattle that were more tsetse-resistant.

19. G. D. H. Bell, "Frank Leonard Engledow," *Biographical Memoirs of Fellows of the Royal Society* 32 (December 1986): 197–98.

20. Palladino has shown how plant breeding and crop improvement under John Percival at University College, Reading, very much followed this "alternative" approach, arguing that the Mendelian concept of genetically uniform "pure lines" was not applicable to expressions of "yield" or "strength," characteristics which depended on complex physiological and environmental interactions and

not just heritability. See Paolo Palladino, "Between Craft and Science: Plant Breeding, Mendelian Genetics and British Universities, 1900–1920," *Technology and Culture* 34, no. 2 (1993): 312–18. Engledow's own skepticism, though more muted, is evident in his 1938 contribution to a British Association for the Advancement of Science symposium on practical problems of crop production: "Plant-breeding has gained from genetics new confidence, analytical grasp, and wider horizon; but it remains, despite the common illusion, a blend of art and empiricism . . . In general . . . genetics has not fundamentally improved breeding-methods." See F. L. Engledow, "The Place of Plant Physiology and of Plant-Breeding in the Advancement of British Agriculture," *Empire Journal of Experimental Agriculture* 7 (1939): 146.

21. Masefield, *History of the Colonial Agricultural Service,* 96; Bell, "Frederick Leonard Engledow," 205–6; Simmonds, "Earlier British Contribution to Tropical Agricultural Research," 5.

22. Lynn, who had studied at Cambridge under Engledow, was at the time the senior lecturer in extension methods at the ICTA. See CO 996/8 CAC Papers 1949, CAC 822, C. W. Lynn, *Agricultural Extension and Advisory Work with Special Reference to the Colonies,* Col. no. 241 (London 1949); BLCAS Mss. Brit. Emp. s. 476, Box 3 (27), Charles William Lynn, 1908–.

23. Suzanne Moon, "Empirical Knowledge, Scientific Authority, and Native Development: The Controversy over Sugar/Rice Ecology in the Netherlands East Indies, 1905–1914," *Environment and History* 10, no. 1 (2004): 74–75.

24. Monica van Beusekom, "Disjunctures in Theory and Practice: Making Sense of Change in Agricultural Development at the Office du Niger, 1920–1960," *Journal of African History* 41, no. 1 (2000): 93–96.

25. Helen Tilley, "African Environments and Environmental Sciences: The African Research Survey, Ecological Paradigms and British Colonial Development, 1920–1940," in *Social History and African Environments,* ed. William Beinart and JoAnn McGregor (Oxford: James Currey, 2003), 116.

26. Richard Grove, *Ecology, Climate and Empire,* 5–36.

27. See Fairhead and Leach, "Desiccation and Domination," 40–41; Millington, "Environmental Degradation, Soil Conservation and Agricultural Policies in Sierra Leone," 229–35; S. Ravi Rajan, "Foresters and the Politics of Colonial Agroecology: The Case of Shifting Cultivation and Soil Erosion, 1920–1950," *Studies in History,* n.s., 14, no. 2 (1998): 217–35.

28. See, for example, C. W. Wardlaw, "Virgin Soil Deterioration: The Deterioration of Virgin Soils in the Caribbean Banana Lands," *Tropical Agriculture* 6, no. 9 (1929): 243–48.

29. In addition to Tansley, Chipp, and Elton, the "school" included such prominent figures as Robert Scott Troup and Ray Bourne at the Imperial Forestry Institute and Julian Huxley, Edward Max Nicholson, and Alexander Morris Carr-Saunders in Oxford's Department of Zoology. See Peder Anker, *Imperial Ecology: Environmental Order in the British Empire, 1895–1945* (Cambridge, Mass.: Harvard

University Press, 2001). For the widening influence of "tropical ecology" on colonial scientific discourse generally, see David Anderson, "Organizing Ideas," 10, and *Eroding the Commons: The Politics of Ecology in Baringo, Kenya, 1890–1963* (Athens: Ohio University Press, 2002).

30. Tansley and Chipp's adaptation of such forest survey methods as the use of traverse lines to plant ecology seems to have been particularly important. See A. G. Tansley and T. F. Chipp, eds., *Aims and Methods in the Study of Vegetation* (London: British Empire Vegetation Committee, 1926); Anker, *Imperial Ecology,* 35–40.

31. William M. Adams, *Green Development: Environment and Sustainability in the Third World* (London: Routledge, 1990), 26–27.

32. I credit Helen Tilley for drawing my attention to this crucial point.

33. What was really needed, Stevenson suggested, was a proper topographical survey to carefully study the actual possibilities of production. Based on his own research, he believed such a survey would reveal the much more sobering truth that the colony was living on its "natural" capital, treating its forests as a mine, with the result that tree populations were steadily declining. See BLCAS Mss. W. Ind. s. 28, Duncan Stevenson, "The Value of Forestry."

34. Clement Gillman, "Southwest Tanganyika Territory," *Geographical Journal* 69, no. 2 (1927): 97–126.

35. BLCAS Mss. Brit. Emp. s. 457, Geoffrey Milne, 1898–1942, Papers Relating to Soil Science in East Africa, West Indies, and America, Box 2, "Biographical Note by Milne's Wife."

36. BLCAS Mss. Brit. Emp. s. 457, Geoffrey Milne, Box 16, Soil Papers by and Correspondence with G. Milne, 1932, 1940–41, "Soil Survey in East Africa: Some Needed Development" (1940).

37. Geoffrey Milne, "Soil Reconnaissance Journey through Parts of Tanganyika Territory, December 1935 to February 1936," *Journal of Ecology* 35, no. 1–2 (1947): 194.

38. See Geoffrey Milne, "Some Aspects of Modern Practice in Soil Survey," *East African Agricultural Journal* 5, no. 4 (1940): 1–7.

39. See Geoffrey Milne, "Soil Conservation—The Research Side," *East African Agricultural Journal* 6, no. 1 (1940): 26–31; Milne, "Some Aspects of Modern Practice in Soil Survey"; Milne, "Soil Reconnaissance Journey through Parts of Tanganyika Territory."

40. Geoffrey Milne, "Soil and Vegetation," *East African Agricultural Journal* 5, no. 4 (1940): 294–98; Milne, "Soil and Vegetation," *Tropical Agriculture* 17, no. 5 (1940): 95–97.

41. Frederick Hardy, "Soil Deterioration in the Tanganyikan Uplands: A Lesson in Tropical Agricultural Exploitation," *Tropical Agriculture* 15, no. 4 (1938): 79–81.

42. Milne, "Soil Reconnaissance Journey through Parts of Tanganyika Territory," 250–51. Milne was not alone in his condemnation. Gillman called it the "de-

stroy your Land Policy," and, according to Iliffe, some agricultural staff left their provinces, viewing it as a blatant policy of exploitation. See John Iliffe, *A Modern History of Tanganyika* (Cambridge: Cambridge University Press, 1979), 349.

43. BLCAS Mss. Brit. Emp. s. 457, Geoffrey Milne, Box 28, Milne, "A Defence of Shamba Burning: Wood-Ash as Manure," *East Africa* 29 (December 1932).

44. In fact, Trapnell and Milne corresponded and exchanged work with each other in the 1930s. See BLCAS Mss. Brit. Emp. s. 457, Geoffrey Milne, Box 28, Letter to Milne, Amani, from Trapnell, Department of Agriculture, Northern Rhodesia, 3/8/37. For Trapnell and Allan's approach, see C. G. Trapnell and J. N. Clothier, *The Soils, Vegetation and Agricultural Systems of North-Western Rhodesia* (Lusaka: Government Printer, 1937); C. G. Trapnell, *The Soils, Vegetation and Agricultural Systems of North-Eastern Rhodesia* (Lusaka: Government Printer, 1943); William Allan, *Studies in African Land Usage in Northern Rhodesia,* Rhodes-Livingstone Papers 15 (London: Oxford University Press, 1949).

45. From there, measurements of cultivable and uncultivable area could be made, and from them the percentage of cultivable land could be estimated and compared with the various vegetation-soil types to determine the mean cultivable percentage for each type. Next, it was necessary to determine how each vegetation-soil type was actually used under the traditional system of agriculture in order to estimate its method of land usage or "cultivation factor." Then it was a relatively simple matter to calculate the land carrying capacity for any single soil type, which was the area required per head of population under a given system of land usage. See Allan, *Studies in African Land Usage in Northern Rhodesia,* 1–17.

46. BLCAS Mss. Brit. Emp. s. 476, Box 3 (21), C. E. Johnson, 1911–.

47. See Richard Allen, "The Slender, Sweet Thread: Sugar, Capital and Dependency in Mauritius, 1860–1936," *Journal of Imperial and Commonwealth History* 16, no. 2 (1988): 192–94; Michael Havinden and David Meredith, *Colonialism and Development: Britain and Its Tropical Colonies, 1850–1960* (London: Routledge, 1993), 174–83; A. G. Hopkins, *An Economic History of West Africa* (London: Longman, 1973), 254–67; John Forbes Munro, *Africa and the International Economy, 1800–1960* (London: J. M. Dent, 1976), 150–70.

48. See, for example, David Anderson and David Throup, "Africans and Agricultural Production in Colonial Kenya: The Myth of the War as a Watershed," *Journal of African History* 26, no. 4 (1985): 329–30.

49. C. V. Brayne, "The Problem of Peasant Agriculture in Ceylon," *Ceylon Economic Journal* 6 (December 1934): 34–46; B. H. Farmer, *Pioneer Peasant Colonization in Ceylon: A Study in Asian Agrarian Problems* (London: Oxford University Press, 1957); Swarna Jayaweera, "Land Policy and Peasant Colonization, 1914–1948," in *History of Ceylon,* vol. 3, *From the Beginning of the 19th Century to 1948,* ed. K. M. De Silva (Peradeniya: University of Ceylon Press Board, 1973), 446–60.

50. C. Y. Shephard, "Co-operation and Peasant Development in the Tropics," *Tropical Agriculture* 6, no. 2 (1929): 39–42; C. C. Parisinos, C. Y. Shephard, and A. L. Jolly, "Peasant Agriculture: An Economic Survey of the Las Lomas District,

Trinidad," *Tropical Agriculture* 21, no. 5 (1944): 84–98; C. Y. Shephard, "Peasant Agriculture in the Leeward and Windward Islands," *Tropical Agriculture* 24, no. 4–6 (1947): 61–71; GB, *PP,* Cmd. 3517 (1930), "Report of the West Indian Sugar Commission."

51. For Tanganyika, see John Iliffe, *A Modern History of Tanganyika* (Cambridge: Cambridge University Press, 1979), 349. For Kenya, see John Lonsdale, "The Depression and the Second World War in the Transformation of Kenya," in *Africa and the Second World War,* ed. David Killingray and Richard Rathbone (London: Macmillan, 1986), 103–11.

52. The increasing concern about soil erosion in the late 1920s and 1930s and the emergence of a new colonial conservationist discourse have been extensively documented in the secondary literature. Among the most important historical studies on which this paragraph is based, see David Anderson, "Depression, Dust Bowl, Demography and Drought: The Colonial State and Soil Conservation in East Africa during the 1930s," *African Affairs* 83, no. 332 (July 1984): 321–43; Anderson, *Eroding the Commons,* esp. 157–89; William Beinart, "Soil Erosion, Conservationism and Ideas about Development: A Southern African Exploration, 1900–1960," *Journal of Southern African Studies* 11, no. 1 (October 1984): 52–83; Lonsdale, "Depression and the Second World War in the Transformation of Kenya," 114; Mackenzie, *Land, Ecology and Resistance in Kenya,* 89–97, 117–24; John McCracken, "Experts and Expertise in Colonial Malawi," *African Affairs* 81, no. 322 (January 1982): 11–12; Kate Showers, "Soil Erosion in the Kingdom of Lesotho: Origins and Colonial Response, 1830s–1950s," *Journal of Southern African Studies* 15, no. 2 (1989): 263–86; Michael Stocking, "Soil Conservation Policy in Colonial Africa," *Agricultural History* 59, no. 2 (1985): 148–61.

53. See "Interim Report of the South African Drought Investigation Committee" (Cape Town, 1922), reprinted in Michael H. Glantz, ed., *Desertification: Environmental Degradation in and around Arid Lands* (Boulder, Colo.: Westview, 1977), 233–74.

54. W. M. Sykes, *The Sukumaland Development Scheme, 1947–57,* Oxford Development Records Project, Report 7 (Oxford: Rhodes House Library, 1984), 10–25.

55. Iliffe, *Modern History of Tanganyika,* 348.

56. H. H. Storey, *Basic Research in Agriculture: A Brief History of Research at Amani, 1928–1947* (Nairobi: East African Agriculture and Forestry Research Organization, 1950).

57. Fairhead and Leach, "Desiccation and Domination," 38; Grove, *Ecology, Climate and Empire,* 33–34.

58. See GB, *PP,* Cmd. 2744 (1926), "Report by the Honourable W. G. A. Ormsby-Gore on his Visit to West Africa during the Year 1926," 139–53; Fairhead and Leach, "Desiccation and Domination," 41.

59. A. S. Thomas, "The Dry Season in the Gold Coast and Its Relations to Cultivation of Cacao," *Journal of Ecology* 20, no. 2 (1932): 263–69.

60. BLCAS Mss. Brit. Emp. s. 333, Herbert W. Moor, Conservator of Forests, "Deforestation in the Bissa Cocoa Area, Gold Coast," *Malayan Forester,* July 1936.

61. CO 323/1088/7, Report by F. A. Stockdale, Agricultural Adviser to the Secretary of State for the Colonies, on His Visit to West Africa, 1929, 16 January 1930 (CAC 19).

62. E. E. Cheesman, "The Economic Botany of Cocoa: A Critical Survey of the Literature to the End of 1930," *Tropical Agriculture,* supplement, June 1932; D. B. Murray, "Cocoa Research at ICTA," *Tropical Agriculture* 51, no. 4 (1974): 477–79.

63. See "Swollen Shoot of Cacao," *Tropical Agriculture* 15, no. 5 (1938): 110–11; BLCAS Mss. Brit. Emp. s. 333, Herbert W. Moor, Conservator of Forests, Letter to J. N. Oliphant from Moor, Forestry Department, Accra, August 1937.

64. Frederick Hardy and N. Ahmad, "Soil Science at ICTA/UWI, 1922–1972," *Tropical Agriculture* 51, no. 4 (October 1974): 476.

65. At the Imperial Agricultural Research Conference held in Australia in 1932, for example, soil conservation was not even on the agenda. See FCOL CAC, Minutes of the 5th Meeting of CAC, 27 May 1930.

66. See Royal African Society, "Land Usage and Soil Erosion in Africa: A Report of the Speeches at the Dinner of the Royal African Society, 1 December 1937," *Journal of the Royal African Society,* supplement, vol. 37, no. 146 (January 1938): 10.

67. CO 847/10/9, Africa, Soil Erosion, 1937, "Resolution Adopted by the Council of the Royal African Society," 20 October 1937.

68. See Anderson, "Depression, Dust Bowl, Demography, and Drought," 331–33; Thomas Dunlap, *Nature and the English Diaspora: Environment and History in the United States, Canada, Australia and New Zealand* (Cambridge: Cambridge University Press, 1999), 179–89; Donald Worster, *Dust Bowl: The Southern Plains in the 1930s* (Oxford: Oxford University Press, 1979).

69. See E. P. Stebbing, "The Encroaching Sahara: The Threat to the West African Colonies," *Geographical Journal* 85, no. 6 (June 1935): 506–24; Stebbing, "The Threat of the Sahara," *Journal of the Royal African Society,* extra supplement, 25 May 1937, 3–35.

70. Stebbing, "Encroaching Sahara," 510.

71. See Brynmor Jones, "Desiccation and the West African Colonies," *Geographical Journal* 91, no. 5 (May 1938): 401–23.

72. This seems to have been the view of Tansley, Throup, and Stockdale as well, who argued that much of the vegetation of West Africa was undergoing edaphic, not climatic, climaxes. The same carefully observed facts, Stamp noted, were thus capable of more than one interpretation, which pointed to the lesson that observations needed to be carried out diachronically, not just synchronically. See Dudley Stamp, "Southern Margin of the Sahara: Comments on Some Recent Studies on the Question of Desiccation in West Africa," *Geographical Review* 30, no. 2 (1940): 297–300.

73. See E. P. Stebbing, "Africa and Its Intermittent Rainfall: The Role of the Savannah," *Journal of the Royal African Society,* supplement, vol. 37, no. 149 (October 1938): 3–32.

74. G. V. Jacks and R. O. Whyte, *The Rape of the Earth: A World Survey of Soil Erosion* (London: Faber and Faber, 1939), 250.

75. Jeremy Swift, "Desertification: Narratives, Winners and Losers," in *The Lie of the Land: Challenging Received Wisdom on the African Environment,* ed. Melissa Leach and Robin Mearns (Portsmouth, N.H.: Heinemann, 1996), 73–90.

76. R. MacLagan Gorrie, *The Use and Misuse of the Land,* Oxford Forestry Memoirs 19 (Oxford, 1935); Gorrie, "The Problem of Soil Erosion in the British Empire with Special Reference to India," *Journal of the Royal Society of Arts* 86, no. 4471 (1938): 901–29; Hardy, "Soil Deterioration in the Tanganyikan Uplands"; Elspeth Huxley, "The Menace of Soil Erosion," *Journal of the Royal African Society* 36, no. 144 (1937): 357–70; G. V. Jacks and R. O. Whyte, *Erosion and Soil Conservation,* Technical Communication 36 (Harpenden: Imperial Bureau of Soil Science, 1938); E. P. Stebbing, "The Man-Made Desert in Africa: Erosion and Drought," *Journal of the Royal African Society,* supplement, vol. 37, no. 146 (January 1938): 3–40; Frank Stockdale, "Soil Erosion in the Colonial Empire," *Empire Journal of Experimental Agriculture* 5, no. 20 (October 1937): 281–97; H. A. Tempany, G. M. Roddan, and L. Lord, "Soil Erosion and Soil Conservation in the Colonial Empire," *Empire Journal of Experimental Agriculture* 12, no. 47 (1944): 121–53.

77. Top priority was given to Milne's reconnaissance surveys of East African soil types, with the goal of providing basic data to frame a rational policy of land utilization and control of soil erosion in the region. Stockdale fully supported Milne's request for further funding of the station's soil chemistry work, so as to extend the range of soil reconnaissance and surveying beyond Tanganyika. See CO 822/77/11, Visit by Sir Frank Stockdale to East Africa, 1937; CO 822/79/17, East African Agricultural Research Station, Sir Frank Stockdale's Report: Programme of Work, 1937; FCOL CAC, Minutes of the 34th Meeting of CAC, 29 June 1937.

78. For a sense of the paperwork generated over the erosion question in the late 1930s, see CO 323/1620/8, Soil Erosion, Annual Reports, 1937; CO 323/1529/7, Soil Erosion in USA, 1937; CO 533/496/1, Kenya Circular no. 23, "Soil Conservation," from H. G. Pilling, Acting Colonial Secretary, Nairobi, 20 December 1937, forwarding Colin Maher's reports; CO 137/836/8, Jamaica, Department of Science and Agriculture, H. H. Croucher and C. Swabey, *Soil Erosion and Conservation in Jamaica, 1937,* Bulletin no. 17 (Jamaica, 1938); CO 323/1621/4, Conference of Colonial Directors of Agriculture, 1938, Session on Soil Erosion, July 26 1938; CO 852/249/15, Soil Erosion, Annual Reports, 1938; CO 852/249/16, Soil Erosion, Annual Reports, 1939; CO 852/249/17, "Soil Conservation in the Tropics," by Frank Stockdale, prepared for the Netherlands Conference on Tropical Agriculture, June 1939; CO 733/431/17, Soil Erosion in Palestine, enclosure, "Soil Erosion and Conservation Measures, Palestine 1939," G. N. Sale, Conservator of Forests, 23 June 1940.

79. FCOL CAC Minutes of the 36th Meeting of CAC, 7 December 1937; CO 323/1529/7, Draft of Circular Despatch to all Colonies and Protectorates from Ormsby-Gore, 27 January 1938.

80. Libby Robin, "Ecology: A Science of Empire?" in *Ecology and Empire: Environmental History of Settler Societies,* ed. Tom Griffiths and Libby Robin (Edinburgh: Keele University Press, 1997), 70.

81. As early as 1885, the government passed a Forest Ordinance prohibiting the alienation of Crown lands above five thousand feet. See Webb, *Tropical Pioneers,* 139–43.

82. On the importance of soil conservation work in Ceylon, see T. Eden, "The Menace of Soil Erosion," *Tropical Agriculture* 11, no. 4 (1934): 77–79; Jacks and Whyte, *Erosion and Soil Conservation,* 41–45; Frank Stockdale, "Soil Erosion in the Colonial Empire," 292–93; CO 852/249/17, "Soil Conservation in the Tropics"; R. C. Wood, "Agriculture in Ceylon," *Tropical Agriculture* 6, no. 3 (1929): 83.

83. "Soil Erosion," *Tropical Agriculture* 7, no. 1 (1930): 19–20.

84. CO 323/1529/7, Soil Erosion in USA, 1937, "Notes on Soil Conservation Work in America" (CAC 367), by F. A. Stockdale, 17 November 1937; CO 852/249/17, "Soil Conservation in the Tropics." On the Soil Conservation Service, see Sandra S. Batie, "Soil Conservation in the 1980s: A Historical Perspective," *Agricultural History* 59, no. 2 (April 1985): 109–13.

85. The influence of American ideas in Southern and East Africa, especially in settler colonies, has been well documented in the literature. Numerous agricultural and soil erosion officers from the region corresponded with members of the SCS, and even traveled to the United States to observe the operations being carried out under Bennett and the SCS. See McCracken, "Experts and Expertise in Colonial Malawi," 112; Beinart, "Soil Erosion, Conservation and Ideas about Development," 67–69; Anderson, "Depression, Dust Bowl, Demography, and Drought," 333.

86. CO 852/27/3, "Forestry in Relation to Agriculture Under Tropical Conditions," paper presented by Frank Stockdale, British Empire Forestry Conference, South Africa, 1935.

87. FCOL CAC Minutes of the 36th meeting of the CAC, 7 December 1937

88. See CO 323/1529/7, Note by Calder to Stockdale, 30 December 1937; Minute by Stockdale, 30 December 1937; Note from Schuckburgh to Cosmo Parkinson, 4 January 1938.

89. On the Colonial Office, see Ittmann, "Colonial Office and the Population Question," 62–63. For Africa, see Anderson, "Depression, Dust Bowl, Demography and Drought," 328–31; Michael P. Cowen and Robert W. Shenton, "The Origin and Course of Fabian Colonialism in Africa," *Journal of Historical Sociology* 4, no. 2 (1991): 143–74; Megan Vaughan, *The Story of an African Famine: Gender and Famine in Twentieth-Century Malawi* (Cambridge: Cambridge University Press, 1987), 61. On India and Ceylon, see Farmer, *Pioneer Peasant Colonization in Ceylon,* 141–60; John Megaw, "Pressure of Population in India," *Journal of the Royal Society of Arts* 86, no. 4491 (1938): 134–49; K. M. De Silva, *A History of Sri Lanka* (Berkeley: University of California Press, 1981), 407–12. On the West Indies, see R. R. Kuczynski, *West Indian and American Territories,* vol. 3, *Demographic Survey of the British Colonial Empire* (London: Oxford University Press, 1953), 3–25; Howard Johnson, "The British Caribbean from Demobilization to Constitutional Decolonization," in *The Twentieth Century,* ed. Judith Brown and William

Roger Louis, vol. 4, *The Oxford History of the British Empire,* ed. William Roger Louis (Oxford: Oxford University Press, 1999), 605–6.

90. Megan Vaughan, *Curing Their Ills: Colonial Power and African Illness* (Cambridge: Polity Press, 1991), 143.

91. CO 885/33/418, Papers Relating to the Health and Progress of Native Population in Certain Parts of the Empire, 1929–31, "Despatch from S/S/C to Officers Administering the Governments of Kenya, Uganda Protectorate, Tanganyika Territory, Northern Rhodesia and Zanzibar, 8 March 1930."

92. See, for example, Report to the Committee of the Belgian Colonial Congress, *La Question Sociale au Congo Belge* (1924), 68, as quoted in Joseph H. Oldham, "Population and Health in Africa," *International Review of Missions* 15 (1926): 411. For the Pacific region, see Stephen H. Roberts, *Population Problems of the Pacific* (New York: Routledge, 1969 [1927]), 58–85; T. H. Harrison, "The New Hebrides People and Culture," *Geographical Journal* 88, no. 4 (October 1936): 332–41.

93. See, for example, Royal Institute of International Affairs, *The Colonial Problem* (London: Oxford University Press, 1937), 127–39.

94. On demographic trends in India between 1870 and 1940, see B. R. Tomlinson, *The Economy of Modern India, 1860–1970* (Cambridge: Cambridge University Press, 1993); Leela Visaria and Pravin Visaria, "Population (1757–1947)," in *The Cambridge Economic History of India,* vol. 2, *c. 1751–c. 1970,* ed. Dharma Kumar and Meghnad Desai (Cambridge: Cambridge University Press, 1983), 463–532.

95. See Megaw, "Pressure of Population in India."

96. Ittmann, "Colonial Office and the Population Question," 63.

97. Joseph H. Oldham, *White and Black in Africa: A Critical Examination of the Rhodes Lectures of General Smuts* (London: Longmans, Green, 1930), 46–49.

98. J. Merle Davis, *Modern Industry and the African: An Enquiry into the Effect of the Copper Mines of Central Africa upon Native Society and the Work of the Christian Missions* (London: Frank Cass, 1967 [1933]), vii–xxiv, 376–93.

99. Ibid., 34.

100. Much of this and the following paragraph is based on Frederick Cooper, "Reforming Imperialism, 1935–1940," in *Decolonization and African Society: The Labor Question in French and British Africa* (Cambridge: Cambridge University Press, 1996), 57–109. See also Forbes Munro, *Africa and the International Economy,* 168; Hopkins, *Economic History of West Africa,* 254–67; Iliffe, *Modern History of Tanganyika,* 342–56; Phillips, *Enigma of Colonialism,* 136–55; John Sender and Sheila Smith, *The Development of Capitalism in Africa* (London: Methuen, 1986), 46–52.

101. Gavin Kitching, *Class and Economic Change in Kenya: The Making of an African Petite Bourgeoisie* (New Haven, Conn.: Yale University Press, 1980), 82–94.

102. CO 318/433/1, West India Royal Commission, Minute by Sir John Campbell, 23 May 1938, in A. N. Porter and A. J. Stockwell, *British Imperial Policy and Decolonization, 1938–64,* vol. 1, *1938–51* (London: Macmillan, 1987), 82–85.

103. Major G. St. J. Orde-Browne, *Labour Conditions in Northern Rhodesia,* Colonial 150 (London: Colonial Office, 1938); GB, *PP,* Cmd. 6277, IV (1941), 1, Major G. St. J. Orde-Browne, "Labour Conditions in West Africa."

104. Michael Worboys, "The Discovery of Colonial Malnutrition between the Wars," in *Imperial Medicine and Indigenous Societies,* ed. David Arnold (Manchester: Manchester University Press, 1988), 220.

105. See Ittmann, "Colonial Office and the Population Question," 62; Lenore Manderson, *Sickness and the State: Health and Illness in Colonial Malaya, 1870–1940* (Cambridge: Cambridge University Press, 1996), 201; Vaughan, *Curing Their Ills,* 143.

106. See John Boyd Orr and John L. Gilks, "The Nutritional Condition of the East Africa Native," *Lancet,* 12 March 1927, 1:560–62.

107. The subcommittee worked closely with the Rowett Research Institute, which received nearly £7,500 in grants from the Empire Marketing Board for the project. In 1930 the CCR, including the Dietetics Committee, was subsumed into the new Economic Advisory Council. EMB grants enabled the research to continue until the board's demise in 1933. See Roy MacLeod and Kay Andrews, "The Committee of Civil Research: Scientific Advice for Economic Development, 1925–1930," *Minerva* 7, no. 2 (1969): 680–705.

108. John Boyd Orr and John L. Gilks, *Studies in Nutrition: The Physique and Health of Two African Tribes* (London: HMSO, 1931), 49.

109. See A. I. Richards, *Hunger and Work in a Savage Tribe* (London: Routledge, 1932); A. I. Richards and E. M. Widdowson, "A Dietary Study of North-Eastern Rhodesia," in "Problems of African Native Diet," special issue, *Africa: Journal of the International African Institute* 9, no. 2 (5 April 1936): 166–96; A. I. Richards, *Land, Labour and Diet in Northern Rhodesia: An Economic Study of the Bemba Tribe* (London: Routledge, 1939).

110. Richards and Widdowson, "Dietary Study of North-Eastern Rhodesia"; BLCAS Mss. Afr. s. 1420, D. M. Wright, "Outline of Nutrition Work in Northern Rhodesia," Memo Submitted by the Executive Secretary and Chief Public Relations Officer to the National Food and Nutrition Commission, 1960.

111. Richards tended both to overstate the importance of male labor supply and to downplay the increasing interdependence of cash and kinship links in the subsistence strategies of rural households. Similar to Stebbings's predictions of desiccation in West Africa, her warning of a pending collapse in the traditional Bemba redistributive system would turn out to be largely misplaced, with relations of kinship and reciprocity becoming, if anything, even more crucial as individual households became increasingly reliant on cash income for survival. See Henrietta L. Moore and Megan Vaughan, "Relishing Porridge: The Gender Politics of Food," in *Cutting Down Trees: Gender, Nutrition and Agricultural Change in the Northern Province of Zambia, 1890–1990* (London: James Currey, 1994), 46–78.

112. Raymond Firth, "The Sociological Study of Native Diet," *Africa* 7, no. 4 (October 1934): 401–14.

4. Charles Jeffries, *The Colonial Office* (London: Allen and Unwin, 1956), 166.

5. On the impact of the Depression, see Philip Williamson, *National Crisis and National Government: British Politics, the Economy and Empire, 1926–1932* (London: Cambridge University Press, 1992).

6. For the impact of the Depression on British colonial Africa, see A. G. Hopkins, *An Economic History of West Africa* (London: Longman, 1973), 254–67; John Forbes Munro, *Africa and the International Economy, 1800–1960* (London: J. M. Dent, 1976), 150–70; John Iliffe, *A Modern History of Tanganyika* (Cambridge: Cambridge University Press, 1979), chap. 11; David Anderson, "Depression, Dust Bowl, Demography, and Drought: The Colonial State and Soil Conservation in East Africa during the 1930s," *African Affairs* 83, no. 332 (July 1984): 321–43; John Lonsdale, "The Depression and the Second World War in the Transformation of Kenya," in *Africa and the Second World War,* ed. David Killingray and Richard Rathbone (London: Macmillan, 1986), 103–11; Anne Phillips, *The Enigma of Colonialism: British Policy in West Africa* (London: James Currey, 1989), 140; Frederick Cooper, *Decolonization and African Society,* 58–65.

7. See Hopkins, *Economic History of West Africa,* 258.

8. On the significance of the West Indian disturbances, see Johnson, "West Indies and the Conversion of the British Official Classes to the Development Idea," 55–56; Ken Post, *Arise Ye Starlings: The Jamaican Labour Rebellion of 1938 and Its Aftermath* (The Hague: Nijhoff, 1978).

9. William Kelleher Storey, "Small-Scale Sugar Cane Farmers and Biotechnology in Mauritius: The 'Uba' Riots of 1937," *Agricultural History* 69, 2 (1995): 163–76.

10. David Fieldhouse has estimated that that the empire's share of British exports rose from 37.2 percent in the late 1920s to 41.3 percent between 1934 and 1938, while the empire's share of British imports increased from 32.9 percent to 41.2 percent. See Fieldhouse, "The Metropolitan Economics of Empire," in *Oxford History of the British Empire,* vol. 4, *The Twentieth Century,* ed. Judith Brown and William Roger Louis (Oxford: Oxford University Press, 1999), 100–101; Niall Ferguson, *Empire: The Rise and Demise of the British World Order and the Lessons for Global Power* (New York: Basic Books, 2002), 321.

11. George Orwell, *The Road to Wigan Pier* (London: Gollancz, 1937).

12. On Huxley and Carr-Saunders, see Peder Anker, *Imperial Ecology: Environmental Order in the British Empire, 1895–1945* (Cambridge, Mass.: Harvard University Press, 2002), 417–509; Karl Ittmann, "The Colonial Office and the Population Question in the British Empire, 1918–62," *Journal of Imperial and Commonwealth History* 27, no. 3 (1999): 57–65.

13. Hall expressed his concerns in the Heath Clark Lectures at the University of London in 1935. See A. Daniel Hall, *The Improvement of Native Agriculture in Relation to Population and Public Health* (London: Oxford University Press, 1936).

14. Hall's book was also favorably received by those working on nutrition such as Audrey Richards and E. B. Worthington. See A. I. Richards, review of *The Improvement of Native Agriculture in Relation to Population and Public Health,* by

A. Daniel Hall, *Africa: Journal of the International African Institute* 9, no. 4 (1936): 558–59; E. B. Worthington, "On the Food and Nutrition of African Natives," in "Problems of African Native Diet," special issue, *Africa: Journal of the International African Institute* 9, no. 2 (5 April 1936): 150.

15. See GB, CO, *Report and Proceedings of the Conference of Colonial Directors of Agriculture, July 1938,* Colonial no. 156 (London, 1938); CO 323/1621/4, Conference of Colonial Directors of Agriculture, Conference Papers, 1938.

16. See, for example, CO 323/1693/2, *The Colonial Empire in 1938/39,* Agriculture, Sir F. Stockdale and Dr. Tempany, chap. 5, "The Development of Natural Resources"; CO 852/397/1 CAC, Papers and Minutes, 1942, "Co-ordination in Land Utilisation Policy and Rural Development, Prepared by the Agricultural, Animal Health and Forestry Advisers," by H. A. Tempany, J. Smith, and W. A. Robertson, 12 March 1942 (CAC 600); CO 852/397/2, "Principles of Agricultural Policy in the Colonial Empire," 3 March 1943 (CAC 639).

17. Engledow held a very similar position. Even as early as 1927 he observed that, so far, West Africans had lived very well. They benefited from wonderful growing conditions and virgin soils, there was no competition for their products, and they had enjoyed a period of crop immunity to plant diseases that marks the beginnings of agriculture in all regions. However, these conditions were coming to an end as population pressure rose, soil exhaustion set in, and competition was felt from plantation industries in Southeast Asia. Given this, he felt it was essential to introduce more intensive agricultural methods through legislation and education before a crisis occurred. See BLCAS Mss. Brit. Emp. s. 373 (4), Professor F. L. Engledow, Journal of Tour of Nigeria, 1927.

18. CO 323/1621/4, Conference of Colonial Directors of Agriculture, Conference Papers, 1938, C.P. 4, Papers on Land Settlement, Tanganyika, Mr. R. D. Linton, Agricultural Officer.

19. CO 323/1619/3, Report of the Meeting of the Sub-Committee Appointed to Consider the Proposal of Professor F. L. Engledow as to the Desirability of Colonial Agricultural and Veterinary Departments Being Asked to Formulate Their Policy for a Period of Years, 24 June 1938 (CAC 412).

20. FCOL CAC, Minutes of the 38th Meeting of CAC, 28 June 1938; CO 323/1702/5, Draft Circular Despatch from S/S/C, Malcolm MacDonald, to all Colonies and Protectorates, 23 February 1939; CO 323/1757/6, Agricultural Policy in the Colonial Empire, Sub-Committee to Consider Memo by Professor Engledow, 1940.

21. CO 852/397/1 CAC, Papers and Minutes, 1942, Note by Prof. Engledow on Agricultural Policy in Relation to Colonial Departments and the Work of CAC.

22. CO 323/1354/4 ACEC, *Memorandum on the Education of African Communities,* Colonial no. 103 (London: HMSO, 1935); CO 885/41/446, Miscellaneous, Minutes of the 55th Meeting of ACEC, 4 October 1934.

23. R. J. Mason, "Education of Rural Communities," quoted in A. R. Thompson, "The Adaptation of Education to African Society in Tanganyika under British Rule" (Ph.D. thesis, University of London, 1968), 197.

24. CO 1045/1376, Miscellaneous Papers on Community Development, 1954–55, Stanley Milburn, *Methods and Techniques of Community Development in the United Kingdom Dependent and Trust Territories: A Study,* United Nations Series on Community Organization and Development (New York: United Nations, 1954).

25. On William McLean's career as a CO adviser, see BLCAS Mss. Perham 687/1, Promotion of Social and Economic Development, McLean's Memo, 1943–44, rev. 1963; BLCAS Mss. Brit. Emp. s. 332, Arthur Creech Jones Papers, 7/5 ff. 19–21, McLean to Creech Jones, 5 August 1947.

26. CO 885/41/446, Minutes of the 51st and 56th Meetings of the ACEC, 26 April and 22 November 1934. Dr. McLean's penchant for systematic planning and coordination did not begin in response to Oldham's memo. McLean was a former colonial town planner who had earned a Ph.D. in Engineering from the University of Glasgow, writing his thesis on the broader application of the principles of town planning to regional, national, and international development planning. McLean conceptualized development as largely a technical problem, involving the application of the principles of planning and coordination to the systematic "opening up" of undeveloped or partially developed territories. He felt that a new kind of professional, the "development planner," was needed to collect and collate the necessary knowledge and prepare national schemes. See BLCAS Mss. Perham, Margery Perham Papers, 687/1 ff. 41, Biography on William McLean; FCOL, Miscellaneous Pamphlets, no. 311, folio (1929), "Principles of the Regional Development Planning of Colonies and Imperial Development Planning," by Dr. W. H. McLean.

27. CO 323/1279/9, Colonial Development Fund: Coordination of Schemes, Memorandum on the Need for Coordination, by Dr. W. H. McLean, 28 June 1934; CO 323/1406/18, Advisory Committees in the Colonial Office, Question of Coordination, Memorandum by Dr. W. H. McLean, 26 November 1935.

28. CO 323/1406/18, Minute by R. V. Vernon, 10 December 1935.

29. In the end, it was decided that closer coordination could be best achieved by circulating the agendas for all meetings of the three committees to the various specialist advisers in advance. The advisers, if they deemed it necessary, could ask for a joint meeting to consider questions of common interest. See CO 323/1406/18, Letter to McLean from Sir Cecil Bottomley (for Sir John Shuckburgh), 25 April 1936.

30. CO 323/1415/6, Activities of the Social and Economic Development Sub-Committee, 1937, Minute to Calder, Bottomley, and Shuckburgh, from C. G. Eastwood, 12 October 1937; CO 885/41/446, Minutes of the 72nd and 78th Meetings of the ACEC, 26 November 1936, 22 July 1937.

31. The subcommittee included a number of prominent ACEC members, including its joint secretaries, Vischer and Mayhew; Dr. Philippa Esdaile, reader in biology, University of London; and Dr. James Dougall, former principal of the Kabete Jeanes School and successor to Oldham as secretary of the International Missionary Council. It also enlisted the help of Clauson, Stockdale, and Dr. Ambrose Thomas Stanton, the CO's chief medical adviser.

32. It also closely paralleled the recurrent debate over infant mortality and child welfare in Britain, where the teaching of domestic science and mothercraft in schools, the appointment of lady health visitors, and the establishment of Infant Welfare Centers had been part of the campaign for the education and improvement of "motherhood" since the early 1900s. See Anna Davin, "Imperialism and Motherhood," in *Tensions of Empire: Colonial Cultures in a Bourgeois World,* ed. Frederick Cooper and Ann Laura Stoler (Berkeley: University of California Press, 1997), 87–151.

33. CO 885/67/472, Miscellaneous, "Certain Aspects of the Welfare of Women and Children in the Colonies," by Mary G. Blacklock, Liverpool School of Tropical Medicine, 19 June 1936.

34. See CO 323/1354/6 ACEC, "Memorandum on the Teaching of Domestic Science in England and its Application to Work in the Colonies," by Philippa C. Esdaile, 31 October 1936.

35. For the influence of Blacklock's memo, see Lenore Manderson, *Sickness and the State: Health and Illness in Colonial Malaya, 1870–1940* (Cambridge: Cambridge University Press, 1996), 201.

36. CO 885/41/446, Minutes of the 70th Meeting of the ACEC, 23 July 1936.

37. CO 323/1415/6, "Memorandum Showing the Importance of Programmes Covering All Social and Economic Development Activities," 1937.

38. CO 323/1415/6, "Memorandum on Community Education and Social and Economic Development Programmes in Rural Areas," 1937.

39. Antagonism toward the McLean subcommittee remained strong among the CO's permanent staff. C. G. Eastwood complained he was acting unwillingly as secretary of the committee, and suggested that he be replaced by the appointment of someone else. Calder expressed dismay at the number of despatches originating from the ACEC, noting that the Blacklock despatch had already produced thirty long and complicated replies which he felt should just be "put by." Judging by the response, the advisory committees' work had struck a nerve in the colonies and was raising contentious issues which officials in Whitehall wished would simply go away. It is also clear, however, that attitudes were beginning to change. The new head of the General Department, John Shuckburgh, was prepared to associate himself with the recommendations coming out of the ACEC and other advisory committees, since in his view they were an inevitable outcome of policy and could not be reversed. Cecil Bottomley noted that the new subcommittee on social and economic development arose out of a perceived need for coordination among the various advisory committees, and must be regarded as permanent. See CO 323/1415/6, Minute to Calder, Bottomley, and Shuckburgh from C. G. Eastwood, 12 October 1937; Minute by Calder, 1 November 1937; Minute by W. C. Bottomley, 9 November 1937; Minute by John Shuckburgh to Parkinson, 12 November 1937.

40. GB, *PP,* Cmd. 6175, X (1940), 25, *Statement of Policy on Colonial Development and Welfare.*

41. David Anderson, "Organising Ideas: British Colonialism and African Rural Development," unpublished paper, 1993, 16–17; Pearce, *Turning Point in Africa,* 20.

42. Preliminary discussions began in 1943 with a report on "Mass Education in African Society" prepared by Clarke and Read themselves. This was, in many respects, an extension of the 1935 *Memorandum on the Education of African Communities* and the work of the Sub-Committee on Social and Economic Development. In May 1944 the ACEC decided to form an Adult and Mass Education Sub-Committee with the education adviser, Christopher Cox, as chair. See CO 879/148/1171, "Mass Education of African Colonial Peoples," Memorandum Prepared by Sir Fred Clarke, Director of the Institute of Education, University of London, December 1943.

43. In choosing the term "mass education," the subcommittee explained that "much of the work which is going on in the Colonies at present, and which is in fact community education, is being organised and carried out under the name of welfare work . . . We think there is justification for introducing [the new term] at this stage, for the ideas suggested in the name are the keynote of the new advance." See GB, CO, *Mass Education in African Society,* Report of the Adult and Mass Education Sub-Committee of the ACEC, Colonial no. 186 (London, 1944).

44. GB, CO, ACEC, *Report of the Sub-Committee on Education for Citizenship in Africa,* Colonial no. 216 (London, 1948).

45. CO 847/53/2, Africa, Community Development Organization, 1950, "Community Development in Africa." See also Ursula Hicks, *Development from Below: Local Government and Finance in Developing Countries of the Commonwealth* (Oxford: Clarendon Press, 1961), 520; I. C. Jackson, *Advance in Africa: A Study of Community Development in Eastern Nigeria* (London: Oxford University Press, 1956).

46. W. E. F. Ward, one of the founders of Achimota College in the Gold Coast, and from 1945 deputy education adviser at the CO, also served as one of the UK's representatives at the United Nations and at UNESCO conferences. See Clive Whitehead, "The Advisory Committee on Education in the (British) Colonies, 1924–1961," *Paedagogica Historica* 27, no. 3 (1991): 410. For Huxley's role in the founding of UNESCO, see Anker, *Imperial Ecology.*

47. The West Indies Development and Welfare Organization was known locally as the "Stockdale Circus." It was unique in the sense that it was a regional body, responsible to the secretary of state for the colonies and not to any of the West Indian governments, which resulted in considerable resistance from local authorities and political leaders who viewed the whole project with great suspicion. As Hewitt-Myring noted in his detailed account of the organization, "The general reception of the 'men on the spot' was that this was a gang of busybodies—experts, possibly in their own fields, but almost certainly ignorant of local conditions—who would poke their noses into other people's business . . . and tell them what they ought to do." See BLCAS Mss. W. Ind. 33, P. Hewitt-Myring, "A History of Development and Welfare in the West Indies," 1962, 13.

48. GB, CO, *Development and Welfare in the West Indies, 1943–44: Report by Sir Frank Stockdale,* Colonial no. 189 (London, 1945), 12–13.

49. GB, Cmd. 6607, VI (1944–45), *West India Royal Commission Report,* December 1939, 245.

50. GB, Cmd. 6608, F. L. Engledow, *Report on Agriculture, Fisheries, Forestry and Veterinary Matters,* Supplement to the West India Royal Commission Report, June 1945, 2.

51. FCOL CAC, Minutes of the 46th Meeting of CAC, 2 April 1940.

52. FCOL CAC, Minutes of the 44th Meeting of CAC, 17 October 1939.

53. See CO 852/397/2 CAC, Papers and Minutes, 1943, "Principles of Agricultural Policy in the Colonial Empire" (CAC 639), 31 March 1943.

54. See CO 852/397/4, CAC, Sub-Committee to Consider a Memorandum Prepared by Prof. F. Engledow, 1943, "Principles of Agricultural Policy in the Colonial Empire" (CAC 671), and Extracts from 58th Meeting of CAC, 1 February 1944; CO 996/4, CACAAH&F, Papers, 1945 (CAC 713), Memo on Colonial Agricultural Policy (London, 1945); CO 852/397/3, CAC, Papers and Minutes, 1943, "Principles of Forest Policy in the Colonial Empire" (CAC 656), by W. A. Robertson, 16 September 1943. The CAC was reorganized and extended in 1943 to include colonial forestry, and W. A. Robertson was appointed as forestry adviser. Several new members were also added and a separate standing committee on forestry established, with A. E. Rambaut, the former conservator of forests in Malaya, serving as joint secretary of the council. See FCOL CAC, Minutes of the 53rd and 55th Meetings of CAC, 4 December 1942, 6 April 1943; CO 852/391/1, CAC, Papers and Minutes, 1942, "The Enlargement of the Function of the Council to Include Forestry, 19 Nov. 1942" (CAC 624); CO 852/396/15, CAC, Enlargement of Functions, 1942–43.

55. See CO 852/397/1, CAC, Papers and Minutes, 1942, "Coordination in Land Utilization Policy and Rural Development," Prepared by the Agricultural, Animal Health and Forestry Advisers, H. Tempany, J. Smith, W. A. Robertson, 12 March 1942, and "Note by Prof. Engledow on Agricultural Policy in Relation to Colonial Departments and the Work of the CAC"; FCOL CAC, Minutes of the 51st Meeting of CAC, 12 May 1942.

56. Jose Harris, "Political Thought and the Welfare State, 1870–1940: An Intellectual Framework for British Social Policy," *Past and Present* 135, no. 1 (May 1992): 116–41.

57. Lee and Petter, *Colonial Office, War and Development Policy,* 76; Michael Cowen and Nicholas Westcott, "British Imperial Economic Policy during the War," in *Africa and the Second World War,* ed. David Killingray and Richard Rathbone (London: Macmillan, 1986), 30; CO 852/482/6, no. 11, Circular Despatch from Lord Moyne to Colonial Governments, 5 June 1941, in S. R. Ashton and S. E. Stockwell, eds., *Imperial Policy and Colonial Practice, 1925–1945,* British Documents on the End of Empire, ser. A, vol. 1 (London: HMSO, 1996), 127–33; Keith Jeffery, "The Second World War," in *The Twentieth Century,* ed. Judith Brown and

History of the British Empire, ed. William Roger Louis (Oxford: Oxford University Press, 1999), 479–80.

7. Frederick Cooper, "Imperial Plans," in *Decolonization and African Society: The Labor Question in French and British Africa* (Cambridge: Cambridge University Press, 1996), 176–224; Monica van Beusekom and Dorothy Hodgson, "Lessons Learned? Development Experiences in the Late Colonial Period," *Journal of African History* 41, no. 1 (2000): 31; James Fairhead and Melissa Leach, "Desiccation and Domination: Science and Struggles over Environment and Development in Colonial Guinea," *Journal of African History* 41, no. 1 (2000): 44–45.

8. The planning mission included A. J. Wakefield, former director of agriculture for Tanganyika, David Martin, plantation manager for the United Africa Company, and John Rosa, who acted as its financial consultant. See BLCAS Mss. Brit. Emp. s. 476, ODRP, Food and Cash Crops Collection, Box 7 (48), Sir Roger Swynnerton, Letter to Anthony Kirk-Greene, 15 March 1983.

9. BLCAS Mss. Afr. s. 352, A. J. Wakefield Papers, 1923–1949, "The Groundnut Scheme," Paper Given by Wakefield to First Conference of the British Society of Soil Science, 1947; BLCAS Mss. Afr. s. 352, A. J. Wakefield Papers, "Soil Erosion," Broadcast Talk by Wakefield, 17 December 1946.

10. See Alan Wood, *The Groundnut Affair* (London: The Bodley Head, 1950), app., "Summary of the Wakefield Report," 252–56; Jan Hogendorn and K. M. Scott, "Very Large-Scale Agricultural Projects: The Lessons of the East African Groundnut Scheme," in *Imperialism, Colonialism and Hunger: East and Central Africa,* ed. Robert Rotberg (Lexington, Mass.: Lexington Books, 1983), 167–98.

11. See Wood, *Groundnut Affair.* See also CO 691/215/3, East Africa, Tanganyika, Overseas Food Corporation, "Report on a Visit by Frank Sykes and Geoffrey Clay, Agricultural Adviser to the S/S/C, to the Three Areas of the Groundnut Scheme in Tanganyika between 19 February and 3 March 1951."

12. CO 967/67, East Africa, Groundnut Scheme, Correspondence between Sir Edward Twinning, Governor, and Sir Thomas Lloyd, Permanent Undersecretary of State, about the Scheme, 1950, Letter dated 9 January 1950.

13. CO 822/237, East Africa, Criticisms of Policy of the OFC, 1951, "Land Survey in the East African Groundnut Scheme," by Dr. John Phillips, Chief Agricultural Adviser, OFC, Extract from *Nature* 64, no. 4254 (2 June 1951).

14. CO 822/237, "Resolutions of the Association of Scientific Workers," Enclosure in Letter from General Secretary T. Ainley to James Griffiths, S/S/C, 6 July 1951.

15. John Phillips, *Agriculture and Ecology in Africa: A Study of Actual and Potential Development South of the Sahara* (London: Faber and Faber, 1959), 358.

16. "[T]he one fundamental and unanswerable criticism of the groundnut scheme," as Alan Woods concluded, "is that it attempted large-scale production before trying out a prototype plan; all other mistakes, all the misfortunes we have faithfully chronicled, sprang from this." Woods, *Groundnut Affair,* 242. See also Phillips, *Agriculture and Ecology in Africa,* 348.

17. BLCAS, ODRP, Mss. Afr. s. 1872, Medicine and Public Health in British Tropical Africa, Box VI (21), W. R. Burkitt, Aide-Mémoire.

18. For a good overview of the scheme, see W. M. Sykes, *The Sukumaland Development Scheme, 1947–57,* Oxford Development Records Project, Report 7 (Oxford: Rhodes House Library, 1984).

19. John Ford, *The Role of the Trypanosomiases in African Ecology: A Study of the Tsetse Fly Problem* (Oxford: Clarendon Press, 1971), 196–97.

20. For a detailed record of the investigations undertaken by Malcolm, Rounce, and other officers at Ukiriguru and Lubaga in the 1930s, see N. V. Rounce, *The Agriculture of the Cultivation Steppe of the Lake, Western and Central Provinces* (Capetown: Longmans, Green, 1949); D. W. Malcolm, *Sukumaland: An African People and Their Country* (London: Oxford University Press, 1953).

21. See BLCAS Mss. Afr. s. John Ford Papers, Reports and Memoranda on Land Reclamation from Tsetse Fly in Sukumaland, 1938–1958, "The Development, Expansion and Rehabilitation of Sukumaland," by N. V. Rounce, Senior AO, 1947; N. V. Rounce, "Technical Considerations in the Economic Development of Sukumaland," *Empire Journal of Experimental Agriculture* 19, no. 76 (1951): 253–65.

22. Most of the revenues came internally, from the local Native Authority, or else from the government's Agricultural Development Fund, which derived almost entirely from a cess on cotton. Since the overwhelming bulk of cotton exports came from Lake Province, the Fund amounted to a tax on the Sukuma in all but name. Only £70,000 was received from the CD&W Fund. See Peter McLoughlin, *An Economic History of Sukumaland, Tanzania, to 1964: Field Notes and Analysis* (Fredericton, N.B., Canada: n.p., 1971), 13.

23. Rounce, *Agriculture of the Cultivation Steppe,* 19.

24. McLoughlin, *Economic History of Sukumaland,* 26.

25. John C. de Wilde, *Experiences with Agricultural Development in Tropical Africa* (Baltimore: Johns Hopkins University Press, 1967), 2:415–32.

26. Ford, *Role of the Trypanosomiases in African Ecology,* 217–36.

27. Rounce, *Agriculture of the Cultivation Steppe,* 8.

28. One agricultural officer noted, "During the 1950s, and after the inception of the Scheme, there were signs that the larger farmers tended to concentrate on cash crops, particularly cotton, because they had the means to buy food with the cash received from their cotton. Some of these farmers purchased tractors and equipment with the help of loans so that they could expand the area under cultivation for cotton for themselves, and hired their machinery to others for the same purpose." Quoted in Sykes, *Sukumaland Development Scheme,* 64. See also de Wilde, *Experiences with Agricultural Development in Tropical Africa,* 426–31.

29. McLoughlin, *Economic History of Sukumaland,* 27–31.

30. See Lionel Cliffe's seminal piece, "Nationalism and the Reaction to Enforced Agricultural Change in Tanganyika during the Colonial Period," in *Socialism in Tanzania: An Interdisciplinary Reader,* ed. Lionel Cliffe and John Saul (Nairobi: EAPH, 1972), 17–24.

31. Low and Lonsdale, "Introduction: Towards the New Order," 14.

32. Dorothy Hodgson, "Taking Stock: State Control, Ethnic Identity and Pastoralist Development in Tanganyika, 1948–1958," *Journal of African History* 41, no. 1 (2000): 55–78.

33. McLoughlin, *Economic History of Sukumaland,* 27–28.

34. Swynnerton was born in Tanganyika, where his father was in charge of the Tsetse Research and Survey Department, and spent the first part of his career as an agricultural officer in Tanganyika before becoming assistant director of agriculture, field services, in Kenya in 1951. See BLCAS Mss. Brit. Emp. s. 476, Box 7 (48), Sir Roger Swynnerton.

35. See J. G. M. King, "Mixed Farming in Northern Nigeria, Part I: Origin and Present Conditions," *Empire Journal of Experimental Agriculture* 7, no. 27 (1939): 271–85.

36. BLCAS Mss. Brit. Emp. s. 476, Box 3 (22), Richard W. Kettlewell, 1910–, "Report for the ODRP."

37. BLCAS Mss. Brit. Emp. s. 476, Box 3 (21), C. E. Johnson, 1911–; Mac Dixon-Fyle, "Agricultural Improvement and Political Protest on the Tonga Plateau, Northern Rhodesia," *Journal of African History* 18, no. 4 (1977): 579–96.

38. Roger Tangri, "From the Politics of Union to Mass Nationalism: The Nyasaland African Congress, 1944–1959," in *From Nyasaland to Malawi: Studies in Colonial History,* ed. Roderick J. MacDonald (Nairobi: East African Publishing House, 1975), 267–81.

39. BLCAS Mss. Brit. Emp. s. 476, Box 4 (31), Adrian Frank Posnette, 1914–.

40. CO 554/141/1, West African Visits by Dr. Tempany, 1944, "Report on a Visit to the Gold Coast, January 12th to 26th, 1943, by Dr. Harold Tempany, Agricultural Adviser to the S/S/C."

41. GB, CO, Commission of Enquiry into Disturbances in the Gold Coast, *Minutes of Evidence,* Col. no. 231 (1948), as quoted in Dennis Austin, *Politics in Ghana, 1946–1960* (London: Oxford University Press, 1964), 60.

42. See Austin, *Politics in Ghana,* 58–77; Francis K. Danquah, "Rural Discontent and Decolonization in Ghana, 1945–1951" *Agricultural History* 68, no. 1 (1994): 1–19.

43. See, for example, Lionel Cliffe, "Nationalism and the Reaction to Enforced Agricultural Change"; Low and Lonsdale, "Introduction: Towards the New Order"; David Throup, *The Economic and Social Origins of Mau Mau, 1945–53* (London: James Currey, 1987); Steven Feierman, *Peasant Intellectuals: Anthropology and History in Tanzania* (Madison: University of Wisconsin Press, 1990); Danquah, "Rural Discontent and Decolonization in Ghana"; Hodgson, "Taking Stock"; David Anderson, *Eroding the Commons: The Politics of Ecology in Baringo, Kenya, 1890–1963* (Athens: Ohio University Press, 2002). For southern Africa, where similar soil conservation and rural "betterment" schemes were introduced in this period, see William Beinart, "Soil Erosion, Conservationism and Ideas about Development: A Southern African Exploration, 1900–1960," *Journal of Southern African Studies* 11, no. 1 (1984): 52–83; Michael Drinkwater, "Technical

Development and Peasant Impoverishment: Land Use Policy in Zimbabwe's Midland Province," *Journal of Southern African Studies* 15, no. 2 (1989): 287–305; Leonard Leslie Bessant, "Coercive Development: Land Shortage, Forced Labour, and Colonial Development in the Chiweshe Reserve, Colonial Zimbabwe, 1938–1946," *International Journal of African Historical Studies* 25, no. 1 (1992): 39–65.

44. Joanna Lewis, *Empire State-Building: War and Welfare in Kenya, 1925–52* (Oxford: James Currey, 2000), 13–20.

45. See, especially, Frederick Cooper and Ann Laura Stoler, eds., *Tensions of Empire: Colonial Cultures in a Bourgeois World* (Berkeley: University of California Press, 1997).

46. CO 967/10, Private Office Papers, Minutes Dealing with Working Arrangements in the Colonial Office, 1937–1947, Minute by G. F. Seel, 28 August 1942.

47. CO 967/10, Private Office Papers, Minute by A. J. Dawe, 10 May 1940.

48. BLCAS Mss. Perham, Box 718/2 ff 132, Reply from Phillip Mitchell, Governor of Kenya, to no. 46 of 19 May 1949, S/S/C Despatch on "Education for Citizenship in Africa."

49. CO 852/401/16, Proposed Establishment of Territorial Agricultural Councils, Minute by Sidney Caine (?), 18 January 1943.

50. See BLCAS Mss. Afr. s., J. R. Mackie Papers (1), Papers on Nigerian Agriculture 1939–1945, "Colonial Development and Welfare"; BLCAS Mss. Afr. s. 2124, J. R. Mackie Papers, Letters, 1941–48. See especially "Letter to S/S/C from Mackie explaining his reasons for resigning."

51. BLCAS Mss. Brit. Emp. s. 476, Box 1 (2), Anthony Blair Rains, 1925–, Letter to Anthony Kirk-Greene, Development Records Project, 9 August 1983.

52. Abdul Raufu Mustapha, "Colonialism and Environmental Perception in Northern Nigeria," *Oxford Development Studies* 31, no. 4 (2003): 408.

53. BLCAS Mss. Brit. Emp. s. 476, Box 3 (29), Tom Alan Phillips, 1915–.

54. BLCAS Mss. Afr. s. 352, A. J. Wakefield Papers 1923–1949, "Random Notes on the Colonial System," 1946.

55. See BLCAS Mss. Perham, Box 685, Colonial Policy: Experts, Committees and Groups, (2) ff. 1, Record of Discussion at Carlton Hotel, 6 October 1939.

56. GB, *PP*, Cmd. 6608 (June 1945), West India Royal Commission, *Report on Agriculture, Fisheries, Forestry and Veterinary Matters*, by F. L. Engledow, supplement, 4.

57. H. Vine, "Experiments on the Maintenance of Soil Fertility at Ibadan, Nigeria, 1922–1955," *Empire Journal of Experimental Agriculture* 21, no. 82 (1953): 65–85.

58. BLCAS Mss. Brit. Emp. s. 476, Box 7 (55), Paul Tuley, 1929–.

59. Colin Maher, "Soil Conservation in Kenya Colony, Part I," *Empire Journal of Experimental Agriculture* 18, no. 71 (1950): 137–49; C. C. Webster, "The Ley and Soil Fertility in Britain and Kenya," *East African Agricultural Journal* 20, no. 2 (1954): 71–74; BLCAS Mss. Brit. Emp. s. 476, Box 7 (61), Cyril Charles Webster, 1909–.

60. O. T. Faulkner and C. Y. Shephard, "Mixed Farming: The Basis of a System for West Indian Peasants," *Tropical Agriculture* 20, no. 7 (1943): 136–42.

61. W. S. Martin, "Grass Covers in Their Relation to Soil Structure," *Empire Journal of Experimental Agriculture* 12, no. 47 (1944): 21–33.

62. J. K. Robertson, "Mixed or Multiple Cropping in Native Agricultural Practice," *East African Agricultural Journal* 4 (1941): 228–32.

63. Murray Lunan, "Mound Cultivation in Ufipa, Tanganyika," *East African Agricultural Journal* 16, no. 2 (1950): 88–89.

64. Mustapha, "Colonialism and Environmental Perception in Northern Nigeria," 409–10.

65. CO 996/4 CACAAH&F, Papers, 1945 (CAC 735), Report of the Meeting of the Committee of Agriculture, 31 October 1945; CO 996/5 CACAAH&F, Papers 1946 (CAC 746), Report of the Meeting of the Committee of Agriculture, 11 June 1946.

66. CO 996/7 CACAAH&F, Papers 1948 (CAC 772), Report of the Joint Meeting of the Committee of Agriculture and Committee of Animal Health of the CAC, 6 January 1948.

67. CO 996/6 CACAAH&F, Papers 1947 (CAC 766), Observations by Engledow for the 70th Meeting of the CAC, Item 2, "Mechanized Cultivation and Agricultural Engineering Research" (CAC 758).

68. CO 852/588/2, no. 1, [Colonial Development and Welfare], Memorandum by S. Caine, 12 August 1943, in S. R. Ashton and S. E. Stockwell, eds., *Imperial Policy and Colonial Practice, 1925–1945,* British Documents on the End of Empire, ser. A, vol. 1 (London: HMSO, 1996), 166–72.

69. Both Julian Huxley and William Macmillan, for example, were arguing by 1940 that colonial "backwardness" was largely a natural condition, reflecting the fundamental poverty of the tropical environment itself and only indirectly affected by outside forces. What was required was *more,* not less, imperial state intervention and capital to overcome these inherent deficiencies. W. Arthur Lewis would push these arguments even further, laying down in the process many of the basic principles of what would become known as "Development Economics." See Julian Huxley, "The Future of Colonies," *Fortnightly,* n.s., 148 (August 1940): 120–30; William Macmillan, "The Real Colonial Question," *Fortnightly,* n.s., 148 (December 1940): 548–57.

70. For the importance of the "Caine memorandum," see J. M. Lee and Martin Petter, *The Colonial Office, War and Development Policy: Organisation and the Planning of a Metropolitan Initiative, 1939–1945* (London: Institute of Commonwealth Studies, 1982), 168–75.

71. See Frederick Cooper's insightful discussion of the CO's internal debate over production and raising colonial living standards. Cooper, *Decolonization and African Society,* 114–20.

72. CO 852/803/3, Cmd. 7032, International Trade Distribution, from Washington to the Foreign Office, 28 January 1947, "Summary of the Report of the FAO Preparatory Commission on World Food Proposals."

73. Michael Cowen, "The British State and Agrarian Accumulation in Kenya," in *Industry and Accumulation in Africa,* ed. Martin Fransman (London: Heinemann, 1982), 142–69.

74. CO 996/7 CACAAH&F, Papers 1948 (CAC 774), Note on the General Economic Policy of the Overseas Resources Development Bill, Speech by Rees-Williams, 6 November 1948.

75. PREM 8/923, Norman Brook to Clement Atlee, 14 January 1948, quoted in Cowen and Shenton, "Origin and Course of Fabian Colonialism in Africa," 144.

76. CO 996/8 CACAAH&F, Papers 1949 (CAC 812), "The Organizational Development of Peasant Agricultural Production in the Colonies," Talk by Geoffrey F. Clay, Agricultural Adviser to the S/S/C, Given at the Opening of the Second Course at the Summer School, St. John's College, Cambridge, 11 September 1948.

77. "Report on Discussion on Fertilizer Experiments in East Africa," *East African Agricultural Journal* 15, no. 2 (1949): 61–68; Bernard Keen and D. W. Duthie, "Crop Responses to Fertilizers and Manures in East Africa," *East African Agricultural Journal* 19, no. 1 (1953): 19–32; Vine, "Experiments on the Maintenance of Soil Fertility at Ibadan," 76–79, 84.

78. BLCAS Mss. Brit. Emp. s. 476, Box 1 (4), D. H. Brown, 1903–.

79. Maher, "Soil Conservation in Kenya Colony, Part I"; BLCAS Mss. Brit. Emp. s. 476, Box 3 (29), Tom Alan Phillips, 1915–.

80. BLCAS Mss. Brit. Emp. s. 476, Box 1 (6), Donald Vincent Chambers, 1925–.

81. BLCAS Mss. Brit. Emp. s. 476, Box 7 (48), Sir Roger Swynnerton.

82. BLCAS Mss. Brit. Emp. s. 476, Box 3 (17), Charles W. S. Hartley, 1911–, Letter to Anthony Kirk-Greene, 4 March 1983.

83. C. W. S. Hartley, "Establishment of New Rice Areas in Malaya," *World Crops*, May 1951, 174.

84. An agricultural conference held in Dodoma, Tanganyika, in 1951 reported that it could not be accepted that mechanization was the best agricultural method for opening up unoccupied areas. If it was to be used at all, it should by preceded by adequate research and approached with extreme caution. See Co691/215/14, Tanganyika, Agricultural Policy, 1951, Minutes of Agricultural Development Schemes Conference, Dodoma, 14–17 March 1951. See also comments by various agricultural officers: BLCAS Mss. Brit. Emp. s. 476, Box 1 (8), Reginald Child, 1903–; Box 4 (35), Peter Harold Rosher, 1927–; Box 4 (37), John A. Sandys, 1914–; Box 4 (40), M. F. H. Selby, 1916–; Box 7 (45), Anthony John Smyth, 1927–; Box 7 (57), Robert L. Waddell, 1930–.

85. CO 996/8 CACAAH&F, Papers 1949 (CAC 838) "Report on the Problems of Mechanization of Native Agriculture in Tropical African Colonies,"—Comments by Professor Engledow, 24 October 1949.

86. J. E. Mayne, "Progress in the Mechanization of Farming in the Colonial Territories," *Tropical Agriculture* 31, no. 3 (1954): 178–87; 32, no. 2 (1955): 95–99; 33, no. 4 (1956): 272–77.

87. CO 852/397/4, "Principles of Agricultural Policy in the Colonial Empire" (CAC 671), 4.

88. CO 852/310/38, Visit by Colin Maher to U.S.A. to Study Soil Conservation, 1940–41.

89. Malcolm, *Sukumaland,* 113.

90. FCOL CAC, Minutes of the 46th Meeting of the CAC, 2 April 1940.

91. See, for example, C. Y. Shephard, "Peasant Agriculture in the Leeward and Windward Islands," *Tropical Agriculture* 24, no. 4–6 (1947): 61–71.

92. FCOL CAC, Minutes of the 54th Meeting of the CAC, 21 December 1942.

93. See GB, *PP,* C. 8655, L (1898), 1–176, *Report of the West India Royal Commission.*

94. William Macmillan, *Warning from the West Indies: A Tract for Africa and Empire* (London: Faber and Faber, 1936), 84.

95. Ibid., 91–93.

96. See, for example, C. C. Parisinos, C. Y. Shephard, and A. L. Jolly, "Peasant Agriculture: An Economic Survey of the Las Lomas District, Trinidad," *Tropical Agriculture* 21, no. 5 (1944): 84–98; Alan Pim, *Colonial Agricultural Production: The Contribution Made by Native Peasants and by Foreign Enterprise* (London: Oxford University Press, 1946); C. Y. Shephard, "Peasant Agriculture in the Leeward and Windward Islands."

97. Mona Macmillan, *Champion of Africa: The Second Phase of the Work of W. M. Macmillan, 1934–1974* (London: Swindon Press, 1985), 76.

98. Pim, *Colonial Agricultural Production,* 109.

99. Anne Phillips, *The Enigma of Colonialism: British Policy in West Africa* (London: James Currey, 1989), 89.

100. GB, CO, *Development and Welfare in the West Indies, 1943–4: Report by Sir Frank Stockdale (Comptroller for Development and Welfare in the West Indies),* Col. no. 189 (London, 1945), 31.

101. See CO 323/1621/4, Conference of Colonial Directors of Agriculture, 1938, Session on Land Settlement, Thursday July 28th, "Memo by Director of Agriculture, Jamaica."

102. See, for example, CO 852/310/38, Visit by Colin Mayer to USA to Study Soil Conservation, 1940–41; FCOL CAC, Minutes of the 54th Meeting of CAC, 21 December 1942; B. H. Farmer, *Pioneer Peasant Colonization in Ceylon: A Study in Asian Agrarian Problems* (London: Oxford University Press, 1953), 158–60.

103. CO 852/119/16, Land Settlement with Special Reference to Work of the Land Settlement Association in the UK, Memo Prepared by H. Tempany, 1940.

104. BLCAS Mss. Afr. s. 1425, E. B. Worthington Papers, Box 2, Colonial Office, "Notes on Some Agricultural Development Schemes in the British Colonial Territories," October 1955, Trinidad—Land Settlement Development Schemes; BLCAS Mss. W. Ind. s. 33, P. Hewitt-Myring, "History of Development and Welfare in the West Indies," 1962, 41–44.

105. BLCAS Mss. Afr. s. 352, A. J. Wakefield Papers, Press Clippings on Agricultural Policy Committee Report, 1942–43, "Wakefield Report Is Our 'Beveridge Plan,'" 4 April 1943; "Wakefield Report Promises New Life for Countryside," by a Woman Correspondent, 6 April 1943.

106. CO 1045/156, West Indies, Correspondence with Sir Frank Stockdale, Barbados, Letters 1941–45, "Extract from Agricultural Policy Committee."

107. James Wright, “Lucky Hill Community Project,” *Tropical Agriculture* 24, nos. 10–12 (1947): 142.

108. See, for example, A. L. Jolly, “An Economic Survey on the La Pastora Land Settlement, Trinidad,” *Tropical Agriculture* 28, no. 7 (1946): 117–22.

109. BLCAS Mss. Brit. Emp. s. 476, Box 2 (15), Prof. Hedley John Gooding, H. J. Gooding and D. B. Williams, “Food Crop Research,” *Farmer* (Kingston, Jamaica) 67:136–42.

110. CO 1042/312, Jamaican Land Settlement, 1954, “Observations on Wakefield's Report on Agricultural Development in Jamaica,” Commissioner of Lands, 10 June 1942.

111. As Wakefield protested to Stockdale, Benham's recommendations flew in the face of the principles laid down by the CAC, the West Indies Royal Commission, and the UN's FAO conference, and would stultify the West Indies Development and Welfare Organization's efforts to promote mixed farming and local food production. See CO 1042/311, WID&WO, Jamaica, Agricultural Policy Committee, Letter from Wakefield to Stockdale, 15 February 1945, “Summary of the Benham Report.”

112. CO 852/1342/5, CACAAH&F, Papers 1951, Conference of Heads of Departments of Agriculture, British Caribbean, Jamaica, 24–26 May 1950.

113. CO 852/1342/5, CACAAH&F, Papers 1951 (CAC 888), “Land Settlement in the West Indies,” Memo by Geoffrey Clay, 1 March 1951.

114. George Kay, *Changing Patterns of Settlement and Land Use in the Eastern Province of Northern Rhodesia* (Hull, England: University of Hull Publications, 1967), 41–45; BLCAS Mss. Brit. Emp. s. 476, Box 3 (21), C. E. Johnson, 1911–.

115. Henrietta Moore and Megan Vaughan, *Cutting Down Trees: Gender, Nutrition and Agricultural Change in the Northern Province of Zambia, 1890–1990* (London: James Currey, 1994), 116.

116. “Planned Group Farming in Nyanza Province, Kenya,” *Tropical Agriculture* 28, nos. 7–12 (1951): 153–57; Eric Clayton, *Agrarian Development in Peasant Economies: Some Lessons from Kenya* (New York: Macmillan, 1964), 26–28; Anne Thurston, *The Intensification of Smallholder Agriculture in Kenya: The Genesis and Implementation of the Swynnerton Plan,* Report 6 (Oxford: Oxford Development Records Project, 1984), 16; Robert M. Maxon, *Going Their Separate Ways: Agrarian Transformation in Kenya, 1930–1950* (Madison, N.J.: Fairleigh Dickinson University Press, 2003), 250–57.

117. This account is based on Thurston, *Intensification of Smallholder Agriculture in Kenya,* 16–20; Throup, *Economic and Social Origins of Mau Mau,* 140–59; Bruce Berman, *Control and Crisis in Colonial Kenya: The Dialectic of Domination* (London: James Currey, 1990), 274–79; A. Fiona D. MacKenzie, *Land, Ecology and Resistance in Kenya, 1880–1952* (Portsmouth, N.H.: Heinemann, 1998), 156–67.

118. Moon first expressed his opposition to the use of traditional authorities in soil conservation measures in Western Kenya, where he was a senior agricultural officer in North Kavirondo until 1946. Moon was subsequently promoted to

provincial agricultural officer for Central Province, where his actions helped provoke the widespread resistance in Murang'a. See Throup, *Economic and Social Origins of Mau Mau,* 142; Maxon, *Going Their Separate* Ways, 208–18.

119. In addition to Storrar's farm planning methods, the plan also drew heavily on Brown's ecological surveys of Central and Nyanza Provinces, which divided the regions into different ecological zones based on the association of different indigenous plant communities with certain kinds of soils and rainfall. See Thurston, *Intensification of Smallholder Agriculture in Kenya,* 35–40.

120. For details, see ibid., 45–54; Judith Heyer, "Agricultural Development Policy in Kenya from the Colonial Period to 1975," in *Rural Development in Tropical Africa,* ed. Judith Heyer, Pepe Roberts, and Gavin Williams (New York: St. Martin's, 1981), 101–7.

121. Berman, *Control and Crisis in Colonial Kenya,* 366–71.

122. Clayton, *Agrarian Development in Peasant Economies,* 24–26; Nicholas Ekutu Makana, "Changing Patterns of Indigenous Economic Systems: Agrarian Change and Rural Transformation in Bungoma District, 1930–1960" (Ph.D. diss., West Virginia University, 2005), 274–91.

123. The African Farming Improvement Scheme was begun in the Tonga maize areas of Southern Province, Northern Rhodesia, in 1946 (and revised in 1949), but was challenged by opposition from farmers who believed the real object of the scheme was to improve the land for European settlement. The scheme was extended to areas of Central Province in 1951 and Eastern Province in 1953. By 1959 there were some 1,600 "improved farmers." The Nyasaland Master Farmers Scheme was launched in 1950 and revised in 1955. By 1956 there were 180 Master Farmers, about half of whom were in Central Province. See Kay, *Changing Patterns of Settlement and Land Use in the Eastern Province of Northern Rhodesia,* 39–51; George Kay, "Agricultural Progress in Zambia," in *Environment and Land Use in Africa,* ed. M. F. Thomas and G. W. Whittington (London: Methuen, 1969), 495–523; BLCAS Mss. Brit. Emp. s. 373, Frank L. Engledow Papers (14), Notes and Maps, Nyasaland, 20 September–14 October 1957, "Agricultural Adviser's Newsletter no. 1 of 1958, Appendix I: The Nyasaland Master Farmer's Scheme, Departmental Circular no. 3 of 1957"; Owen J. M. Kalinga, "The Master Farmers' Scheme in Nyasaland, 1950–1962: A Study of a Failed Attempt to Create a 'Yeoman' Class," *African Affairs* 92, no. 368 (July 1993): 367–87.

124. Heyer, "Agricultural Development Policy in Kenya," 102–6; Kalinga, "Master Farmers' Scheme in Nyasaland," 382, 385–86.

125. Kalinga, "Master Farmers' Scheme in Nyasaland," 379–80.

126. Kay, "Agricultural Progress in Zambia," 495–523; Moore and Vaughan, *Cutting Down Trees,* 135–37.

127. Frank Samuel, "Economic Potential of Colonial Africa," *Tropical Agriculture* (1951): 150.

128. Anthony Kirk-Greene, *On Crown Service: A History of HM Colonial and Overseas Civil Services, 1837–1997* (London: I. B. Tauris, 1999), 67–69.

129. See CO 1029/83, Future Work of the CACAAH&F, 1954–56, Minutes of a Meeting between Geoffrey Clay, Frank Engledow, Geoffrey Nye, and R. V. Vernon on the future of CAC.

130. See Clive Whitehead, "The Advisory Committee on Education in the (British) Colonies, 1924–1961," *Paedagogica Historica* 27, no. 3 (1991): 415–19.

131. For the above quotations, see BLCAS Mss. Brit. Emp. s. 476, Box 1 (6), Donald Vincent Chambers, 1925–; Box 3 (27), Charles William Lynn, 1908–; Box 7 (45), Anthony John Smyth, 1927–.

Conclusion

1. Doug Porter, Bryant Allen, and Gaye Thompson, *Development in Practice: Paved with Good Intentions* (London: Routledge, 1991), xv–xvii.

2. Arturo Escobar, *Encountering Development: The Making and Unmaking of the Third World* (Princeton, N.J.: Princeton University Press, 1995), 24; Philip McMichael, "Instituting the Development Project," in *Development and Social Change: A Global Perspective,* 2nd ed. (Thousand Oaks, Calif.: Pine Forge Press, 2000), 3–42.

3. Colin Leys, *The Rise and Fall of Development Theory* (London: James Currey, 1996), 5.

4. David B. Moore, "Development Discourse as Hegemony: Towards an Ideological History, 1945–1995," in *Debating Development Discourse: Institutional and Popular Perspectives,* ed. David B. Moore and Gerald G. Schmitz (London: Macmillan, 1995), 22.

5. See, for example, *Proceedings of the International Congress on Population and World Resources in Relation to the Family [Cheltenham, August 1948]* (London: H. K. Lewis, 1948); Fairfield Osborn, *Our Plundered Planet* (Boston: Little, Brown, 1948); William Vogt, *Road to Survival* (London: Gollancz, 1949); Julian Huxley, "The World's Greatest Problem," *World Review,* January 1950; Paul K. Hatt, *World Population and Future Resources: Proceedings of the Second Centennial Academic Conference of Northwestern University [1951]* (New York: American Book Co., 1952); Elmer Pendell, *Population on the Loose* (New York: W. Funk, 1951); Frank McDougall, "Food and Population," *International Conciliation,* December 1952; United Nations, Department of Social and Economic Affairs, *The Determinants and Consequences of Population Change* (New York, 1953); Dudley Stamp, *Our Undeveloped World* (London: Faber and Faber, 1953); E. John Russell, *World Population and World Food Supplies* (London: Allen and Unwin, 1954).

6. David M. Anderson, "Organising Ideas: British Colonialism and African Rural Development" (unpublished paper, Social Science Research Council Workshop on Social Science and Development, 1993), 18.

7. William M. Adams, *Green Development: Environment and Sustainability in the Third World* (London: Routledge, 1990), 20–22.

8. See, especially, Mathew Connelly, "The Failure of Progress: Algeria and the Crisis of the Colonial World," in *A Diplomatic Revolution: Algeria's Fight for*

Independence and the Origins of the Post–Cold War Era (Oxford: Oxford University Press, 2002)17–38.

9. See Harold Wilson, *The War on World Poverty: An Appeal to the Conscience of Mankind* (London: Gollancz, 1953).

10. Frederick Cooper, *Africa since 1940: The Past of the Present* (Cambridge: Cambridge University Press, 2002), 88.

11. James Ferguson, *The Anti-Politics Machine: "Development," Depoliticization, and Bureaucratic Power in Lesotho* (Cambridge: Cambridge University Press, 1991), 254–56.

12. See, for example, Melissa Leach and Robin Mearns, "Environmental Change and Policy: Challenging Received Wisdom in Africa," in *The Lie of the Land: Challenging Received Wisdom on the African Environment,* ed. Melissa Leach and Robin Mearns (Portsmouth, N.H.: Heinemann, 1996), 1–33; Gavin Williams, "Studying Development and Explaining Policies," *Oxford Development Studies* 31, no. 1 (2003): 37–58.

13. FCOL CAC, Minutes of the 56th Meeting of CAC, 6 July 1943; CO 996/4 CACAAH&F, Papers, 1945, "UN Food Conference and Colonial Nutrition Policy, December 1943."

14. See John Boyd Orr and David Lubbock, *The White Man's Dilemma: Food and the Future* (London: Allen and Unwin, 1953), chaps. 10 and 11. Boyd Orr won the Nobel Peace Prize in 1949 for his efforts at the FAO to foster peace through international scientific cooperation.

15. CO 879/121, Confidential Print, African no. 1100, Part II, Memoranda and Reports, 1923–29, "Memorandum Prepared by Professor Julian Huxley and Dr. W. K. Spencer for the Advisory Committee on Education in Tropical Africa," and "Draft Memorandum: Biology in Tropical Africa."

16. Peder Anker, *Imperial Ecology: Environmental Order in the British Empire, 1895–1945* (Cambridge, Mass.: Harvard University Press, 2002), 230–33.

17. BLCAS Mss. Brit. Emp. s. 476, Box 7 (47), Alexander Storrar, 1921–.

18. BLCAS Mss. Brit. Emp. s. 476, Box 2 (12), Archibald P. S. Forbes, 1913–.

19. BLCAS Mss. Brit. Emp. s. 476, Box 7 (45), Anthony John Smyth, 1927–.

20. See BLCAS Mss. Afr. s. 1872, Medicine and Public Health in British Tropical Africa, Box 18 (75), Robert Samuel Hennessey; Box 6 (28A), R. L. Cheverton.

21. BLCAS Mss. Brit. Emp. s. 476, Box 3 (21), C. E. Johnson, 1911–.

22. BLCAS Mss. Brit. Emp. s. 476, Box 1 (2), Anthony Blair Rains, 1925–.

23. In addition to Johnson and Rains, Charles Lynn, director of agriculture and later chairman of the Natural Resources Board in Northern Rhodesia, served as a consultant for the ODA in St. Helena, Mauritius, Botswana, Antigua, and Lesotho. Robert Waddell, an agricultural officer in Nigeria in the 1950s, worked as a project leader for the ODA on the Cotton Research and Development Project, Thailand, then as an agricultural adviser at the ODA's Southeast Asian Development Division, Bangkok, and finally at the Central and South America Office, London; Reginald Child, former director of the Ceylon Coconut Research Institute in the 1930s

and 1940s and the Tea Research Institute of East Africa in Kenya in the 1950s, advised the ODM as a coconut specialist between 1961 and 1974; Tom Alan Phillips, a twenty-year veteran of the Nigerian agricultural department, went on to work for the Commonwealth Development Corporation (the successor to the Colonial Development Corporation) as a senior agricultural adviser, and in the 1970s became a member of its executive management board. See BLCAS Mss. Brit. Emp. s. 476, Box 3 (27), Charles William Lynn, 1908–; Box 7 (57), Robert L. Waddell, 1930–; Box 1 (8), Reginald Childs, 1903–; Box 3 (29), Tom Alan Phillips, 1915–.

24. Among those who joined Hunting were Robert Campbell, a land development officer in Fiji in the late 1950s and 1960s; P. G. Thompson, a senior agricultural officer, also in Fiji, between 1959 and 1974; Donald V. Chambers, a senior agricultural research officer in Tanganyika from 1951 to 1961; and Richard Kettlewell, director of agriculture and later minister of lands and surveys, Nyasaland, from 1936 to 1962. Thompson would become Director of HTS in 1975. See BLCAS Mss. Brit. Emp. s. 476, Box 1 (5), Robert Owen Campbell, 1934–; Box 1 (6), Donald Vincent Chambers, 1925–; Box 3 (22), Richard Kettlewell, 1910–; Box 7 (49), P. G. Thompson.

25. Charles Hartley, for example, who, as noted earlier, worked in Malaya and Nigeria, reinvented himself on retirement as a consultant with the ODM, the World Bank, the Ecuadorian and Colombian governments, and finally for a private plantation company in Malaysia. Sir Charles Pereira, the deputy director of the East African Agricultural and Forestry Research Organization (EAAFRO) (1957–61) and then director of the Agricultural Research Council of Central Africa (1961–67), would later become a leading soil scientist and consultant in tropical land use and hydrology working for UNESCO, FAO, World Bank, UNDP, and the International Development Research Council in Ottawa, Canada. M. A. Rosenquist, oil palm botanist in Malaya in the 1950s, accepted a post as agricultural research assistant for the Cameroun Development Corporation, where he was involved in negotiations for the first World Bank loan to the country. See BLCAS Mss. Brit. Emp. s. 476, Box 3 (17), Charles W. S. Hartley, 1911–; Box 3 (28), Sir Charles Pereira, 1913–; Box 4 (34), M. A. Rosenquist.

26. BLCAS Mss. Brit. Emp. s. 476, Box 4 (34), M. A. Rosenquist.

27. BLCAS Mss. Brit. Emp. s. 476, Box 7 (48), Sir Roger Swynnerton.

28. BLCAS Mss. Brit. Emp. s. 476, Box 7 (52), T. W. Tinsley, 1924–; Box 7(61), Cyril Charles Webster, 1909–. See, for example, T. W. Tinsley, "The Ecology of Cocoa Viruses: The Role of Wild Hosts in the Incidence of Swollen Shoot Virus in West Africa," *Journal of Ecology* 8, no. 2 (1971): 491–95.

29. C. C. Webster and P. N. Wilson, eds., *Agriculture in the Tropics*, 3rd ed. (London: Blackwell Science, 1998).

30. See J. W. Purseglove, *Tropical Crops*, 4 vols. (London: Longman, 1968–72); M. H. Arnold, ed., *Agricultural Research for Development: The Namulonge Contribution* (Cambridge: Cambridge University Press, 1976).

31. See William Allan, *The African Husbandman* (New York: Barnes and Noble, 1965); John Ford, *The Role of the Trypanosomiases in African Ecology: A Study of the*

Tsetse Fly Problem (Oxford: Clarendon Press, 1971); H. C. Pereira, *Land Use and Water Resources in Temperate and Tropical Climates* (London: Cambridge University Press, 1973); John Phillips, *Agriculture and Ecology in Africa: A Study of Actual and Potential Development South of the Sahara* (London: Faber and Faber, 1959); E. W. Russell, *Soils and Soil Fertility in Tropical Pastures* (London: Faber and Faber, 1966); A. H. Savile, *Extension in Rural Communities: A Manual for Agricultural and Home Extension Workers* (Oxford: Oxford University Press, 1965).

32. BLCAS Mss. Afr. s. 1420, D. M. Wright, Nutrition in Northern Rhodesia, 1930–1960.

33. BLCAS Mss. W. Ind. s. 61 (1), Professor Frederick Hardy, *A History of Soil Science at the ICTA, Trinidad, West Indies, 1922–1956*, vol. 2, *Pedology and Soil Technology.*

34. BLCAS Mss. Afr. 352, A. J. Wakefield Papers, District Agricultural Officer in Tanganyika, Northern Rhodesia, and Trinidad and Tobago, 1923–1949, Extract from Memo Regarding Application of Technical Assistance, for the FAO Technical Assistance Program, September 1949.

35. BLCAS Mss. Ind. Ocn. s. 267, R. G. Heath Papers, Letter to Heath from J. F. Dekker, Plant Production and Protection Division, FAO, 25 August 1960.

36. BLCAS Mss. Ind. Ocn. s. 267, R. G. Heath Papers (3), "Report on Crop Production Possibilities under Conditions of Irrigation in the Lower Volta Flood Plain Areas," 1961.

37. See, for example, BLCAS Mss. Brit. Emp. s. 476, Box 1 (7), Donald Vincent Chambers, 1925–; Box 2 (12), Archibald P. S. Forbes, 1913–.

38. CO 323/162/4, Conference of Colonial Directors of Agriculture, 1938, Session on Soil Erosion, 26 July 1938, Nigeria, Memo by Chief Conservator of Forests.

39. It is a stunning example of what Michael P. Cowen and Robert W. Shenton have called an agrarian doctrine of development: "Agrarian doctrine consists of proposals, usually associated with official policy, to undertake agrarian schemes of development based on small-farm, household production. The intention is to compensate for mass unemployment, urban poverty and the threat of rural emigration." See "Agrarian Doctrines of Development," *Journal of Peasant Studies* 25, no. 2 (1998): 49–75; and 25, no. 3 (1998): 31–62.

40. CO 993/1, "Native Land Tenure in Africa: Report of an Informal Committee," Hailey as Chair, 1945.

41. BLCAS Mss. Afr. s. 1872, Box 27 (110B), Dr. T. A. M. Nash, Medical Entomologist, Sleeping Sickness Service, Nigeria, 1937–1956, "The Anchau Settlement Scheme," *Farm and Forest* 2, no. 2 (October 1941).

42. Ferguson, *Anti-Politics Machine*, xiv–xv.

43. See J. H. Oldham, *White and Black in Africa: A Critical Examination of the Rhodes Lectures of General Smuts* (London: Longmans, Green, 1930).

44. Frederick Cooper, *Decolonization and African Society: The Labour Question in French and British Africa* (Cambridge: Cambridge University Press, 1996), 68–69.

45. See Robert W. Shenton, "The Labour Question in Late Colonial Africa," review of *Decolonization and African Society,* by Frederick Cooper, *Canadian Journal of African Studies* 32, no. 1 (1998): 174–80. See also Joanna Lewis, *Empire State-Building: War and Welfare in Kenya, 1925–52* (Oxford: James Currey, 2000), 125n2.

46. GB, CO, *Development and Welfare in the West Indies, 1943–4: Report by Sir Frank Stockdale (Comptroller for Development and Welfare in the West Indies),* Col. no. 189 (London, 1945), 12–13.

47. Ibid., 3.

48. See Terrence Farrell, "Arthur Lewis and the Case for Caribbean Industrialization," *Social and Economic Studies* 29, no. 4 (December 1980): 52–75; Barbara Ingham, "Shaping Opinion on Development Policy: Economists at the Colonial Office during World War II," *History of Political Economy* 24, no. 3 (1992): 689–710.

49. W. Arthur Lewis, "Developing Colonial Agriculture," *Tropical Agriculture* 27, nos. 4–6 (1950): 73.

50. CO 852/586/9, [Colonial Economic Advisory Committee], Minute by W. A. Lewis Explaining His Resignation as Secretary, 30 November 1944, in S. R. Ashton and S. E. Stockwell, eds., *Imperial Policy and Colonial Practice, 1925–1945,* British Documents on the End of Empire, ser. A, vol. 1 (London: HMSO, 1996), 206–11.

51. See Farrell, "Arthur Lewis and the Case for Caribbean Industrialization," 60–63.

52. Tetteh Kofi, "Arthur Lewis and West African Economic Development," *Social and Economic Studies* 29, no. 4 (December 1980): 202–27.

53. Harry Cleaver, "The Contradictions of the Green Revolution," *American Economic Review* 62, no. 2 (May 1972): 177–86; Robert Anderson, "The Origins of the International Rice Research Institute," *Minerva* 29, no. 1 (Spring 1991): 61–89.

54. Lester R. Brown, *Seeds of Change: The Green Revolution and Development in the 1970s* (New York: Praeger, 1970), chap. 7.

55. Cleaver, "Contradictions of the Green Revolution"; George Blyn, "The Green Revolution Revisited," *Economic Development and Cultural Change* 31, no. 4 (1983): 705–25.

56. See, for example, Bruce F. Johnston and John Cownie, "The Seed-Fertilizer Revolution and Labor Force Absorption," *American Economic Review* 59, no. 4 (September 1969): 569–82; J. P. Grant, "Marginal Men: The Global Employment Crisis," *Foreign Affairs,* October 1971.

57. J. R. McNeill, *Something New under the Sun: An Environmental History of the Twentieth-Century World* (New York: Norton, 2000), 219–26.

58. See E. B. Worthington, "The Nile Catchment: Technological Change and Aquatic Biology," 189–205, John Phillips, "Problems in the Use of Chemical Fertilizers," 549–66, and E. W. Russell, "The Impact of Technological Developments on Soils in East Africa," 567–76, in *The Careless Technology: Ecology and International Development,* ed. M. Taqhi Farvar and John P. Milton (New York: Natural History Press, 1972).

59. Adams, *Green Development,* 38.

60. BLCAS Mss. Brit. Emp. s. 476, Box 1 (2), Anthony Blair Rains, 1925–.

61. BLCAS Mss. Brit. Emp. s. 476, Box 7 (55), Paul Tuley, 1929–.

62. BLCAS Mss. Brit. Emp. s. 476, Box 7 (57), Robert L. Waddell, 1930–. Numerous other cases could also be cited. See, for example, BLCAS Mss. Brit. Emp. s. 476, Box 2 (12), Archibald P. S. Forbes, 1913–; Box 7 (48), Sir Roger Swynnerton; N. W. Simmonds, "The Earlier British Contribution to Tropical Agricultural Research," *Tropical Agricultural Association Newsletter* 11, no. 2 (June 1991): 2–7.

63. United Nations, *Integrated Approaches to Development in Africa: Social Welfare Services in Africa* (New York, 1971).

64. See World Bank, *Population Growth and Policies in Sub-Saharan Africa* (Washington, D.C., 1986); and *Sub-Saharan Africa: From Crisis to Sustainable Growth* (Washington, D.C., 1989).

65. See, for example, Gavin Williams, "The World Bank and the Peasant Problem," in *Rural Development in Tropical Africa,* ed. Judith Heyer, Pepe Roberts, and Gavin Williams New York: St. Martin's Press, 1981), 16–51; Robert Chambers, *Rural Development: Putting the Last First* (London: Longmans, 1983); G. H. R. Chipande, "Smallholder Agriculture as a Rural Development Strategy: The Case of Malawi" (Ph.D. thesis, University of Glasgow, 1983), chap. 3; Porter, Allen, and Thompson, *Development in Practice.*

66. John W. Mellor, "Ending Hunger: An Implementable Program for Self-Reliant Growth," in *The World Food Crisis: Food Security in Comparative Perspective,* ed. J. I. Hans Bakker (Toronto: Canadian Scholar's Press, 1990), 489–90.

67. For excellent overviews of this process, see Eric Helleiner, "From Bretton Woods to Global Finance: A World Turned Upside Down," in *Political Economy and the Changing Global Order,* ed. Richard Stubbs and Geoffrey R. D. Underhill (Toronto: McClelland and Stewart, 1994), 163–75; McMichael, *Development and Social Change,* 129–50.

68. Leys, *Rise and Fall of Development Theory,* 19–25.

69. The major exceptions to this trend were the Newly Industrializing Countries (NICs) of South East Asia (Singapore, Taiwan, South Korea, and Hong Kong), which together accounted for almost half of all exports of manufactures from the third world between 1980 and 1992, along with China and India, which also grew faster in the 1980s. See P. W. Preston, *Development Theory: An Introduction* (Cambridge, Mass.: Blackwell, 1996), chaps. 13 and 14.

70. Perhaps the most forceful of the neoliberal critiques was P. T. Bauer's *Equality, the Third World and Economic Delusion* (London: Methuen, 1981). The best overview of the development debate in the 1980s remains John Toye, *Dilemmas of Development: Reflections on the Counter-Revolution in Development Economics* (Oxford: Blackwell, 1987).

71. McMichael, *Development and Social Change,* 149–65.

72. Helleiner, "From Bretton Woods to Global Finance," 170.

73. According to the United Nations, more than one hundred countries experienced declining living standards in the 1980s and 1990s. See United Nations,

United Nations Development Report (New York, 1997). In the case of Latin America, as Joseph Stiglitz notes, overall growth for the 1990s was just over half what it was in the 1950s, 1960s, and 1970s, and recent performance has been, if anything, even less satisfactory. See Joseph Stiglitz, "Development Policies in a World of Globalization," paper presented at the 50th anniversary of the Brazilian Economic and Social Development Bank (BNDES), Rio de Janeiro, 12–13 September 2002.

74. Benedict Anderson, "The New World Disorder," *New Left Review* 193 (May/June 1992): 3–14.

75. See, for example, Robert H. Bates, *Beyond the Miracle of the Market: The Political Economy of Agrarian Development in Kenya* (Cambridge: Cambridge University Press, 1989); Pranab Bardhan, ed., *The Economic Theory of Agrarian Institutions* (Oxford: Clarendon Press, 1989); Joseph Stiglitz, *The Economic Role of the State* (Oxford: Blackwell, 1989); John Harriss, Janet Hunter, and Colin M. Lewis, eds., *The New Institutional Economics and Third World Development* (London: Routledge, 1995).

76. See Moore, "Development Discourse as Hegemony," 9; Gavin Williams, "Modernizing Malthus: The World Bank, Population Control and the African Environment," in *Power of Development,* ed. Jonathan Crush (London: Routledge, 1995), 158–75.

77. Joseph Stiglitz, "Failure of the Fund: Rethinking the IMF Response," *Harvard International Review* 23, no. 2 (Summer 2001): 14.

78. Jeffrey Sachs, "Global Capitalism: Making It Work," *Economist,* 12 September 1998.

79. A. J. McMichael, *Planetary Overload: Global Environmental Change and the Health of the Human Species* (Cambridge: Cambridge University Press, 1993); Richard Douthwaite, "Is It Possible to Build a Sustainable World?" in *Critical Development Theory: Contributions to a New Paradigm,* ed. Ronaldo Munck and Denis O'Hearn (London: Zed Books, 1999), 157–77.

80. See F. A. Marglin and S. A. Marglin, *Dominating Knowledge: Development, Culture, and Resistance* (Oxford: Clarendon Press, 1990); Wolfgang Sachs, ed., *The Development Dictionary: A Guide to Knowledge as Power* (London: Zed Books, 1992); Richard Norgaard, *Development Betrayed: The End of Progress and a Coevolutionary Revisioning of the Future* (London: Routledge, 1994); Ozay Mehmet, *Westernizing the Third World: The Eurocentricity of Economic Development Theories* (London: Routledge, 1995); Escobar, *Encountering Development.*

81. Arturo Escobar, "Imagining a Post-Development Era," in *Power of Development,* ed. Jonathan Crush (London: Routledge, 1995), 211–27; Majid Rahnemâ, "Towards Post-Development," in *The Post-Development Reader,* ed. Majid Rahnema and Victoria Bawtree (London: Zed Books, 1997), 377–404; Vandana Shiva, "The Living Democracy Movement: Alternatives to the Bankruptcy of Globalization," in *Another World Is Possible: Popular Alternatives to Globalization at the World Social Forum,* ed. William F. Fisher and Thomas Ponniah (London: Zed Books, 2003), 115–24; McMichael, *Development and Social Change,* 293–95, 302–3.

82. See Rahnema, "Towards Post-Development," 384, 388, 391–92, 394.

83. McMichael, *Development and Social Change,* 295–97.

84. Ibid., 306.

85. Michael Ignatieff, "The Burden," *New York Times Magazine,* 5 January 2003, 50.

Bibliography

Archival Sources

Bodleian Library of Commonwealth and African Studies (BLCAS), Rhodes House, Oxford

Mss W. Ind. s. 58	G. K. Argles papers
Mss W. Ind. s. 42	A. C. Barnes papers
Mss Brit. Emp. t. 1	R. O. Barnes papers
Mss Brit. Emp. s. 332	Arthur Creech Jones papers
Mss Brit. Emp. s. 373	F. L. Engledow papers
Mss Brit. Emp. s. 319	N. B. Favell papers
Mss Ind. Ocn. s. 267	R. G. Heath papers
Mss Afr. s. 1402	J. B. Kirk papers
Mss Afr. s. 1739	J. G. M. King papers
Mss W. Ind. s. 61(1)	Professor Frederick Hardy papers
Mss W. Ind. s. 33	P. Hewitt-Myring papers
Mss W. Ind. s. 55	B. J. Silk papers
Mss Brit. Emp. s. 481	W. C. Lester-Smith papers
Mss Lugard	Frederick Lugard papers
Mss Brit. Emp. s. 76	Frederick Lugard papers
Mss Afr. s. 1471	Colin Maher papers
Mss Afr. s. 873	J. R. Mackie papers
Mss Brit. Emp. s. 457	Geoffrey Milne papers
Mss Brit. Emp. s. 33	H. W. Moor papers
Mss Perham	Margery Perham papers
Mss W. Ind. s. 28	Duncan Stevenson papers
Mss Afr. s. 352	A. J. Wakefield papers
Mss Afr. s. 1425	E. B. Worthington papers
Mss Afr. s. 1420	D. M. Wright papers
Mss Brit. Emp. s. 476	Oxford Development Records Project (ODRP), Food and Cash Crop Collection

Mss Afr. s. 1872 — Oxford Development Records Project (ODRP), Medicine and Public Health in British Tropical Africa Collection

Foreign and Commonwealth Office Library (FCOL), London

Colonial Advisory Council on Agriculture and Animal Health (CAC), Minutes of the Meetings, 1929–1943
East African Pamphlets
Malaria Pamphlets
Miscellaneous Pamphlets

National Archives, Kew, London

Series CAB 58/199 Series CAB 58/206 Series CAB 58/208	Economic Advisory Council: Papers of the Committee on Nutrition in the Colonial Empire, 1936–9
Series CO 273/630/1	Visit of Sir Frank Stockdale to Malay Straits States, 1938
Series CO 295/565/14	Visit by Engledow to Trinidad, 1928
Series CO 318	West Indies: Original Correspondence, 1940–1945
Series CO 323	General Department/Division: Original Correspondence, 1920–1952
Series CO 533	Kenya: Original Correspondence, 1930–1950
Series CO 554	West Africa: Original Correspondence, 1944
Series CO 691	Tanganyika: Original Correspondence, 1945–1951
Series CO 822	East Africa: Original Correspondence, 1930–1951
Series CO 847	Africa: Original Correspondence, 1937–1950
Series CO 852	Economic Department: Original Correspondence, 1935–1952
Series CO 859	Social Services Department: Original Correspondence, 1939–1945
Series CO 879	Confidential Print: Africa, 1895–1945
Series CO 884	Confidential Print: West Indies, 1921
Series CO 885	Confidential Print: Miscellaneous, 1899–1945
Series CO 908	Colonial Agricultural Research Committee (CAR) Papers, 1950–1
Series CO 927	Research Department: Original Correspondence, 1944–1950
Series CO 967/2b	Private Office Papers: Proposed Committee on Appointment of Technical Advisers to Secretary of State Amery, 1926
Series CO 967/67	Private Office Papers: Groundnut Scheme: Correspondence b/w Sir Edward Twinning, Governor and Sir Thomas Lloyd, Permanent U/S/S about the Scheme, 1950.

Series CO 993 — Colonial Land Tenure Advisory Panel Papers, 1945
Series CO 996 — CAC Papers, 1945–1951
Series CO 1029/83 — Future Work of the CAC, 1954–6
Series CO 1042 — WIDWO, Jamaica: Agricultural Policy, 1944–1954
Series CO 1045 — West Indies: Imperial College of Tropical Agriculture, 1941–1945

Sir Ronald Ross Archives (RA), London School of Hygiene and Tropical Medicine, London

Rnum series 14
Rnum series 24
Rnum series 28
Rnum series 32
Rnum series 49
Rnum series 51

Joint International Missionary Council/Conference of British Missionary Societies Archives (IMC/CBMS Archives), Yale Divinity College Library, Yale University

Box 204 — Africa General: International Institute of African Languages and Cultures
Box 218 — Africa General Education: Policy
Box 219 — Africa General Education: Advisory Committee on Education in the Colonies

Official Publications

UK Government

Colonial Office, Colonial Reports

Report and Proceedings of the Conference of Colonial Directors of Agriculture. Colonial no. 67. London: HMSO, 1931.

Labour Conditions in Northern Rhodesia, Report by Major G. St. J. Orde Brown. Colonial no. 150. London: HMSO, 1938.

Report and Proceedings of the Conference of Colonial Directors of Agriculture. Colonial no. 156. London: HMSO, 1938.

Mass Education in African Society: Report of the Adult and Mass Education Sub-Committee of ACEC. Colonial no. 186. London: HMSO, 1943.

Development and Welfare in the West Indies, 1943–44: Report by Sir Frank Stockdale. Colonial no. 189. London: HMSO, 1945.

Report of the Sub-Committee on Education for Citizenship in Africa, ACEC. Colonial no. 216. London: HMSO, 1948.

Parliamentary Command Papers (GB, *PP*)

Report of the West India Royal Commission, with subsidiary report by D. Morris, Assistant Director of the RBG, Kew. C. 8655, L, 1–176. 1897.

Report by Professor W. J. Simpson on Sanitary Matters in Various West Africa Colonies and the Outbreak of Plague in the Gold Coast. Cd. 4718, LXI. June 1909.

Report of the Departmental Committee on the West African Medical Staff. Cd. 4720, LXI. July 1909.

Report of the Northern Nigeria Lands Committee. Cd. 5102, XLIV. 1910.

Correspondence Relating to the Recent Outbreak of Yellow Fever in West Africa. Cd. 5581, LII. March 1911.

Report of the Machinery of Government Committee. Cd. 9230, XII. 1918.

Report of the Committee on the Staffing of the Agricultural Departments in the Colonies. Cmd. 730, XII, 253. June 1920.

Report of the Committee on the Staffing of the Veterinary Departments in the Colonies and Protectorates. Cmd. 920, XII, 279. August 1920.

Report of the Departmental Committee Appointed to Enquire into the Colonial Medical Service. Cmd. 939, XII, 267. September 1920.

Education Policy in British Tropical Africa. Memorandum Submitted to the Secretary of State for the Colonies by the Advisory Committee on Native Education in the British Tropical African Dependencies. Cmd. 2374. March 1925.

Report of the East Africa Commission. Cmd. 2387. April 1925.

Report by the Rt. Hon. W.G.A. Ormsby-Gore on His Visit to West Africa during the Year 1926. Cmd. 2744, IX, 211. 1927.

Agricultural Research and Administration on the Non-Self-Governing Dependencies: Report of a Committee Appointed by the Secretary of State for the Colonies. Cmd. 2825, VII, 563. March 1927.

Report of the Committee on Colonial Scientific and Research Services. Colonial Office Conference, Summary of Proceedings. Cmd. 2883. 1927.

Colonial Office Conference, 1927; Appendices. Cmd. 2884, VII, 834. 1927.

Colonial Agricultural Service: Report of a Committee Appointed by the Secretary of State for the Colonies. Cmd. 3049, VII, 547. March 1928.

Report by the Rt. Hon. W.G.A. Ormsby-Gore on His Visit to Malaya, Ceylon and Java during the Year 1928. Cmd. 3235, V, 791. 1928–29.

Report of the Committee on the System of Appointments in the Colonial Services. Cmd. 3554. April 1930.

Report of the West Indian Sugar Commission. Cmd. 3517. 1930.

Nutrition in the Colonial Empire, Report of the Economic Advisory Council, Part I. Cmd. 6050, 6951, X, 55. 1938–39.

Statement of Policy on Colonial Development and Welfare. Cmd. 6175, X. 1940.

Report on the Social Insurance and Allied Services (Beveridge Report). Cmd. 6404. 1942.

West India Royal Commission Report, Dec. 1939. Cmd. 6607, VI, 245. 1944–45.

Report on Agriculture, Fisheries, Forestry and Veterinary Matters, by F. L. Engledow. Cmd. 6608. June 1945.

International Organizations

League of Nations. *The Problem of Nutrition.* 4 vols. Geneva, 1936.

———. *Nutrition: Final Report of the Mixed Committee of the League of Nations on the Relation of Nutrition to Health, Agriculture and Economic Policy.* Geneva, 1937.

United Nations. *Methods and Techniques of Community Development in the United Kingdom Dependent and Trust Territories: A Study.* United Nations Series on Community Organization and Development. New York, 1954.

———. *Integrated Approaches to Development in Africa: Social Welfare Services in Africa.* New York, 1971.

———. *United Nations Development Report.* New York, 1997.

———. Department of Social and Economic Affairs. *The Determinants and Consequences of Population Change.* New York, 1953.

World Bank. *Population Growth and Policies in Sub-Saharan Africa.* Washington, D.C., 1986.

———. *Sub-Saharan Africa: From Crisis to Sustainable Growth.* Washington, D.C., 1989.

Secondary Sources

Books

Adams, William M. *Green Development: Environment and Sustainability in the Third World.* London: Routledge, 1990.

Adas, Michael. *Machines as the Measure of Men: Science, Technology, and Ideologies of Western Dominance.* Ithaca: Cornell University Press, 1989.

Allan, William. *The African Husbandman.* New York: Barnes and Noble, 1965.

———. *Studies in African Land Usage in Northern Rhodesia.* Rhodes-Livingstone Papers 15. London: Oxford University Press, 1949.

Alter, Peter. *The Reluctant Patron: Science and the State in Britain, 1850–1920.* Oxford: Berg, 1987.

Amery, Julian. *The Life of Joseph Chamberlain.* Vol. 4, *1901–1903: At the Height of His Power.* London: Macmillan, 1951.

Amery, Leopold S. *My Political Life.* 3 vols. London: Hutchinson, 1953–55.

Anderson, David. *Eroding the Commons: The Politics of Ecology in Baringo, Kenya, 1890–1963.* Athens: Ohio University Press, 2002.

Anderson, David, and Richard Grove, eds. *Conservation in Africa: People, Policies and Practice.* Cambridge: Cambridge University Press, 1987.

Anker, Peder. *Imperial Ecology: Environmental Order in the British Empire, 1895–1945.* Cambridge, Mass.: Harvard University Press, 2002.

Arnold, David, ed. *Colonizing the Body: State Medicine and Epidemic Disease in Nineteenth-Century India.* Berkeley: University of California Press, 1993.

———. *Imperial Medicine and Indigenous Societies.* Manchester: Manchester University Press, 1988.

Arnold, M. H., ed. *Agricultural Research for Development: The Namulonge Contribution.* Cambridge: Cambridge University Press, 1976.

Ashcroft, Bill, Gareth Griffiths, and Helen Tiffin. *The Empire Writes Back: Theory and Practice in Post-Colonial Literatures.* London: Routledge, 1989.

Ashton, S. R., and S. E. Stockwell, eds. *Imperial Policy and Colonial Practice, 1925–1945.* British Documents on the End of Empire, ser. A, vol. 1. London: HMSO, 1996.

Austin, Dennis. *Politics in Ghana, 1946–1960.* London: Oxford University Press, 1964.

Baber, Zaheer. *The Science of Empire: Scientific Knowledge, Civilization, and Colonial Rule in India.* Albany: State University of New York Press, 1996.

Balfour, Andrew, and Henry Harold Scott. *Health Problems of the Empire: Past, Present and Future.* London: W. Collins, 1924.

Bardhan, Pranab, ed. *The Economic Theory of Agrarian Institutions.* Oxford: Clarendon Press, 1989.

Bates, Robert H. *Beyond the Miracle of the Market: The Political Economy of Agrarian Development in Kenya.* Cambridge: Cambridge University Press, 1989.

Bauer, P. T. *Equality, the Third World and Economic Delusion.* London: Methuen, 1981.

Bayly, C. A. *Imperial Meridian: The British Empire and the World, 1780–1830.* London: Longman, 1989.

———. *Indian Society and the Making of the British Empire.* Vol. 2, *New Cambridge History of India.* Cambridge: Cambridge University Press, 1988.

Beloff, Max. *Imperial Sunset.* 2 vols. London: Macmillan, 1982–89.

Bennett, George, ed. *The Concept of Empire: Burke to Attlee, 1774–1947.* London: A. and C. Black, 1953.

Berman, Bruce. *Control and Crisis in Colonial Kenya: The Dialectic of Domination.* London: James Currey, 1990.

Berman, Bruce, and John Lonsdale. *Unhappy Valley: Conflict in Kenya and Africa.* 2 vols. Athens: Ohio University Press, 1992.

Birkett, Deborah. *Mary Kingsley: Imperial Adventuress.* London: Palgrave Macmillan, 1992.

Boyd Orr, John. *Food, Health and Income: Report on a Survey of Diet in Relation to Income.* London: Macmillan, 1936.

Boyd Orr, John, and John L. Gilks. *Studies in Nutrition: The Physique and Health of Two African Tribes.* London: HMSO, 1931.

Boyd Orr, John, and David Lubbock. *The White Man's Dilemma: Food and the Future.* London: Allen and Unwin, 1953.

Brockway, Lucile H. *Science and Colonial Expansion: The Role of the British Royal Botanic Gardens.* London: Academic Press, 1979.

Brown, Lester R. *Seeds of Change: The Green Revolution and Development in the 1970s.* New York: Praeger, 1970.

Bruce, Charles. *The Broad Stone of Empire.* Vol. 1. London: Macmillan, 1910.

Buell, Raymond L. *The Native Problem in Africa.* 2 vols. New York: Macmillan, 1928.

Callender, Ann. *How Shall We Govern India? A Controversy among British Administrators, 1800–1882.* New York: Garland, 1987.

Cell, John W. *By Kenya Possessed: The Correspondence of Norman Leys and J. H. Oldham, 1918–1926.* Chicago: University of Chicago Press, 1976.

Chamberlain, Joseph. *Foreign and Colonial Speeches.* London: George Routledge and Sons, 1897.

Chambers, Robert. *Rural Development: Putting the Last First.* London: Longmans, 1983.

Chanock, Martin. *Unconsummated Union: Britain, Rhodesia and South Africa, 1900–45.* Manchester: Manchester University Press, 1977.

Church, Archibald G. *East Africa, a New Dominion: A Crucial Experiment in Tropical Development and Its Significance to the British Empire.* Westport, Conn.: Negro Universities Press, 1970 [1927].

Clatworthy, Frederick James. *The Formulation of British Colonial Education Policy, 1923–1948.* Ann Arbor: University of Michigan School of Education, 1971.

Clayton, Eric. *Agrarian Development in Peasant Economies: Some Lessons from Kenya.* New York: Macmillan, 1964.

Cline, Catherine Ann. *E. D. Morel, 1873–1924: The Strategies of Protest.* Belfast: Blackstaff, 1980.

Colley, Linda. *Britons: Forging the Nation, 1707–1837.* New Haven: Yale University Press, 1992.

Conklin, Alice. *A Mission to Civilize: The Republican Idea of Empire in France and West Africa, 1895–1930.* Berkeley: University of California Press, 1997.

Connelly, Matthew. *A Diplomatic Revolution: Algeria's Fight for Independence and the Origins of the Post–Cold War Era.* Oxford: Oxford University Press, 2002.

Constantine, Stephen. *The Making of British Colonial Development Policy, 1914–1940.* London: Frank Cass, 1984.

Cooper, Frederick. *Africa since 1940: The Past of the Present.* Cambridge: Cambridge University Press, 2002.

———. *Decolonization and African Society: The Labor Question in French and British Africa.* Cambridge: Cambridge University Press, 1996.

———. *From Slaves to Squatters: Plantation Labor and Agriculture in Zanzibar and Coastal Kenya, 1890–1925.* New Haven: Yale University Press, 1980.

Cooper, Frederick, and Randall Packard, eds. *International Development and the Social Sciences: Essays on the History and Politics of Knowledge.* Berkeley: University of California Press, 1997.

Cooper, Frederick, and Ann Laura Stoler, eds. *Tensions of Empire: Colonial Cultures in a Bourgeois World.* Berkeley: University of California Press, 1997.

Cowen, Michael P., and Robert W. Shenton. *Doctrines of Development.* New York: Routledge, 1996.

Cranefield, Paul. *Science and Empire: East Coast Fever in Rhodesia and the Transvaal.* Cambridge: Cambridge University Press, 1991.

Cronin, James E. *The Politics of State Expansion: War, State and Society in Twentieth-Century Britain.* London: Routledge, 1991.

Cronon, William. *Changes in the Land: Indians, Colonists, and the Ecology of New England.* New York: Hill and Wang, 1983.

Crosby, Alfred W. *Ecological Imperialism: The Biological Expansion of Europe, 900–1900.* Cambridge: Cambridge University Press, 1986.

Crush, Jonathan, ed. *Power of Development.* London: Routledge, 1995.

Curtin, Philip D. *The Image of Africa: British Ideas and Action, 1780–1850.* Madison: University of Wisconsin Press, 1964.

Dale, H. E. *Daniel Hall: Pioneer in Scientific Agriculture.* London: John Murray, 1956.

Davis, J. Merle. *Modern Industry and the African: An Enquiry into the Effect of the Copper Mines of Central Africa upon Native Society and the Work of the Christian Missions.* London: Frank Cass, 1967 [1933].

Denoon, Donald, *Settler Capitalism: The Dynamics of Dependent Development in the Southern Hemisphere.* Oxford: Oxford University Press, 1983.

Denoon, Donald, et al., eds. *The Cambridge History of the Pacific Islanders.* Cambridge: Cambridge University Press, 1997.

De Silva, K. M. *A History of Sri Lanka.* Berkeley: University of California Press, 1981.

de Wilde, John C. *Experiences with Agricultural Development in Tropical Africa.* 2 vols. Baltimore: Johns Hopkins University Press, 1967.

Diamond, Alan, ed. *The Victorian Achievement of Sir Henry Maine: A Centennial Reappraisal.* Cambridge: Cambridge University Press, 1991.

Dilke, Charles Wentworth. *Problems of Greater Britain.* 2 vols. London: Macmillan, 1890.

Dirks, Nicholas B., ed. *Colonialism and Culture.* Ann Arbor: University of Michigan Press, 1992.

Drayton, Richard. *Nature's Government: Science, Imperial Britain, and the "Improvement" of the World.* New Haven: Yale University Press, 2000.

Drummond, Ian M. *British Economic Policy and the Empire, 1919–1939.* London: Allen and Unwin, 1972.

Dunlap, Thomas. *Nature and the English Diaspora: Environment and History in the United States, Canada, Australia and New Zealand.* Cambridge: Cambridge University Press, 1999.

Escobar, Arturo. *Encountering Development: The Making and Unmaking of the Third World.* Princeton: Princeton University Press, 1995.

Farley, John. *Bilharzia: A History of Imperial Tropical Medicine.* Cambridge: Cambridge University Press, 1991.

Farmer, B. H. *Pioneer Peasant Colonization in Ceylon: A Study in Asian Agrarian Problems.* London: Oxford University Press, 1957.

Farvar, M. Taqhi, and John P. Milton, eds. *The Careless Technology: Ecology and International Development.* New York: Natural History Press, 1972.

Faulkner, O. T., and J. R. Mackie. *West African Agriculture.* Cambridge: Cambridge University Press, 1933.

Feierman, Steven. *Peasant Intellectuals: Anthropology and History in Tanzania.* Madison: University of Wisconsin Press, 1990.

Ferguson, James. *The Anti-Politics Machine: "Development," Depoliticization, and Bureaucratic Power in Lesotho.* Cambridge: Cambridge University Press, 1990.

Ferguson, Niall. *Empire: The Rise and Demise of the British World Order and the Lessons for Global Power.* New York: Basic Books, 2002.

Fiddes, George V. *The Dominions and Colonial Offices.* London: G. P. Putnam's Sons, 1926.

Forbes Munro, John. *Africa and the International Economy, 1800–1960.* London: J. M. Dent, 1976.

Ford, John. *The Role of the Trypanosomiases in African Ecology: A Study of the Tsetse Fly Problem.* Oxford: Clarendon Press, 1971.

Fry, Geoffrey Kingdon. *The Growth of Government: The Development of Ideas about the Role of the State and the Machinery and Functions of Government in Britain since 1780.* London: Frank Cass, 1979.

Furse, Ralph. *Aucuparius: Recollections of a Recruiting Officer.* Oxford: Clarendon Press, 1962.

Ghosh, Suresh Chandra. *Dalhousie in India, 1848–56: A Study of His Social Policy as Governor-General.* New Delhi: Munshiram Manoharlal, 1975.

Glantz, Michael H., ed. *Desertification: Environmental Degradation in and around Arid Lands.* Boulder, Colo.: Westview, 1977.

Gorrie, R. M. *The Use and Misuse of the Land.* Oxford Forestry Memoirs 19. Oxford, 1935.

Green, E. H. H. *The Crisis of Conservatism: The Politics, Economics and Ideology of the British Conservative Party, 1880–1914.* London: Routledge, 1995.

Griffiths, Tom, and Libby Robin, eds. *Ecology and Empire: Environmental History of Settler Societies.* Edinburgh: Keele University Press, 1997.

Grigg, Edward W. M. (Lord Altrincham). *Kenya's Opportunity: Memories, Hopes and Ideas.* London: Faber and Faber, 1955.

Grove, Richard. *Ecology, Climate and Empire: Colonialism and Global Environmental History, 1400–1940.* Cambridge: Cambridge University Press, 1997.

———. *Green Imperialism: Colonial Expansion, Tropical Island Edens, and the Origins of Environmentalism, 1600–1860.* Cambridge: Cambridge University Press, 1995.

Hailey, Malcolm. *An African Survey: A Study of Problems Arising in Africa South of the Sahara.* London: Oxford University Press, 1938.

Hall, A. Daniel. *The Improvement of Native Agriculture in Relation to Population and Public Health.* London: Oxford University Press, 1936.

Hall, Henry L. *The Colonial Office: A History.* London: Longmans, 1937.

Hargreaves, John D. *Prelude to the Partition of West Africa.* London: Macmillan, 1963.
Harrison, Gordon. *Mosquitoes, Malaria and Man: A History of the Hostilities since 1880.* London: John Murray, 1978.
Harrison, Mark. *Public Health in British India: Anglo-Indian Preventive Medicine, 1859–1914.* Cambridge: Cambridge University Press, 1994.
Harriss, John, Janet Hunter, and Colin M. Lewis, eds. *The New Institutional Economics and Third World Development.* London: Routledge, 1995.
Hatt, Paul K. *World Population and Future Resources: Proceedings of the Second Centennial Academic Conference of Northwestern University [1951].* New York: American Book Co., 1952.
Havinden, Michael, and David Meredith. *Colonialism and Development: Britain and Its Tropical Colonies, 1850–1960.* London: Routledge, 1993.
Headrick, Daniel R. *The Tools of Empire: Technology and European Imperialism in the Nineteenth Century.* New York: Oxford University Press, 1981.
Hetherington, Penelope. *British Paternalism and Africa, 1920–1940.* London: Frank Cass, 1978.
Hicks, Ursula. *Development from Below: Local Government and Finance in Developing Countries of the Commonwealth.* Oxford: Clarendon Press, 1961.
Hill, Christopher. *The English Bible and the Seventeenth-Century Revolution.* London: Allen Lane, 1993.
Hobson, J. A. *Imperialism: A Study.* Ann Arbor: University of Michigan Press, 1965 [1902].
Hopkins, A. G. *An Economic History of West Africa.* London: Longman, 1973.
Howe, Stephen. *Anticolonialism in British Politics: The Left and the End of Empire, 1918–1964.* Oxford: Clarendon Press, 1993.
Hyam, Ronald. *Elgin and Churchill at the Colonial Office, 1905–1908.* London: Macmillan, 1968.
Iliffe, John. *A Modern History of Tanganyika.* Cambridge: Cambridge University Press, 1979.
Jacks, G. V., and R. O. Whyte. *Erosion and Soil Conservation.* Technical Communication 36. Harpenden: Imperial Bureau of Soil Science, 1938.
———. *The Rape of the Earth: A World Survey of Soil Erosion.* London: Faber and Faber, 1939.
Jackson, I. C. *Advance in Africa: A Study of Community Development in Eastern Nigeria.* London: Oxford University Press, 1956.
Jeffries, Charles. *The Colonial Empire and Its Civil Service.* Cambridge: Cambridge University Press, 1938.
———. *The Colonial Office.* London: Allen and Unwin, 1956.
Kay, Geoffrey B. *The Political Economy of Colonialism in Ghana: A Collection of Documents and Statistics, 1900–1960.* Cambridge: Cambridge University Press, 1972.
Kay, George. *Changing Patterns of Settlement and Land Use in the Eastern Province of Northern Rhodesia.* Hull: University of Hull Publications, 1967.

Keen, B. A. *The East African Agriculture and Forestry Research Organization: Its Origins and Objects.* Nairobi, 1948.

Kesner, Richard. *Economic Control and Colonial Development: Crown Colony Financial Management in the Age of Joseph Chamberlain.* Oxford: Clio, 1981.

Kidd, Benjamin. *The Control of the Tropics.* New York: Macmillan, 1898.

King, Kenneth James. *Pan-Africanism and Education: A Study of Race Philanthropy and Education in the Southern States of America and East Africa.* Oxford: Clarendon Press, 1971.

Kingsley, Mary H. *Travels in West Africa: Congo Français, Corisco and Cameroons.* 3rd ed. London: Frank Cass, 1965 [1897].

———. *West African Studies.* London: Frank Cass, 1964 [1899].

Kirk-Greene, Anthony. *On Crown Service: A History of HM Colonial and Overseas Civil Services, 1837–1997.* London: I. B. Tauris, 1999.

Kitching, Gavin. *Class and Economic Change in Kenya: The Making of an African Petite Bourgeoisie.* New Haven: Yale University Press, 1980.

Koponen, Juhani. *Development for Exploitation: German Colonial Policies in Mainland Tanzania, 1884–1914.* Helsinki: Finnish Historical Society, 1994.

Kubicek, Robert V. *The Administration of Imperialism: Joseph Chamberlain at the Colonial Office.* Durham, N.C.: Duke University Press, 1969.

Kuczynski, R. R. *West Indian and American Territories.* Vol. 3, *Demographic Survey of the British Colonial Empire.* London: Oxford University Press, 1953.

Kuklick, Henrika. *The Savage Within: The Social History of British Anthropology, 1885–1945.* Cambridge: Cambridge University Press, 1991.

Kupperman, Karen. *Settling with the Indians: The Meeting of English and Indian Cultures in America, 1580–1640.* Totowa, N.J.: Rowman and Littlefield, 1980.

Kuster, Sybille. *Neither Cultural Imperialism nor Precious Gift of Civilization: African Education in Colonial Zimbabwe, 1890–1962.* Munster: Lit Verlag, 1994.

La-Anyane, Seth. *Ghana Agriculture: Its Economic Development from Early Times to the Middle of the Twentieth Century.* London: Oxford University Press, 1963.

Langan, Mary, and Bill Schwartz, eds. *Crises in the British State, 1880–1930.* London: Hutchinson, 1985.

Lee, J. M. *Colonial Development and Good Government: A Study of the Ideas Expressed by the British Official Classes in Planning Decolonization, 1939–1964.* Oxford: Clarendon Press, 1967.

Lee, J. M., and Martin Petter. *The Colonial Office, War and Development Policy: Organisation and the Planning of a Metropolitan Initiative, 1939–1945.* London: Institute of Commonwealth Studies, 1982.

Lewis, Joanna. *Empire State-Building: War and Welfare in Kenya, 1925–52.* Oxford: James Currey, 2000.

Leys, Colin. *The Rise and Fall of Development Theory.* London: James Currey, 1996.

Leys, Norman. *Kenya.* 3rd ed. London: Hogarth, 1926.

Locke, John. *Second Treatise of Government.* Indianapolis, Ind.: Hackett, 1980 [1690].

Louis, William Roger. *In the Name of God, Go! Leo Amery and the British Empire in the Age of Churchill.* New York: Norton, 1992.

Lugard, Frederick D. *The Dual Mandate in British Tropical Africa.* London: Frank Cass, 1965 [1922].

———. *The Rise of Our East African Empire.* 2 vols. London: Frank Cass, 1968 [1893].

MacDonald, Ramsay. *Labour and the Empire.* Edinburgh: Ballantyre, Hanson, 1907.

MacDonald, Robert H. *The Language of Empire: Myths and Metaphors of Popular Imperialism, 1880–1918.* Manchester: Manchester University Press, 1994.

Mackenzie, A. Fiona D. *Land, Ecology and Resistance in Kenya, 1880–1952.* Portsmouth, N.H.: Heinemann, 1998.

MacKenzie, John. *Orientalism: History, Theory and the Arts.* Manchester: Manchester University Press, 1995.

MacLeod, Roy, ed. *Government and Expertise: Specialists, Administrators and Professionals, 1860–1919.* Cambridge: Cambridge University Press, 1988.

MacLeod, Roy, and Milton Lewis, eds. *Disease, Medicine and Empire: Perspectives on Western Medicine and the Experience of European Expansion.* London: Routledge, 1988.

Macmillan, Mona. *Champion of Africa: The Second Phase of the Work of W. M. Macmillan, 1934–1974.* London: Swindon Press, 1985.

Macmillan, William. *Warning from the West Indies: A Tract for Africa and Empire.* London: Faber and Faber, 1936.

Maine, Henry Sumner. *Village Communities in the East and West.* 2nd ed. London: John Murray, 1872.

Malcolm, D. W. *Sukumaland: An African People and Their Country.* London: Oxford University Press, 1953.

Mamdani, Mahmood. *Citizen and Subject: Contemporary Africa and the Legacy of Late Colonialism.* Princeton: Princeton University Press, 1996.

Manderson, Lenore. *Sickness and the State: Health and Illness in Colonial Malaya, 1870–1940.* Cambridge: Cambridge University Press, 1996.

Manson-Bahr, Philip. *History of the School of Tropical Medicine in London, 1899–1949.* London: H. K. Lewis, 1956.

Marglin, F. A., and S. A. Marglin. *Dominating Knowledge: Development, Culture, and Resistance.* Oxford: Clarendon Press, 1990.

Masefield, G. B. *A History of the Colonial Agricultural Service.* Oxford: Clarendon Press, 1972.

———. *A Short History of Agriculture in the British Colonies.* Oxford: Clarendon Press, 1950.

Maxon, Robert M., *Going Their Separate Ways: Agrarian Transformation in Kenya, 1930–1950.* London: Associated University Presses, 2003.

Maycock, Willoughby. *With Mr. Chamberlain in the United States and Canada, 1887–88.* Toronto: Bell and Cockburn, 1914.

McClintock, Anne. *Imperial Leather: Race, Gender and Sexuality in the Colonial Conquest.* London: Routledge, 1995.

McCracken, Donal P. *Gardens of Empire: Botanical Institutions of the Victorian British Empire.* London: Leicester University Press, 1997.

McKendrick, Neil, John Brewer, and J. H. Plumb. *The Birth of a Consumer Society: The Commercialization of Eighteenth-Century England.* Bloomington: Indiana University Press, 1982.

McLoughlin, Peter. *An Economic History of Sukumaland, Tanzania, to 1964: Field Notes and Analysis.* Fredericton, N.B., Canada, 1971.

McMichael, A. J. *Planetary Overload: Global Environmental Change and the Health of the Human Species.* Cambridge: Cambridge University Press, 1993.

McMichael, Philip. *Development and Social Change: A Global Perspective.* 2nd ed. Thousand Oaks, Calif.: Pine Forge Press, 2000.

McNeill, J. R. *Something New under the Sun: An Environmental History of the Twentieth-Century World.* New York: Norton, 2000.

Mehmet, Ozay. *Westernizing the Third World: The Eurocentricity of Economic Development Theories.* London: Routledge, 1995.

Metcalf, Thomas R. *The Aftermath of Revolt: India, 1857–1870.* Princeton: Princeton University Press, 1964.

———. *Ideologies of the Raj.* New Cambridge History of India 3.4. Cambridge: Cambridge University Press, 1994.

Mills, Richard Charles. *The Colonization of Australia (1829–42): The Wakefield Experiment in Empire Building.* London: Sidgwick and Jackson, 1915.

Moore, Henrietta L., and Megan Vaughan. *Cutting Down Trees: Gender, Nutrition and Agricultural Change in the Northern Province of Zambia, 1890–1990.* London: James Currey, 1994.

Morel, Edmund Dene. *Affairs of West Africa.* London: Heinemann, 1902.

Morgan, D. J. *The Official History of Colonial Development.* 5 vols. London: Macmillan, 1980.

Munck, Ronaldo, and Denis O'Hearn, eds. *Critical Development Theory: Contributions to a New Paradigm.* London: Zed Books, 1999.

Nandy, Ashis. *The Intimate Enemy: Loss and Recovery of Self under Colonialism.* Oxford: Oxford University Press, 1983.

Nimocks, Walter. *Milner's Young Men: The "Kindergarten" in Edwardian Imperial Affairs.* Durham, N.C.: Duke University Press, 1968.

Norgaard, Richard. *Development Betrayed: The End of Progress and a Coevolutionary Revisioning of the Future.* London: Routledge, 1994.

Nye, Edwin R., and Mary E. Gibson. *Ronald Ross: Malariologist and Polymath: A Biography.* London: Macmillan, 1997.

Oldham, Joseph H. *White and Black in Africa: A Critical Examination of the Rhodes Lectures of General Smuts.* London: Longmans, Green, 1930.

Oliver, Roland. *The Missionary Factor in East Africa.* London: Longmans, Green. 1952.

Orde-Browne, Major G. St. J. *Labour Conditions in Northern Rhodesia.* Colonial 150. London: Colonial Office, 1938.

Osborn, Fairfield. *Our Plundered Planet.* Boston: Little, Brown, 1948.

Pagden, Anthony. *Lords of All the World: Ideologies of Empire in Spain, Britain and France, c. 1500–c. 1800.* New Haven: Yale University Press, 1995.

Parkinson, Cosmo. *The Colonial Office from Within, 1909–1945.* London: Faber and Faber, 1945.

Pearce, Robert D. *The Turning Point in Africa: British Colonial Policy, 1938–48.* London: Frank Cass, 1982.

Pendell, Elmer. *Population on the Loose.* New York: W. Funk, 1951.

Pereira, H. C. *Land Use and Water Resources in Temperate and Tropical Climates.* London: Cambridge University Press, 1973.

Phillips, Anne. *The Enigma of Colonialism: British Policy in West Africa.* London: James Currey, 1989.

Phillips, John. *Agriculture and Ecology in Africa: A Study of Actual and Potential Development South of the Sahara.* London: Faber and Faber, 1959.

Pim, Alan. *Colonial Agricultural Production: The Contribution Made by Native Peasants and by Foreign Enterprise.* London: Oxford University Press, 1946.

Porter, A. N., and A. J. Stockwell. *British Imperial Policy and Decolonization, 1938–64,* vol. 1, *1938–51.* London: Macmillan, 1987.

Porter, Bernard. *Critics of Empire: British Radical Attitudes to Colonialism in Africa, 1895–1914.* London: Macmillan, 1968.

Porter, Doug, Bryant Allen, and Gaye Thompson. *Development in Practice: Paved with Good Intentions.* London: Routledge, 1991.

Porter, Roy. *The Creation of the Modern World: The Untold Story of the British Enlightenment.* New York: W. W. Norton, 2000.

Post, Ken. *Arise Ye Starlings: The Jamaican Labour Rebellion of 1938 and Its Aftermath.* The Hague: Nijhoff, 1978.

Power, Helen J. *Tropical Medicine in the Twentieth Century: A History of the Liverpool School of Tropical Medicine, 1898–1990.* London: Kegan Paul International, 1999.

Preston, P. W. *Development Theory: An Introduction.* Cambridge, Mass.: Blackwell, 1996.

Proceedings of the International Congress on Population and World Resources in Relation to the Family, August 1948. London: H. K. Lewis, 1948.

Purseglove, J. W. *Tropical Crops.* 4 vols. London: Longman, 1968–72.

Rahnema, Majid, and Victoria Bawtree, eds. *The Post-Development Reader.* London: Zed Books, 1997.

Ricardo, David. *The Works and Correspondence of David Ricardo.* 11 vols. Cambridge: Cambridge University Press, 1951–73.

Rich, Paul B. *Race and Empire in British Politics.* Cambridge: Cambridge University Press, 1986.

Richards, A. I. *Hunger and Work in a Savage Tribe.* London: Routledge, 1932.

———. *Land, Labour and Diet in Northern Rhodesia: An Economic Study of the Bemba Tribe.* London: Routledge, 1939.

Richey, J. A., ed. *Selections from Educational Records, Part II: 1840–1859.* Calcutta: Bureau of Education, 1922.

Rist, Gilbert. *The History of Development: From Western Origins to Global Faith.* London: Zed Books, 1997.

Roberts, Stephen H. *Population Problems of the Pacific.* New York: Routledge, 1969 [1927].

Robinson, Ronald, and John Gallagher with Alice Denny. *Africa and the Victorians: The Official Mind of Imperialism.* 2nd ed. London: Macmillan, 1981.

Ross, Ronald. *Memoirs, with a Full Account of the Great Malaria Problem and Its Solution.* London: John Murray, 1923.

Ross, Ronald, H. E. Annett, and E. E. Austen. *Report of the Malarial Expedition of the Liverpool School of Tropical Medicine and Medical Parasitology.* London: University Press of Liverpool, 1900.

Rosselli, John. *Lord William Bentinck: The Making of a Liberal Imperialist, 1774–1839.* Delhi: Thomson Press, 1974.

Rounce, N. V. *The Agriculture of the Cultivation Steppe of the Lake, Western and Central Provinces.* Cape Town: Longmans, Green, 1949.

Royal Institute of International Affairs. *The Colonial Problem.* London: Oxford University Press, 1937.

Russell, E. John. *A History of Agricultural Science in Great Britain, 1600–1954.* London: Allen and Unwin, 1966.

———. *World Population and World Food Supplies.* London: Allen and Unwin, 1954.

Russell, E. W. *Soils and Soil Fertility in Tropical Pastures.* London: Faber and Faber, 1966.

Sachs, Wolfgang, ed. *The Development Dictionary: A Guide to Knowledge as Power.* London: Zed Books, 1992.

Said, Edward. *Orientalism.* New York: Vintage Books, 1979.

Savage, Gail. *The Social Construction of Expertise: The English Civil Service and Its Influence, 1919–1939.* Pittsburgh: University of Pittsburgh Press, 1996.

Savile, A. H. *Extension in Rural Communities: A Manual for Agricultural and Home Extension Workers.* Oxford: Oxford University Press, 1965.

Schuster, George. *Private Work and Public Causes.* Cowbridge, Wales: Brown and Sons, 1979.

Schuurman, Frans J., ed. *Beyond the Impasse: New Directions in Development Theory.* London: Zed Books, 1993.

Scott, Henry Harold. *A History of Tropical Medicine.* 2 vols. London: Edward Arnold, 1939.

Searle, G. R. *The Quest for National Efficiency: A Study in British Politics and Political Thought, 1899–1914.* Berkeley: University of California Press, 1971.

Seed, Patricia. *American Pentimento: The Invention of Indians and the Pursuit of Riches.* Minneapolis: University of Minnesota Press, 2001.

Semmel, Bernard. *Imperialism and Social Reform: English Social-Imperial Thought 1895–1914*. London: Allen and Unwin, 1960.
Sender, John, and Sheila Smith. *The Development of Capitalism in Africa*. London: Methuen, 1986.
Shenton, Robert W. *The Development of Capitalism in Northern Nigeria*. Toronto: Toronto University Press 1986.
Sivonen, Seppo. *White Collar or Hoe Handle: African Education under British Colonial Policy, 1920–1945*. Helsinki: Suomen Historiallinen Seura, 1995.
Smuts, Jan Christian. *Africa and Some World Problems*. Oxford: Clarendon Press, 1930.
Spitzer, Leo. *The Creoles of Sierra Leone: Responses to Colonialism, 1870–1945*. Madison: University of Wisconsin Press, 1974.
Spurr, David. *The Rhetoric of Empire: Colonial Discourse in Journalism, Travel Writing, and Imperial Administration*. Durham, N.C.: Duke University Press, 1993.
Stamp, Dudley. *Our Undeveloped World*. London: Faber and Faber, 1953.
Stedman Jones, Gareth. *Outcast London: A Study in the Relationship between Classes in Victorian Society*. London: Oxford University Press, 1971.
Stepan, Nancy. *The Idea of Race in Science: Great Britain, 1800–1960*. London: Macmillan, 1982.
Stephens, J. W. W., and S. R. Christophers. *Reports of the Royal Society Malaria Commission*. 3rd ser. London: Harrison, 1900.
Stiglitz, Joseph. *The Economic Role of the State*. Oxford: Blackwell, 1989.
Stockdale, F. A., T. Perch, and H. F. Macmillan. *The Royal Botanic Gardens, Peradeniya, Ceylon, 1822–1922*. Colombo: H. W. Cave, 1922.
Stokes, Eric. *The English Utilitarians and India*. Oxford: Clarendon Press, 1959.
Storey, H. H. *Basic Research in Agriculture: A Brief History of Research at Amani, 1928–1947*. Nairobi: East African Agriculture and Forestry Research Organization, 1950.
Storey, William Kelleher. *Science and Power in Colonial Mauritius*. Rochester: University of Rochester Press, 1997.
Sykes, W. M. *The Sukumaland Development Scheme, 1947–57*. Oxford Development Records Project, Report 7. Oxford: Rhodes House Library, 1984.
Tansley, A. G., and T. F. Chipp, eds. *Aims and Methods in the Study of Vegetation*. London: British Empire Vegetation Committee, 1926.
Thomas, Nicholas. *Colonialism's Culture: Anthropology, Travel and Government*. Princeton: Princeton University Press, 1994.
Throup, David. *The Economic and Social Origins of Mau Mau, 1945–53*. London: James Currey, 1987.
Thurston, Anne. *The Intensification of Smallholder Agriculture in Kenya: The Genesis and Implementation of the Swynnerton Plan*. Oxford Development Records Project, Report 6. Oxford: Rhodes House Library, 1984.

———. *Sources for Colonial Studies in the Public Record Office.* Vol. 1, *Records of the Colonial Office, Dominions Office, Commonwealth Relations Office and Commonwealth Office.* London: HMSO, 1995.

Tomlinson, B. R. *The Economy of Modern India, 1860–1970.* Cambridge: Cambridge University Press, 1993.

Toye, John. *Dilemmas of Development: Reflections on the Counter-Revolution in Development Economics.* Oxford: Blackwell, 1987.

Trapnell, C. G. *The Soils, Vegetation and Agricultural Systems of North-Eastern Rhodesia.* Lusaka: Government Printer, 1943.

Trapnell, C. G., and J. G. Clothier. *The Soils, Vegetation and Agricultural Systems of North-Western Rhodesia.* Lusaka: Government Printer, 1937.

van Beusekom, Monica M. *Negotiating Development: African Farmers and Colonial Experts at the Office du Niger, 1920–1960.* Oxford: James Currey, 2002.

Vaughan, Megan. *Curing Their Ills: Colonial Power and African Illness.* Cambridge: Polity Press, 1991.

———. *The Story of an African Famine: Gender and Famine in Twentieth-Century Malawi.* Cambridge: Cambridge University Press, 1987.

Vogt, William. *Road to Survival.* London: Gollancz, 1949.

Watson, Malcolm. *African Highway: The Battle for Health in Central Africa.* London: John Murray, 1953.

———. *Rural Sanitation in the Tropics: Being Notes and Observations in the Malay Archipelago, Panama and Other Lands.* London: J. Murray, 1915.

Webb, James L. A., Jr. *Tropical Pioneers: Human Agency and Ecological Change in the Highlands of Sri Lanka, 1800–1900.* Athens: Ohio University Press, 2002.

Webster, C. C., and P. N. Wilson, eds. *Agriculture in the Tropics.* 3rd ed. London: Blackwell Science, 1998.

Williamson, Philip. *National Crisis and National Government: British Politics, the Economy and Empire, 1926–1932.* London: Cambridge University Press, 1992.

Willis, J. C. *Agriculture in the Tropics: An Elemental Treatise.* Cambridge: Cambridge University Press, 1909.

Wilson, Harold. *The War on World Poverty: An Appeal to the Conscience of Mankind.* London: Gollancz, 1953.

Wood, Alan. *The Groundnut Affair.* London: Bodley Head, 1950.

Worster, Donald. *Dust Bowl: The Southern Plains in the 1930s.* Oxford: Oxford University Press, 1979.

Worthington, E. B. *A Development Plan for Uganda.* Entebbe: Uganda Protectorate Press, 1949.

———. *The Ecological Century: A Personal Appraisal.* Oxford: Clarendon Press, 1983.

———. *Science in Africa: A Review of Scientific Research Relating to Tropical and Southern Africa.* Oxford: Oxford University Press, 1938.

Young, Robert. *White Mythologies: Writing History and the West.* London: Routledge, 1990.

Articles, Chapters

Allen, Richard P. "The Slender, Sweet Thread: Sugar, Capital and Dependency in Mauritius, 1860–1936." *Journal of Imperial and Commonwealth History* 16, no. 2 (1988): 177–200.

Amery, Leopold S., and W. Ormsby-Gore. "Problems and Development in Africa." *Journal of the African Society* 28, no. 112 (July 1929): 325–39.

Anderson, Benedict. "The New World Disorder." *New Left Review* 193 (May/June 1992): 3–14.

Anderson, David. "Depression, Dust Bowl, Demography, and Drought: The Colonial State and Soil Conservation in East Africa during the 1930s." *African Affairs* 83, no. 332 (July 1984): 321–43.

———. "Managing the Forest: The Conservation History of Lembus, Kenya, 1904–1963." In *Conservation in Africa: People, Policies and Practice,* edited by David Anderson and Richard Grove, 249–68. Cambridge: Cambridge University Press, 1987.

Anderson, David, and David Throup. "Africans and Agricultural Production in Colonial Kenya: The Myth of the War as a Watershed." *Journal of African History* 26, no. 4 (1985): 327–45.

Anderson, Robert. "The Origins of the International Rice Research Institute." *Minerva* 29, no. 1 (Spring 1991): 61–89.

Arndt, H. W. "Economic Development: A Semantic History." *Economic Development and Cultural Change* 29, no. 3 (1981): 457–66.

Arnold, David. "India's Place in the Tropical World, 1770–1930." *Journal of Imperial and Commonwealth History* 26, no. 1 (1998): 1–21.

———. "Introduction: Disease, Medicine and Empire." In *Imperial Medicine and Indigenous Societies,* edited by David Arnold, 1–27. Manchester: Manchester University Press, 1988.

———. "White Colonization and Labour in Nineteenth-Century India." *Journal of Imperial and Commonwealth History* 11, no. 2 (1983): 133–58.

Aspinall, Algernon. "The Imperial College of Tropical Agriculture." *Tropical Agriculture* 11, no. 2 (1934): 40–43.

Baden-Powell, George. "The Development of Tropical Africa." In *Proceedings of the Royal Colonial Institute,* vol. 27, 217–55. London: Royal Colonial Institute, 1896.

Baker, R. A., and R. A. Bayliss. "William John Ritchie Simpson: Public Health and Tropical Medicine." *Medical History* 31, no. 4 (1987): 450–65.

Barron, T. J. "Science and the Nineteenth-Century Ceylon Coffee Planters." *Journal of Imperial and Commonwealth History* 16, no. 1 (1987): 15–23.

Batie, Sandra S. "Soil Conservation in the 1980s: A Historical Perspective." *Agricultural History* 59, no. 2 (April 1985): 107–23.

Bayly, C. A. "Maine and Change in Nineteenth-Century India." In *The Victorian Achievement of Sir Henry Maine: A Centennial Reappraisal,* edited by Alan Diamond, 389–97. Cambridge: Cambridge University Press, 1991.

Beinart, William. "Introduction: The Politics of Colonial Conservation." *Journal of Southern African Studies* 15, no. 2 (1989): 143–62.

———. "Soil Erosion, Conservationism and Ideas about Development: A Southern African Exploration, 1900–1960." *Journal of Southern African Studies* 11, no. 1 (October 1984): 52–83.

Bell, G. D. H. "Frank Leonard Engledow." *Biographical Memoirs of Fellows of the Royal Society* 32 (December 1986): 188–219.

Berman, Bruce, and John Lonsdale. "Coping with the Contradictions: The Development of the Colonial State in Kenya, 1895–1914." *Journal of African History* 20, no. 4 (1979): 487–505.

———. "Crises of Accumulation, Coercion and the Colonial State: The Development of the Labour Control System in Kenya, 1919–1929." *Canadian Journal of African Studies* 14, no. 1 (1980): 55–81.

Berman, Edward H. "American Influence on African Education: The Role of the Phelps-Stokes Fund's Education Commissions." *Comparative Education Review* 15, no. 2 (1971): 132–45.

Bessant, Leonard Leslie. "Coercive Development: Land Shortage, Forced Labour, and Colonial Development in the Chiweshe Reserve, Colonial Zimbabwe, 1938–1946." *International Journal of African Historical Studies* 25, no. 1 (1992): 39–65.

Birkett, Deborah. "West Africa's Mary Kingsley." *History Today* 37 (May 1987): 10–16.

Blyn, George. "The Green Revolution Revisited." *Economic Development and Cultural Change* 31, no. 4 (1983): 705–25.

Boyce, Rubert. "The Colonization of Africa." *Journal of the African Society* 10, no. 40 (July 1911): 392–97.

Boyd Orr, John, and John L. Gilks. "The Nutritional Condition of the East Africa Native." *Lancet,* 12 March 1927, 1:560–62.

Brayne, C. V. "The Problem of Peasant Agriculture in Ceylon." *Ceylon Economic Journal* 6 (December 1934): 34–46.

Burkill, H. M. "Murray Ross Henderson, 1899–1983, and Some Notes on the Administration of Botanical Research in Malaya." *Journal of the Malaysian Branch of the Royal Asiatic Society* 56, no. 2 (1983): 87–101.

Burnet, E., and W. R. Aykroyd. "Nutrition and Public Health." *Quarterly Bulletin of the Health Organisation of the League of Nations* 4, no. 2 (June 1935): 323–495.

Catanach, I. J. "Plague and the Tensions of Empire: India, 1896–1918." In *Imperial Medicine and Indigenous Societies,* edited by David Arnold, 149–71. Manchester: Manchester University Press, 1988.

Cell, John W. "Hailey and the Making of the African Survey." *African Affairs* 88, no. 353 (1989): 481–505.

Cheesman, E. E. "The Economic Botany of Cocoa: A Critical Survey of the Literature to the End of 1930." *Tropical Agriculture,* supplement, June 1932.

Cheesman, E. E., J. B. Hutchinson, G. G. Gianetti, et al. "Special Issue Commemorating the 50th Anniversary of the Founding of the ICTA." *Tropical Agriculture,* 51, no. 3 (July 1974): 457–500.

Church, Archibald G. "Colonial Development and the Scientific Worker." *Nature* 124 (21 September 1929): 433–34.

———. "The Inter-Relations of East African Territories." *Geographical Journal* 67, no. 3 (May 1926): 215–31.

——— "Mr. Ormsby-Gore and Tropical Development." *Nature* 123 (21 January 1929): 27.

———. "Science and Administration in East Africa." *Nature* 115 (10 January 1925): 37–39.

Churchill, Winston. "The Development of Africa." *Journal of the Royal African Society* 6, no. 23 (April 1907): 291–96.

Cleaver, Harry. "The Contradictions of the Green Revolution." *American Economic Review* 62, no. 2 (May 1972): 177–86.

Cliffe, Lionel. "Nationalism and the Reaction to Enforced Agricultural Change in Tanganyika during the Colonial Period." In *Socialism in Tanzania: An Interdisciplinary Reader,* edited by Lionel Cliffe and John Saul, 17–24. Nairobi: EAPH, 1972.

Connelly, Matthew. "Taking Off the Cold War Lens: Visions of North-South Conflict during the Algerian War of Independence." *American Historical Review* 105, no. 3 (2000): 739–69.

Cooper, Frederick. "Conflict and Connection: Rethinking Colonial African History." *American Historical Review* 99, no. 5 (1994): 1516–45.

Cowen, Michael P. "The British State and Agrarian Accumulation in Kenya." In *Industry and Accumulation in Africa,* edited by Martin Fransman, 142–69. London: Heinemann, 1982.

Cowen, Michael P., and Nicholas Westcott. "British Imperial Economic Policy during the War." In *Africa and the Second World War,* edited by David Killingray and Richard Rathbone, 20–67. London: Macmillan, 1986.

Cowen, Michael P., and Robert W. Shenton. "Agrarian Doctrines of Development." *Journal of Peasant Studies* 25, no. 2 (1998): 49–75; and 25, no. 3 (1998): 31–62.

———. "The Invention of Development." In *Power of Development,* edited by Jonathan Crush, 27–43. London: Routledge, 1995.

———. "The Origin and Course of Fabian Colonialism in Africa." *Journal of Historical Sociology* 4, no. 2 (1991): 143–74.

Crowder, Michael. "The First World War and Its Consequences." In *UNESCO General History of Africa,* vol. 7, *Africa Under Colonial Domination, 1880–1935,* edited by A. Adu Boahen, 283–311. Berkeley: University of California Press, 1985.

Curtin, Philip D. "Medical Knowledge and Urban Planning in Tropical Africa." *American Historical Review* 90, no. 3 (1985): 594–613.

Danquah, Francis K. "Rural Discontent and Decolonization in Ghana, 1945–1951." *Agricultural History* 68, no. 1 (1994): 1–19.

Davin, Anna. "Imperialism and Motherhood." In *Tensions of Empire: Colonial Cultures in a Bourgeois World,* edited by Frederick Cooper and Ann Laura Stoler, 87–151. Berkeley: University of California Press, 1997.

Dewey, Clive. "The Influence of Sir Henry Maine on Agrarian Policy in India." In *The Victorian Achievement of Sir Henry Maine: A Centennial Reappraisal,* edited by Alan Diamond, 353–75. Cambridge: Cambridge University Press, 1991.

Dixon-Fyle, Mac. "Agricultural Improvement and Political Protest on the Tonga Plateau, Northern Rhodesia." *Journal of African History* 18, no. 4 (1977): 579–96.

Douthwaite, Richard. "Is It Possible to Build a Sustainable World?" In *Critical Development Theory: Contributions to a New Paradigm,* edited by Ronaldo Munck and Denis O'Hearn, 157–77. London: Zed Books, 1999.

Drinkwater, Michael. "Technical Development and Peasant Impoverishment: Land Use Policy in Zimbabwe's Midland Province." *Journal of Southern African Studies* 15, no. 2 (1989): 287–305.

Duder, C. J. D. "The Settler Response to the Indian Crisis of 1923 in Kenya: Brigadier General Philip Wheatley and 'Direct Action.'" *Journal of Imperial and Commonwealth History* 17, no. 3 (1989): 349–73.

Duffy, Michael. "World-Wide War and British Expansion, 1793–1815." In *The Eighteenth Century,* edited by P. J. Marshall, 184–207. Vol. 2, *The Oxford History of the British Empire,* edited by William Roger Louis. Oxford: Oxford University Press, 1999.

Dumett, Raymond. "The Campaign against Malaria and the Expansion of Scientific Medical and Sanitary Services in British West Africa, 1898–1910." *African Historical Studies* 1, no. 2 (1968): 153–96.

———. "Obstacles to Government-Assisted Agricultural Development in West Africa: Cotton-Growing Experimentation in Ghana in the Early Twentieth Century." *Agricultural History Review* 23, no. 2 (1975): 156–72.

Dumont, Louis. "The 'Village Community' from Munro to Maine." *Contributions to Indian Sociology* 9 (December 1966): 67–89.

Dunstan, Wyndham R. "Some Imperial Aspects of Applied Chemistry." *Bulletin of the Imperial Institute* 4 (1906): 310–19.

Eden, T. "The Menace of Soil Erosion." *Tropical Agriculture* 11, no. 4 (1934): 77–79.

Engledow, F. L. "The Place of Plant Physiology and of Plant-Breeding in the Advancement of British Agriculture." *Empire Journal of Experimental Agriculture* 7, no. 26 (1939): 145–49.

Escobar, Arturo. "Imagining a Post-Development Era." In *Power of Development,* edited by Jonathan Crush, 211–27. London: Routledge, 1995.

Evans, Geoffrey. "Research and Training in Tropical Agriculture." *Journal of the Royal Society of Arts* 87, no. 4499 (1939): 341.

Fairhead, James, and Melissa Leach. "Desiccation and Domination: Science and Struggles over Environment and Development in Colonial Guinea." *Journal of African History* 41, no. 1 (2000): 35–54.

Fage, J. D. "When the African Society Was Founded, Who Were the Africanists?" *African Affairs* 94, no. 376 (1995): 369–81.

Farrell, Terrence. "Arthur Lewis and the Case for Caribbean Industrialization." *Social and Economic Studies* 29, no. 4 (December 1980): 52–75.

Faulkner, O. T., and C. Y. Shephard. "Mixed Farming: The Basis of a System for West Indian Peasants." *Tropical Agriculture* 20, no. 7 (1943): 136–42.

Firth, Raymond: "The Sociological Study of Native Diet." *Africa* 7, no. 4 (1934): 401–14.

Flint, John. "Macmillan as a Critic of Empire: The Impact of an Historian on Colonial Policy." In *Africa and Empire: W. M. Macmillan, Historian and Social Critic,* edited by Hugh Macmillan and Shula Marks, 212–31. London: Gower, 1989, 212–31.

———. "Planned Decolonization and Its Failure in British Africa." *African Affairs* 82, no. 328 (1983): 389–411.

Gianetti, G. G. "The Amani Institute." *Tropical Agriculture* 7, no. 3 (1930): 77.

———. "The Gold Coast Cacao Industry." *Tropical Agriculture* 12, no. 2 (1935): 311–12.

Gillman, Clement. "Southwest Tanganyika Territory." *Geographical Journal* 69, no. 2 (1927): 97–126.

Gilmartin, David. "Scientific Empire and Imperial Science: Colonialism and Irrigation Technology in the Indus Basin." *Journal of Asian Studies* 53, no. 4 (November 1994): 1127–49.

Gorrie, R. M. "The Problem of Soil Erosion in the British Empire with Special Reference to India." *Journal of the Royal Society of Arts* 86, no. 4471 (1938): 901–29.

Grant, J. P. "Marginal Men: The Global Employment Crisis." *Foreign Affairs* 50, no. 1 (1971): 112–24.

Green, R. H., and S. H. Hymer. "Cocoa in the Gold Coast: A Study in the Relations between African Farmers and Agricultural Experts." *Journal of Economic History* 26, no. 3 (1966): 299–319.

Grove, Richard. "Early Themes in African Conservation: The Cape in the Nineteenth Century." In *Conservation in Africa: People, Policies and Practice,* edited by David Anderson and Richard Grove, 21–39. Cambridge: Cambridge University Press, 1987.

Hagberg Wright, C. T. "German Methods of Development in East Africa." *Journal of the Royal African Society* 1, no. 1 (1901): 23–28.

Hailey, Malcolm. "Present Trends in Colonial Policy." *Listener,* 10 August 1939.

Hardy, Frederick. "Soil Deterioration in the Tanganyikan Uplands: A Lesson in Tropical Agricultural Exploitation." *Tropical Agriculture* 15, no. 4 (1938): 79–81.

Hardy, Frederick, and Nazeer Ahmad. "Soil Science at ICTA/UWI, 1922–1972." *Tropical Agriculture* 51, no. 4 (October 1974): 468–73.

Hartley, C. W. S. "Establishment of New Rice Areas in Malaya." *World Crops* 3, no. 5 (1951): 171–75.

Harris, Jose. "Political Thought and the Welfare State, 1870–1940: An Intellectual Framework for British Social Policy." *Past and Present* 135, no. 1 (May 1992): 116–41.

Harrison, Mark. "'The Tender Frame of Man': Disease, Climate and Racial Difference in India and the West Indies, 1760–1860." *Bulletin of the History of Medicine* 70, no. 1 (Spring 1996): 68–93.

Harrison, T. H. "The New Hebrides People and Culture." *Geographical Journal* 88, no. 4 (October 1936): 332–41.

Heely, Dorothy O. "'Informed Opinion' on Tropical Africa in Great Britain 1860–1890." *African Affairs* 68, no. 272 (1969): 195–217.

Helleiner, Eric. "From Bretton Woods to Global Finance: A World Turned Upside Down." In *Political Economy and the Changing Global Order,* edited by Richard Stubbs and Geoffrey R. D. Underhill, 163–75. Toronto: McClelland and Stewart, 1994.

Heuman, Gad. "The British West Indies." In *The Nineteenth Century,* edited by Andrew Porter, 483–90. Vol. 3, *The Oxford History of the British Empire,* edited by William Roger Louis. Oxford: Oxford University Press, 1999.

Heyer, Judith. "Agricultural Development Policy in Kenya from the Colonial Period to 1975." In *Rural Development in Tropical Africa,* edited by Judith Heyer, Pepe Roberts, and Gavin Williams, 101–7. New York: St. Martin's Press, 1981.

Hodgson, Dorothy. "Taking Stock: State Control, Ethnic Identity and Pastoralist Development in Tanganyika, 1948–1958." *Journal of African History* 41, no. 1 (2000): 55–78.

Hogendorn, Jan, and K. M. Scott. "Very Large-Scale Agricultural Projects: The Lessons of the East African Groundnut Scheme." In *Imperialism, Colonialism and Hunger: East and Central Africa,* edited by Robert Rotberg, 167–98. Lexington, Mass.: Lexington Books, 1983.

Holland, Robert. "The British Empire and the Great War, 1914–1918." In *The Twentieth Century,* edited by Judith Brown and William Roger Louis, 114–37. Vol. 4, *The Oxford History of the British Empire,* edited by William Roger Louis. Oxford: Oxford University Press, 1999.

Hutchinson, Joseph. "The Role of the ICTA in Tropical Agriculture." *Tropical Agriculture* 51, no. 4 (October 1974): 459–67.

Huxley, Elspeth. "The Menace of Soil Erosion." *Journal of the Royal African Society* 36, no. 144 (July 1937): 357–70.

Huxley, Julian. "The Future of Colonies." *Fortnightly,* n.s., 148 (August 1940): 120–30.

———. "The World's Greatest Problem." *World Review,* January 1950.

Ignatieff, Michael. "The Burden." *New York Times Magazine,* 5 January 2003.

Ingham, Barbara. "Shaping Opinion on Development Policy: Economists at the Colonial Office during World War II." *History of Political Economy* 24, no. 3 (1992): 689–710.

Ittmann, Karl. "The Colonial Office and the Population Question in the British Empire, 1918–62." *Journal of Imperial and Commonwealth History* 27, no. 3 (1999): 55–81.

Jayaweera, Swarna. "Land Policy and Peasant Colonization, 1914–1948." In *History of Ceylon,* vol. 3, *From the Beginning of the 19th Century to 1948,* edited by K. M. De Silva, 446–60. Peradeniya: University of Ceylon Press Board, 1973.

Jeffery, Keith. "The Second World War." In *The Twentieth Century,* edited by Judith Brown and William Roger Louis, 306–28. Vol. 4, *The Oxford History of the British Empire,* edited by William Roger Louis. Oxford: Oxford University Press, 1999.

Jeffries, Charles. "Recent Social Welfare Developments in British Tropical Africa." *Africa: Journal of the International African Institute* 14, no. 1 (January 1943): 4–11.

Johnson, Gordon. "India and Henry Maine." In *The Victorian Achievement of Sir Henry Maine: A Centennial Reappraisal,* edited by Alan Diamond, 376–88. Cambridge: Cambridge University Press, 1991.

Johnson, Howard. "The British Caribbean from Demobilization to Constitutional Decolonization." In *The Twentieth Century,* edited by Judith Brown and William Roger Louis, 597–622. Vol. 4, *The Oxford History of the British Empire,* edited by William Roger Louis. Oxford: Oxford University Press, 1999.

———. "The West Indies and the Conversion of the British Official Classes to the Development Idea." *Journal of Commonwealth and Comparative Politics* 15, no. 1 (1977): 55–83.

Johnston, Bruce F., and John Cownie. "The Seed-Fertilizer Revolution and Labor Force Absorption." *American Economic Review* 59, no. 4 (September 1969): 569–82.

Jolly, A. L. "An Economic Survey on the La Pastora Land Settlement, Trinidad." *Tropical Agriculture* 28, no. 7 (1946): 117–22.

Jones, Brynmor. "Desiccation and the West African Colonies." *Geographical Journal* 91, no. 5 (May 1938): 401–23.

Kalinga, Owen J. M. "The Master Farmers' Scheme in Nyasaland, 1950–1962: A Study of a Failed Attempt to Create a 'Yeoman' Class." *African Affairs* 92, no. 368 (July 1993): 367–87.

Kay, George. "Agricultural Progress in Zambia." In *Environment and Land Use in Africa,* edited by M. F. Thomas and G. W. Whittington, 495–523. London: Methuen, 1969.

Keen, Bernard, and D. W. Duthie. "Crop Responses to Fertilizers and Manures in East Africa." *East African Agricultural Journal* 19, no. 1 (1953): 19–32.

Kennedy, Dane. "Empire Migration in Post-War Reconstruction: The Role of the Overseas Settlement Committee, 1919–1922." *Albion* 20, no. 2 (1988): 403–19.

———. "Imperial History and Post-Colonial Theory." *Journal of Imperial and Commonwealth History* 24, no. 3 (1996): 345–63.

King, J. G. M. "Mixed Farming in Northern Nigeria, Part I: Origins and Present Conditions." *Empire Journal of Experimental Agriculture* 7, no. 27 (1939): 271–85.

King, Kenneth James. "Africa and the Southern States of the USA: Notes on J. H. Oldham and American Negro Education for Africans." *Journal of African History* 10, no. 4 (1969): 659–77.

Kofi, Tetteh. "Arthur Lewis and West African Economic Development." *Social and Economic Studies* 29, no. 4 (December 1980): 202–27.

Lavin, Deborah. "Margery Perham's Initiation into African Affairs." In *Margery Perham and British Rule in Africa,* edited by Alison Smith and Mary Bull, 45–61. London: Frank Cass, 1991.

Leach, Melissa, and Robin Mearns. "Environmental Change and Policy: Challenging Received Wisdom in Africa." In *The Lie of the Land: Challenging Received Wisdom on the African Environment,* edited by Melissa Leach and Robin Mearns, 1–33. Portsmouth, N.H.: Heinemann, 1996.

Lee, J. M. "The Dissolution of the Empire Marketing Board, 1933: Reflections on a Diary." *Journal of Imperial and Commonwealth History* 1, no. 1 (1973): 49–58.

Lewis, J. E. "'Tropical East Ends' and the Second World War: Some Contradictions in Colonial Office Welfare Initiatives." *Journal of Imperial and Commonwealth History* 28, no. 2 (May 2000): 42–66.

Lewis, W. Arthur. "Developing Colonial Agriculture." *Tropical Agriculture* 27, nos. 4–6 (1950): 63–73.

Lobdell, Richard A. "British Officials and the West Indian Peasantry, 1842–1938." In *Labour in the Caribbean: From Emancipation to Independence,* edited by Malcolm Cross and Gad Heuman, 195–233. London: Macmillan Caribbean, 1988.

Lonsdale, John M. "The Depression and the Second World War in the Transformation of Kenya." In *Africa and the Second World War,* edited by David Killingray and Richard Rathbone, 97–142. London: Macmillan, 1986.

Low, D. A., and J. M. Lonsdale. "Introduction: Towards the New Order, 1945–1963." In *History of East Africa,* vol. 3, edited by D. A. Low and Alison Smith, 1–64. Oxford: Clarendon Press, 1976.

Ludden, David. "India's Development Regime." In *Colonialism and Culture,* edited by Nicholas B. Dirks, 247–87. Ann Arbor: University of Michigan Press, 1992.

Lugard, Frederick D. "Education in Tropical Africa." *Edinburgh Review* 242, no. 493 (July 1925): 1–19.

———. "The International Institute of African Languages and Cultures." *Africa* 1, no. 1 (1928): 3–12.

Lunan, Murray. "Mound Cultivation in Ufipa, Tanganyika." *East African Agricultural Journal,* 16, no. 2 (1950): 88–89.

MacLagan Gorrie, R. "The Problem of Soil Erosion in the British Empire with Special Reference to India." *Journal of the Royal Society of Arts* 86, no. 4471 (29 July 1938): 901–29.

MacLeod, Roy M. "On Visiting the 'Moving Metropolis': Reflections on the Architecture of Imperial Science." In *Scientific Colonialism: A Cross-Cultural Comparison,* edited by Nathan Reingold and Marc Rothenberg, 217–49. Washington: Smithsonian Institution Press, 1987.

MacLeod, Roy M., and E. Kay Andrews. "The Committee of Civil Research: Scientific Advice for Economic Development, 1925–30." *Minerva* 7, no. 2 (1969): 680–705.

———. "The Origins of the DSIR: Reflections on Ideas and Men, 1915–1916." *Public Administration* 48, no. 1 (Spring 1970): 23–48.

Macmillan, William. "The Real Colonial Question." *Fortnightly,* n.s., 148 (December 1940): 548–57.

Maegraith, B. H. "History of the Liverpool School of Tropical Medicine." *Medical History* 16, no. 4 (1972): 354–68.

Maher, Colin. "Soil Conservation in Kenya Colony, Part I." *Empire Journal of Experimental Agriculture* 18, no. 71 (1950): 137–49.

Malinowski, Bronislaw. "Practical Anthropology." *Africa* 2, no. 1 (1929): 22–38.

———. "The Rationalization of Anthropology and Administration." *Africa* 3, no. 4 (October 1930): 405–29.

Marshall, P. J. "Britain without America—A Second Empire?" In *The Eighteenth Century,* edited by P. J. Marshall, 576–95. Vol. 2, *The Oxford History of the British Empire,* edited by William Roger Louis. Oxford: Oxford University Press, 1999.

Martin, W. S. "Grass Covers in Their Relation to Soil Structure." *Empire Journal of Experimental Agriculture* 12, no. 47 (1944): 21–33.

Mayhew, Madeleine. "The 1930s Nutrition Controversy." *Journal of Contemporary History* 23, no. 3 (1988): 445–64.

Mayne, J. E. "Progress in the Mechanization of Farming in the Colonial Territories." *Tropical Agriculture* 31, no. 3 (1954): 178–87; 32, no. 2 (1955): 95–99; 33, no. 4 (1956): 272–77.

McCracken, John. "Experts and Expertise in Colonial Malawi." *African Affairs* 81, no. 322 (January 1982): 101–16.

McDougall, Frank. "Food and Population." *International Conciliation,* no. 486, December 1952: 539–84.

McLean, William H. "The Social and Economic Development of the British Colonial Empire." *Journal of the Royal Society of Arts* 88, no. 4571 (4 October 1940): 871–81; 88, no. 4572 (18 October 1940): 891–914.

Megaw, John. "Pressure of Population in India." *Journal of the Royal Society of Arts* 86, no. 4491 (1938): 134–49.

Mellor, John W. "Ending Hunger: An Implementable Program for Self-Reliant Growth." In *The World Food Crisis: Food Security in Comparative Perspective,* ed. J. I. Hans Bakker, 485–519. Toronto: Canadian Scholar's Press, 1990.

Meredith, David. "The British Government and Colonial Economic Policy, 1919–1939." *Economic History Review* 28, no. 3 (1975): 484–99.

Millington, Andrew. "Environmental Degradation, Soil Conservation and Agricultural Policies in Sierra Leone, 1895–1984." In *Conservation in Africa: People, Policies and Practice,* edited by David Anderson and Richard Grove, 229–48. Cambridge: Cambridge University Press, 1987.

Milne, Geoffrey. "Soil and Vegetation." *East African Agricultural Journal* 5, no. 4 (1940): 294–98.

———. "Soil and Vegetation." *Tropical Agriculture* 17, no. 5 (1940): 95–97.

———. "Soil Conservation—The Research Side." *East African Agricultural Journal* no. 1 (1940): 26–31.

———. "A Soil Reconnaissance Journey through Parts of Tanganyika Territory, December 1935 to February 1936." *Journal of Ecology* 35, no. 1–2 (1947): 192–265.

———. "Some Aspects of Modern Practice in Soil Survey." *East African Agricultural Journal* 6, no. 1 (1940): 1–7.

Moon, Suzanne. "Empirical Knowledge, Scientific Authority, and Native Development: The Controversy over Sugar/Rice Ecology in the Netherlands East Indies, 1905–1914." *Environment and History* 10, no. 1 (2004): 59–81.

Moore, David B. "Development Discourse as Hegemony: Towards an Ideological History, 1945–1995." In *Debating Development Discourse: Institutional and Popular Perspectives,* edited by David B. Moore and Gerald G. Schmitz, 1–53. London: Macmillan, 1995.

Moore, Sally Falk. "Changing Perspectives on a Changing Africa: The Work of Anthropology." In *Africa and the Disciplines: The Contributions of Research in Africa to the Social Sciences and Humanities,* edited by Robert Bates, 3–52. Chicago: University of Chicago Press, 1993.

Murray, D. B. "Cocoa Research at ICTA." *Tropical Agriculture* 51, no. 4 (1974): 477–79.

Mustapha, Abdul Raufu. "Colonialism and Environmental Perception in Northern Nigeria." *Oxford Development Studies* 31, no. 4 (2003): 405–25.

Nowell, William. "The Agricultural Research Station at Amani." *Journal of the Royal Society of Arts* 81, no. 4227 (1933): 1098–1109.

Nworah, K. K. D. "The Liverpool 'Sect' and British West African Policy, 1895–1915." *African Affairs* 70, no. 281 (1970): 349–64.

Olby, Robert. "Social Imperialism and State Support for Agricultural Research in Edwardian Britain." *Annals of Science* 48, no. 6 (1991): 509–26.

Oldham, Joseph H. "Christian Missions and the Education of the Negro." *International Review of Missions* 7 (1918): 242–47.

———. "Population and Health in Africa." *International Review of Missions* 15 (1926): 402–17.

Ormsby-Gore, William. "Agricultural Research." *Journal of the Royal African Society* 33 (January 1934): 18–21.

———. "British West Africa." *United Empire* 18, n.s., 1 (January 1927): 28–41.

———. "Development of Our Tropical Dependencies." Address to the Scottish Geographical Society, Edinburgh, 13 December 1928. *Scottish Geographical Magazine* 14 (May 1929): 129–39.

———. "My Recent Travels in East Africa." *United Empire,* n.s., 16 (1925): 357–64.

Overfield, Richard. "Charles Bessey: The Impact of the 'New Botany' on American Agriculture, 1880–1910." *Technology and Culture* 16, no. 2 (1975): 162–81.

Packard, Randall M. "Visions of Postwar Health and Development and Their Impact on Public Health Interventions in the Developing World." In *International Development and the Social Sciences,* edited by Frederick Cooper and Randall Packard, 93–115. Berkeley: University of California Press, 1997.

Palladino, Paolo. "Between Craft and Science: Plant Breeding, Mendelian Genetics and British Universities, 1900–1920." *Technology and Culture* 34, no. 2 (1993): 300–23.

Parisinos, C. C., C. Y. Shephard, and A. L. Jolly. "Peasant Agriculture: An Economic Survey of the Las Lomas District, Trinidad." *Tropical Agriculture* 21, no. 5 (1944): 84–98.

Pearce, Robert D. "Missionary Education in Colonial Africa: The Critique of Mary Kingsley." *History of Education* 17, no. 4 (1988): 283–94.

Phillips, John. "Problems in the Use of Chemical Fertilizers." In *The Careless Technology: Ecology and International Development,* edited by M. Taqhi Farvar and John P. Milton, 549–66. New York: Natural History Press, 1972.

"Planned Group Farming in Nyanza Province, Kenya." *Tropical Agriculture* 28, nos. 7–12 (1951): 153–57.

Rahnema, Majid. "Towards Post-Development." In *The Post-Development Reader,* edited by Majid Rahnema and Victoria Bawtree, 377–404. London: Zed Books, 1997.

Rajan, S. Ravi. "Foresters and the Politics of Colonial Agroecology: The Case of Shifting Cultivation and Soil Erosion, 1920–1950." *Studies in History,* n.s., 14, no. 2 (1998): 217–35.

"Report on Discussion on Fertilizer Experiments in East Africa." *East African Agricultural Journal* 15, no. 2 (1949): 61–68.

Richards, A. I., and E. M. Widdowson. "A Dietary Study of North-Eastern Rhodesia." In "Problems of African Native Diet," special issue, *Africa* 9, no. 2 (April 1936): 166–96.

Roberts, A. D. "The Imperial Mind." In *The Cambridge History of Africa,* vol. 7, edited by A. D. Roberts, 24–76. Cambridge: Cambridge University Press, 1986.

Robertson, J. K. "Mixed or Multiple Cropping in Native Agricultural Practice." *East African Agricultural Journal* 4 (1941): 228–32.

Robin, Libby. "Ecology: A Science of Empire?" In *Ecology and Empire: Environmental History of Settler Societies,* edited by Tom Griffiths and Libby Robin, 63–75. Edinburgh: Keele University Press, 1997.

Robinson, Kenneth. "Experts, Colonialists, and Africanists, 1895–1960." In *Experts in Africa: Proceedings of a Colloquium at the University of Aberdeen,* edited by J. C. Stone, 55–74. Aberdeen: Aberdeen University African Studies Group, 1980.

Roe, Emery. "'Development Narratives,' or Making the Best of Blueprint Development." *World Development* 19, no. 4 (1991): 287–300.

Rounce, N. V. "Technical Considerations in the Economic Development of Sukumaland." *Empire Journal of Experimental Agriculture* 19, no. 76 (1951): 253–65.

Royal African Society. "Land Usage and Soil Erosion in Africa: A Report of the Speeches at the Dinner of the Royal African Society, 1 December 1937." *Journal of the Royal African Society,* supplement, vol. 37, no. 146 (January 1938): 3–19.

Russell, E. W. "The Impact of Technological Developments on Soils in East Africa." In *The Careless Technology: Ecology and International Development,* edited by M. Taqhi Farvar and John P. Milton, 567–76. New York: Natural History Press, 1972.

Sachs, Jeffrey. "Global Capitalism: Making It Work." *Economist,* 12 September 1998.

Sambon, L. Westenra. "Acclimatization of Europeans in Tropical Lands." *Geographical Journal* 12, no. 6 (1898): 589–99.

Samuel, Frank. "Economic Potential of Colonial Africa." *Tropical Agriculture* 28, nos. 7–12 (1951): 138–50.

Saul, S. B. "The Economic Significance of 'Constructive Imperialism.'" *Journal of Economic History* 17, no. 2 (1957): 173–92.

Self, Robert. "Treasury Control and the Empire Marketing Board: The Rise and Fall of Non-Tariff Preference in Britain, 1924–1933." *Twentieth Century British History* 5, no. 2 (1994): 153–82.

Shenton, Robert W. "The Labour Question in Late Colonial Africa." Review of *Decolonization and African Society,* by Frederick Cooper. *Canadian Journal of African Studies* 32, no. 1 (1998): 174–80.

Shephard, C. Y. "Co-operation and Peasant Development in the Tropics." *Tropical Agriculture* 6, no. 2 (1929): 39–42.

———. "Peasant Agriculture in the Leeward and Windward Islands." *Tropical Agriculture* 24, no. 4–6 (1947): 61–71.

Shiva, Vandana. "The Living Democracy Movement: Alternatives to the Bankruptcy of Globalization." In *Another World Is Possible: Popular Alternatives to Globalization at the World Social Forum,* edited by William F. Fisher and Thomas Ponniah, 115–24. London: Zed Books, 2003.

Showers, Kate. "Soil Erosion in the Kingdom of Lesotho: Origins and Colonial Response, 1830s–1950s." *Journal of Southern African Studies* 15, no. 2 (1989): 263–86.

Simmonds, N. W. "The Earlier British Contribution to Tropical Agricultural Research." *Tropical Agricultural Association Newsletter* 11, no. 2 (June 1991): 2–7.

Simpson, W. J. R. "The Croonian Lectures on Plague." Lecture 4, Royal College of Physicians, London, 27 June 1907. *Lancet,* 27 July 1907, 207–12.

———. "A Presidential Address on Preventive Work in the Tropics." Tropical Section of the Congress of the Royal Institute of Public Health, July, 1904. *Lancet,* 27 August 1904, 2:577–80.

———. "Recent Discoveries Which Have Rendered Antimalarial Sanitation More Precise and Less Costly." Section of Tropical Diseases, British Medical Association. *British Medical Journal,* 19 October 1907, 2:1044–46.

Smith, David, and Malcolm Nicolson. "Nutrition, Education, Ignorance and Income: A Twentieth Century Debate." In *The Science and Culture of Nutrition,*

1840–1940, edited by Harmke Kamminga and Andrew Cunningham, 288–318. Amsterdam: Rodopi, 1995.

"Soil Erosion." *Tropical Agriculture* 7, no. 1 (1930): 19–20.

Spencer, Ian R. G. "The First World War and the Origins of the Dual Policy of Development in Kenya, 1914–1922." *World Development* 9, no. 8 (1981): 735–48.

Stamp, Dudley. "Southern Margin of the Sahara: Comments on Some Recent Studies on the Question of Desiccation in West Africa." *Geographical Review* 30, no. 2 (1940): 297–300.

Stebbing, E. P. "Africa and Its Intermittent Rainfall: The Role of the Savanna." *Journal of the Royal African Society*, supplement, vol. 37, no. 149 (October 1938), 3–32.

———. "The Encroaching Sahara: The Threat to the West African Colonies." *Geographical Journal* 85, no. 6 (June 1935): 506–24.

———. "The Man-Made Desert in Africa: Erosion and Drought." *Journal of the Royal African Society*, supplement, vol. 37, no. 146 (January 1938): 3–40.

———. "The Threat of the Sahara." *Journal of the Royal African Society*, extra supplement, 25 May 1937, 3–35.

Stephens, J. W. W. "The Discussion on the Prophylaxis of Malaria." *British Medical Journal*, 17 September 1904.

Stephens, J. W. W., and S. R. Christophers. "Distribution of *Anopheles* in Sierra Leone, Parts I and II." In *Report to the Malaria Committee of the Royal Society*, 1st ser., 42–74. London: Harrison and Sons, 1900.

———. "The Native as the Prime Agent in the Malarial Infection of Europeans." In *Further Reports to the Malaria Committee of the Royal Society*, 2nd ser., 3–19. London: Harrison and Sons, 1900.

Stiglitz, Joseph. "Failure of the Fund: Rethinking the IMF Response." *Harvard International Review* 23, no. 2 (Summer 2001): 14–18.

Stockdale, Frank A. "Soil Erosion in the Colonial Empire." *Empire Journal of Experimental Agriculture* 5, no. 20 (October 1937): 281–97.

Stocking, Michael. "Soil Conservation Policy in Colonial Africa." *Agricultural History* 59, no. 2 (1985): 148–61.

Stockwell, A. J. "Imperialism and Nationalism in South-East Asia." In *The Twentieth Century*, edited by Judith Brown and William Roger Louis, 465–89. Vol. 4, *The Oxford History of the British Empire*, edited by William Roger Louis. Oxford: Oxford University Press, 1999.

Stokes, Eric. "Milnerism." *Historical Journal* 5, no. 1 (1962): 47–60.

Stoler, Ann Laura. "Rethinking Colonial Categories: European Communities and the Boundaries of Rule." *Comparative Studies in Society and History* 31, no. 1 (1989): 134–61.

Storey, William Kelleher. "Small-Scale Sugar Cane Farmers and Biotechnology in Mauritius: The 'Uba' Riots of 1937." *Agricultural History* 69, no. 2 (1995): 163–76.

Strachan, Henry. "Discussion on the Prophylaxis of Malaria." *British Medical Journal*, 17 September 1904, 637–39.

Sutphen, Mary P. "Not What, but Where: Bubonic Plague and the Reception of Germ Theories in Hong Kong and Calcutta, 1894–1897." *Journal of the History of Medicine and Allied Sciences* 52, no. 1 (January 1997): 81–113.

———. "Striving to Be Separate? Civilian and Military Doctors in Cape Town during the Anglo Boer War." In *War, Medicine and Modernity,* edited by Roger Cooter, Mark Harrison, and Steve Sturdy, 48–64. Gloucestershire: Sutton, 1998.

Swift, Jeremy. "Desertification: Narratives, Winners and Losers." In *The Lie of the Land: Challenging Received Wisdom on the African Environment,* edited by Melissa Leach and Robin Mearns, 73–90. Portsmouth, N.H.: Heinemann, 1996.

"Swollen Shoot of Cacao." *Tropical Agriculture* 15, no. 5 (1938): 110–11.

Tangri, Roger. "From the Politics of Union to Mass Nationalism: The Nyasaland African Congress, 1944–1959." In *From Nyasaland to Malawi: Studies in Colonial History,* edited by Roderick J. MacDonald, 267–81. Nairobi: East African Publishing House, 1975.

Tempany, H. A., G. M. Roddan, and L. Lord. "Soil Erosion and Soil Conservation in the Colonial Empire." *Empire Journal of Experimental Agriculture* 12, no. 47 (1944): 121–53.

Thomas, A. S. "The Dry Season in the Gold Coast and Its Relations to Cultivation of Cacao." *Journal of Ecology* 20, no. 2 (1932): 263–69.

Tilley, Helen. "African Environments and Environmental Sciences: The African Research Survey, Ecological Paradigms and British Colonial Development, 1920–1940." In *Social History and African Environments,* edited by William Beinart and JoAnn McGregor, 109–30. Oxford: James Currey, 2003.

Tinsley, T. W. "The Ecology of Cocoa Viruses: The Role of Wild Hosts in the Incidence of Swollen Shoot Virus in West Africa." *Journal of Ecology* 8, no. 2 (1971): 491–95.

Turner, Frank M. "Public Science in Britain, 1880–1919." *Isis* 71, no. 259 (1980): 598–608.

Turner, John. "'Experts' and Interests: David Lloyd George and the Dilemmas of the Expanding State, 1906–19." In *Government and Expertise: Specialists, Administrators and Professionals, 1860–1919,* edited by Roy M. MacLeod, 203–23. Cambridge: Cambridge University Press, 1988.

van Beusekom, Monica M. "Colonisation Indigène: French Rural Development Ideology at the Office du Niger, 1920–1940." *International Journal of African Historical Studies* 30, no. 2 (1997): 299–323.

———. "Disjunctures in Theory and Practice: Making Sense of Change in Agricultural Development at the Office du Niger, 1920–1960." *Journal of African History* 41, no. 1 (2000): 79–99.

van Beusekom, Monica, and Dorothy Hodgson. "Lessons Learned? Development Experiences in the Late Colonial Period." *Journal of African History* 41, no. 1 (2000): 29–33.

Varcoe, Ian. "Scientists, Government and Organised Research in Great Britain 1914–16: The Early History of the DSIR." *Minerva* 8, no. 2 (April 1970): 192–216.

Vernon, Keith. "Science for the Farmer? Agricultural Research in England 1909–1936." *Twentieth Century British History* 8, no. 3 (1997): 310–33.

Vine, H. "Experiments on the Maintenance of Soil Fertility at Ibadan, Nigeria, 1922–1955." *Empire Journal of Experimental Agriculture* 21, no. 82 (1953): 65–85.

Visaria, Leela, and Pravin Visaria. "Population (1757–1947)." In *The Cambridge Economic History of India,* vol. 2, *c. 1751–c. 1970,* edited by Dharma Kumar and Meghnad Desai, 463–532. Cambridge: Cambridge University Press, 1983.

Wakefield, Edward Gibbon. "The Art of Colonization." In *A Letter from Sydney and Other Writings,* edited by Ernest Rhys. London: J. M. Dent and Sons, 1929.

Wardlaw, C. W. "Virgin Soil Deterioration: The Deterioration of Virgin Soils in the Caribbean Banana Lands." *Tropical Agriculture* 6, no. 9 (1929): 243–48.

Washbrook, D. A. "Law, State and Agrarian Society in Colonial India." *Modern Asian Studies* 15, no. 3 (1981): 649–721.

Watson, Malcolm. "The Effect of Drainage and Other Measures on the Malaria of Klang, Federated Malay States." *Journal of Tropical Medicine* 6, no. 22 (16 November 1903): 349–53; 6, no. 23 (1 December 1903): 368–71.

Weber, Charles W. "The Influence of the Hampton-Tuskegee Model on the Educational Policy of the Permanent Mandates Commission and British Colonial Policy." *Africana Journal* 16 (1994): 66–84.

Webster, C. C. "The Ley and Soil Fertility in Britain and Kenya." *East African Agricultural Journal* 20, no. 2 (1954): 71–74.

Webster, Charles. "Health, Welfare and Unemployment during the Depression." *Past and Present* 109 (November 1985): 204–30.

Weindling, Paul. "The Role of International Organizations in Setting Nutritional Standards in the 1920s and 1930s." In *The Science and Culture of Nutrition, 1840–1940,* edited by Harmke Kamminga and Andrew Cunningham, 319–32. Amsterdam: Rodopi, 1995.

Whitehead, Clive. "The Advisory Committee on Education in the (British) Colonies, 1924–1961." *Paedagogica Historica* 27, no. 3 (1991): 385–421.

———. "British Colonial Education Policy: A Synonym for Cultural Imperialism?" In *"Benefits Bestowed?" Education and British Imperialism,* edited by J. A. Mangan, 211–30. Manchester: Manchester University Press, 1988.

Wickremeratne, L. A. "Economic Development in the Plantation Sector, 1900 to 1947." In *History of Ceylon,* vol. 3, *From the Beginning of the 19th Century to 1948,* edited by K. M. De Silva, 428–45. Peradeniya: University of Ceylon Press Board, 1973.

Williams, Gavin. "Modernizing Malthus: The World Bank, Population Control and the African Environment." In *Power of Development,* edited by Jonathan Crush, 158–75. London: Routledge, 1995.

———. "Studying Development and Explaining Policies." *Oxford Development Studies* 31, no. 1 (2003): 37–58.

———. "The World Bank and the Peasant Problem." In *Rural Development in Tropical Africa,* edited by Judith Heyer, Pepe Roberts, and Gavin Williams, 16–51. New York: St. Martin's Press, 1981.

Williams, Keith. "'A Way Out of Our Troubles': The Politics of Empire Settlement, 1900–1922." In *Emigrants and Empire: British Settlement in the Dominions between the Wars,* edited by Stephen Constantine, 22–44. Manchester: Manchester University Press, 1990.

Wood, R. C. "Agriculture in Ceylon." *Tropical Agriculture* 6, no. 3 (1929): 83.

Worboys, Michael. "British Colonial Science Policy, 1918–1939." In *Les Sciences Coloniales: Figures et Institutions,* edited by Patrick Petitjean, 99–112. Paris: Orstom, 1996.

———. "The Discovery of Colonial Malnutrition between the Wars." In *Imperial Medicine and Indigenous Societies,* edited by David Arnold, 208–25. Manchester: Manchester University Press, 1988.

———. "The Emergence of Tropical Medicine: A Study in the Establishment of a Scientific Specialty." In *Perspectives on the Emergence of Scientific Disciplines,* edited by Gerard Lemaine, Roy MacLeod, Michael Mulkay, and Peter Weingart, 75–98. The Hague: Mouton, 1976.

———. "Germs, Malaria and the Invention of Mansonian Tropical Medicine: From 'Diseases in the Tropics' to 'Tropical Diseases.'" In *Warm Climates and Western Medicine: The Emergence of Tropical Medicine, 1500–1900,* edited by David Arnold, 181–207. Amsterdam: Rodopi, 1996.

———. "Manson, Ross and Colonial Medical Policy: Tropical Medicine in London and Liverpool, 1899–1914." In *Disease, Medicine and Empire: Perspectives on Western Medicine and the Experience of European Expansion,* edited by Roy MacLeod and Milton Lewis, 21–37. London: Routledge, 1988.

Worthington, E. B. "The Nile Catchment: Technological Change and Aquatic Biology." In *The Careless Technology: Ecology and International Development,* edited by M. Taqhi Farvar and John P. Milton, 189–205. New York: Natural History Press, 1972.

———. "On the Food and Nutrition of African Natives." In "Problems of African Native Diet," special issue, *Africa: Journal of the International African Institute* 9, no. 2 (5 April 1936): 150–65.

Wright, James. "Lucky Hill Community Project." *Tropical Agriculture* 24, nos. 10–12 (1947): 142.

Wylie, Diana. "Confrontation over Kenya: The Colonial Office and Its Critics, 1918–1940." *Journal of African History* 18, no. 3 (1977): 427–47.

———. "Norman Leys and McGregor Ross: A Case Study in the Conscience of African Empire 1900–39." *Journal of Imperial and Commonwealth History* 5, no. 3 (1977): 294–309.

Youe, Christopher P. "The Threat of Settler Rebellion and the Imperial Predicament: The Denial of Indian Rights in Kenya, 1923." *Canadian Journal of History* 12, no. 3 (1978): 347–60.

Zimmerman, Andrew. "'What Do You Really Want in German East Africa, *Herr Professor?*' Counterinsurgency and the Science Effect in Colonial Tanzania." *Comparative Studies of Society and History* 48, no. 2 (April 2006): 419–61.

Unpublished Papers, Dissertations, and Theses

Anderson, David M. "Organising Ideas: British Colonialism and African Rural Development." Unpublished paper, Social Science Research Council Workshop on Social Science and Development, 1993.

Chipande, G. H. R. "Smallholder Agriculture as a Rural Development Strategy: The Case of Malawi." Ph.D. thesis, University of Glasgow, 1983.

Dubow, Saul. "The Idea of Race in Early 20th Century South Africa: Some Preliminary Thoughts." African Studies Seminar Paper, University of the Witwatersrand, April 1989.

Gale, Thomas S. "Official Medical Policy in British West Africa, 1870–1930." Ph.D. thesis, School of Oriental and African Studies, University of London, 1972.

Hodge, Joseph M. "Development and Science: British Colonialism and the Rise of the 'Expert,' 1895–1945." Ph.D. thesis, Queen's University at Kingston, Canada, 1999.

Ibokette, Isongesit S. "Contradictions in Colonial Rule: Urban Development in Northern Nigeria, 1900–1940." Ph.D. thesis, Queen's University at Kingston, Canada, 1989.

Joyce, Michael C. W. K. "Education, Culture and Empire: Sir Gordon Guggisberg and Pedagogical Imperatives in the Gold Coast, 1919–27." M.A. thesis, Queen's University at Kingston, Canada, 1994.

Makana, Nicholas Ekutu. "Changing Patterns of Indigenous Economic Systems: Agrarian Change and Rural Transformation in Bungoma District, 1930–1960." Ph.D. thesis, West Virginia University, 2005.

Moon, Suzanne. "Constructing 'Native Development': Technological Change and the Politics of Colonization in the Netherlands East Indies, 1905–1930." Ph.D. diss., Cornell University, 2000.

Nworah, K. K. D. "Humanitarian Pressure-Groups and British Attitudes to West Africa, 1895–1915." Ph.D. diss., University of London, 1966.

Stiglitz, Joseph. "Development Policies in a World of Globalization." Paper presented at the 50th anniversary of the Brazilian Economic and Social Development Bank (BNDES), Rio de Janeiro, 12–13 September 2002.

Thompson, A. R. "The Adaptation of Education to African Society in Tanganyika under British Rule." Ph.D. thesis, University of London, 1968.

Tilley, Helen. "Africa as a 'Living Laboratory': The African Research Survey and the British Colonial Empire: Consolidating Environmental, Medical, and Anthropological Debates, 1920–1940." D.Phil. thesis, Oxford University, 2001.

Worboys, Michael. "Science and British Colonial Imperialism, 1895–1940." Ph.D. thesis, University of Sussex, 1979.

Index